T0211663

Aircraft Communications and Navigation Systems

Introducing the principles of communications and navigation systems, this book is written for anyone pursuing a career in aircraft maintenance engineering or a related aerospace engineering discipline, and in particular will be suitable for those studying for licensed aircraft maintenance engineer status. It systematically addresses the relevant sections (Air Transport Association of America chapters 23/34) of modules 11 and 13 of Part-66 of the European Aviation Safety Agency (EASA) syllabus and is ideal for anyone studying as part of an EASA and FAR-147 approved course in aerospace engineering.

- Delivers the essential principles and knowledge base required by Airframe and Propulsion (A&P) Mechanics for Modules 11 and 13 of the EASA Part-66 syllabus and BTEC National awards in aerospace engineering
- Supports mechanics, technicians and engineers studying for a Part-66 qualification
- Comprehensive and accessible, with self-test questions, exercises and multiple choice questions to enhance learning for both independent and tutor-assisted study
- Additional resources and interactive materials are available at the book's companion website at www.66web.co.uk

This new and updated third edition provides readers with an overview of the latest key technologies that underpin the functioning of safety-critical systems such as those used in flight management, reporting, navigation, and air traffic control.

Mike Tooley has over 30 years' experience of teaching electrical principles, electronics and avionics to engineers and technicians, previously as Head of the Department of Engineering and Vice Principal at Brooklands College, Surrey, UK. He currently works as a consultant and freelance technical author.

David Wyatt has over 45 years' experience in the aviation industry. After a technician apprenticeship in BOAC he progressed as a development engineer at BA. After a period as an avionics engineering lecturer in further education, David went on to become Head of Airworthiness at Gama Aviation. David is the author of several popular books in the Routledge Aircraft Engineering Series.

Aircraft Communications and Navigation Systems
Third edition

Mike Tooley and David Wyatt

Routledge
Taylor & Francis Group

LONDON AND NEW YORK

Cover image: Mike Tooley and David Wyatt

Third edition published 2024
by Routledge
4 Park Square, Milton Park, Abingdon, Oxon OX14 4RN

and by Routledge
605 Third Avenue, New York, NY 10158

Routledge is an imprint of the Taylor & Francis Group, an informa business

© 2024 Mike Tooley and David Wyatt

The right of Mike Tooley and David Wyatt to be identified as authors of this work has been asserted in accordance with sections 77 and 78 of the Copyright, Designs and Patents Act 1988.

All rights reserved. No part of this book may be reprinted or reproduced or utilised in any form or by any electronic, mechanical, or other means, now known or hereafter invented, including photocopying and recording, or in any information storage or retrieval system, without permission in writing from the publishers.

Trademark notice: Product or corporate names may be trademarks or registered trademarks, and are used only for identification and explanation without intent to infringe.

First published 2007
Previous edition published 2017

British Library Cataloguing-in-Publication Data
A catalogue record for this book is available from the British Library

Library of Congress Cataloging-in-Publication Data
Names: Tooley, Michael H., author. | Wyatt, David, 1968- author.
Title: Aircraft communications and navigation systems / Mike Tooley and David Wyatt.
Description: Third edition. | New York, NY : Routledge, 2024. | Includes bibliographical references and index.
Identifiers: LCCN 2023028817 | ISBN 9781032534152 (hardback) | ISBN 9781032518084 (paperback) | ISBN 9781003411932 (ebook) | ISBN 9781032626130 (ebook other)
Subjects: LCSH: Airplanes--Electronic equipment. | Avionics. | Aeronautics--Communication systems.
Classification: LCC TL694 .T664 2024 | DDC 629.135/1--dc23/eng/20230630
LC record available at https://lccn.loc.gov/2023028817

ISBN: 978-1-032-53415-2 (hbk)
ISBN: 978-1-032-51808-4 (pbk)
ISBN: 978-1-003-41193-2 (ebk)
ISBN: 978-1-032-62613-0 (eBook+)

DOI: 10.1201/9781003411932

Typeset in Times New Roman
by SPi Technologies India Pvt Ltd (Straive)

Contents

Preface   x

Acknowledgements   xvi

Online resources   xvi

Chapter 1   **Introduction**   1
   1.1   The radio frequency spectrum   1
   1.2   Electromagnetic waves   3
   1.3   Frequency and wavelength   4
   1.4   The atmosphere   5
   1.5   Radio wave propagation   5
   1.6   The ionosphere   8
   1.7   MUF and LUF   10
   1.8   Silent zone and skip distance   12
   1.9   Space weather   12
   1.10   Satellite communications (SATCOM)   16
   1.11   Communication systems integration and management   19
   1.12   Aircraft communications and reporting system (ACARS)   22
   1.13   Multiple choice questions   24

Chapter 2   **Antennas**   27
   2.1   The isotropic radiator   27
   2.2   The half-wave dipole   28
   2.3   Impedance and radiation resistance   30
   2.4   Radiated power and efficiency   31
   2.5   Antenna gain   31
   2.6   The Yagi beam antenna   32
   2.7   Directional characteristics   34
   2.8   Other practical antennas   36
   2.9   Feeders   40
   2.10   Connectors   44
   2.11   Standing wave ratio   45
   2.12   Waveguide   50
   2.13   Vector network analysis   51
   2.14   Multiple choice questions   55

Chapter 3   **Transmitters and receivers**   58
   3.1   A simple radio system   58
   3.2   Modulation and demodulation   59
   3.3   AM transmitters   60
   3.4   FM transmitters   61
   3.5   Tuned radio frequency receivers   62
   3.6   Superhet receivers   63
   3.7   Selectivity   64
   3.8   Image channel rejection   67
   3.9   Automatic gain control   68

3.10	Double superhet receivers	68
3.11	Digital frequency synthesis	70
3.12	A design example	72
3.13	Software defined radio	73
3.14	Multiple choice questions	79
Chapter 4	**VHF communications**	**81**
4.1	VHF range and propagation	81
4.2	DSB modulation	82
4.3	Channel spacing	83
4.4	Depth of modulation	83
4.5	Compression	84
4.6	Squelch	85
4.7	Data modes	85
4.8	ACARS	88
4.9	VHF radio equipment	90
4.10	Multiple choice questions	92
Chapter 5	**HF communications**	**94**
5.1	HF range and propagation	94
5.2	SSB modulation	95
5.3	SELCAL	97
5.4	HF datalink	97
5.5	HF radio equipment	99
5.6	HF antennas and coupling units	103
5.7	Multiple choice questions	106
Chapter 6	**Flight-deck audio systems**	**107**
6.1	Flight interphone system	107
6.2	Cockpit voice recorder	112
6.3	Multiple choice questions	114
Chapter 7	**Emergency locator transmitters**	**116**
7.1	Types of ELT	116
7.2	Maintenance and testing of ELT	117
7.3	ELT mounting requirements	118
7.4	Typical ELT	119
7.5	Cospas–Sarsat satellites	121
7.6	Multiple choice questions	123
Chapter 8	**Aircraft navigation**	**125**
8.1	The earth and navigation	125
8.2	Dead reckoning	129
8.3	Position fixing	129
8.4	Maps and charts	130
8.5	Navigation terminology	131
8.6	Navigation systems evolution	131
8.7	Navigation systems summary	141
8.8	Multiple choice questions	143
Chapter 9	**Automatic direction finder**	**144**
9.1	Introducing ADF	144
9.2	ADF principles	144
9.3	ADF equipment	145

	9.4	Operational aspects of ADF	151
	9.5	ADF homing	153
	9.6	Multiple choice questions	153
Chapter 10		**VHF omnidirectional range**	**155**
	10.1	VOR principles	155
	10.2	Airborne VOR equipment	159
	10.3	Operational aspects of VOR	164
	10.4	VOR navigation display scenarios	166
	10.5	Multiple choice questions	168
Chapter 11		**Distance measuring equipment**	**170**
	11.1	Radar principles	170
	11.2	DME overview	171
	11.3	DME operation	172
	11.4	Equipment overview	173
	11.5	En route navigation using radio navigation aids	175
	11.6	Multiple choice questions	180
Chapter 12		**Instrument landing system**	**182**
	12.1	ILS overview	182
	12.2	ILS ground equipment	182
	12.3	ILS airborne equipment	186
	12.4	Low range radio altimeter	189
	12.5	ILS approach	190
	12.6	Autoland	190
	12.7	Operational aspects of the ILS	193
	12.8	Multiple choice questions	193
Chapter 13		**Doppler navigation**	**195**
	13.1	The Doppler effect	195
	13.2	Doppler navigation principles	195
	13.3	Airborne equipment overview	199
	13.4	Typical Doppler installations	200
	13.5	Doppler summary	200
	13.6	Other Doppler applications	201
	13.7	Multiple choice questions	201
Chapter 14		**Area navigation**	**203**
	14.1	RNAV overview	203
	14.2	RNAV equipment	208
	14.3	Kalman filters	212
	14.4	Required navigation performance (RNP)	215
	14.5	PBN system errors	217
	14.6	Actual navigation performance (ANP)	217
	14.7	Multiple choice questions	218
Chapter 15		**Inertial navigation systems**	**219**
	15.1	Inertial navigation principles	219
	15.2	System overview	221
	15.3	System description	223
	15.4	Alignment process	230
	15.5	Inertial navigation accuracy	232

	15.6	Inertial navigation summary	232
	15.7	System integration	233
	15.8	Multiple choice questions	233

Chapter 16		**Global navigation satellite systems**	**235**
	16.1	GPS overview	235
	16.2	Principles of wave propagation	235
	16.3	Satellite navigation principles	235
	16.4	GPS segments	236
	16.5	GPS signals	239
	16.6	GNSS Operation	239
	16.7	GNSS evolution	241
	16.8	GNSS augmentation	241
	16.9	GNSS – The future	243
	16.10	Multiple choice questions	244

Chapter 17		**Flight management systems**	**246**
	17.1	FMS overview	246
	17.2	Flight management computer system (FMCS)	246
	17.3	System initialisation	248
	17.4	FMCS operation	250
	17.5	General Aviation FMS	257
	17.6	Four-dimensional (4D) navigation	258
	17.7	Automatic DME tuning	261
	17.8	FMS summary	261
	17.9	Multiple choice questions	262

Chapter 18		**Weather radar**	**264**
	18.1	System overview	264
	18.2	Airborne equipment	265
	18.3	Precipitation and turbulence	269
	18.4	System enhancements	276
	18.5	Lightning detection	277
	18.6	Datalink weather	277
	18.7	Phased array radar	278
	18.8	Multiple choice questions	281

Chapter 19		**Air traffic control systems**	**282**
	19.1	ATC overview	282
	19.2	ATC transponder modes	285
	19.3	Airborne equipment	285
	19.4	System operation	288
	19.5	Automatic dependent surveillance–broadcast (ADS-B)	297
	19.6	Communications, navigation and surveillance/air traffic management (CNS/ATM)	302
	19.7	Single European Sky	305
	19.8	FAA Next generation	305
	19.9	Future Air Navigation Systems (FANS)	306
	19.10	Drones	307
	19.11	Multiple choice questions	307

Chapter 20		**Traffic alert and collision avoidance systems**	**308**
	20.1	Airborne collision avoidance systems (ACAS)	308
	20.2	TCAS overview	309

20.3	TCAS equipment	312
20.4	System operation	314
20.5	ADS-B traffic displays	320
20.6	Traffic advisory system (TAS)	321
20.7	Multiple choice questions	321
Chapter 21	**Electromagnetic compatibility (EMC)**	**323**
21.1	The avionic electromagnetic environment	323
21.2	Effects of EMI	325
21.3	Sources of EMI	325
21.4	Classification of EMI	325
21.5	Some examples of EMI	326
21.6	EMI reduction	329
21.7	Aircraft wiring and cabling	330
21.8	Grounding and bonding	332
21.9	Case Study – The 5G problem	333
21.10	Multiple choice questions	338
Appendix 1	**Abbreviations and acronyms**	**340**
Appendix 2	**Revision papers**	**349**
Appendix 3	**Answers to multiple choice questions**	**355**
Appendix 4	**Decibels**	**360**
Index		**362**

Preface

The first edition of Aircraft Communications and Navigation Systems was published in 2007 as part of a series designed for both independent and tutor-assisted studies. This third edition has been updated with reference to the latest version of EASA's Part 66 syllabus as well as technology updates and requests from readers/reviewers. The book forms part of a series of titles:

- Aircraft Engineering Principles (AEP)
- Aircraft Communications and Navigation Systems (ACNS)
- Aircraft Flight Instruments and Guidance Systems (AFIGS)
- Aircraft Electrical and Electronic Systems (AEES)

ACNS is designed to cover the essential knowledge base required by certifying mechanics, technicians and engineers engaged in engineering maintenance activities on commercial aircraft. In addition, this book should appeal to members of the armed forces and others attending training and educational establishments engaged in aircraft maintenance and related aeronautical engineering programmes (including Foundation Degree courses, BTEC National and Higher National units as well as City and Guilds and NVQ courses).

The book introduces the principles, operation and maintenance of aircraft communications and navigation systems. The aim has been to make the subject material accessible and presented in a form that can be readily assimilated. The book provides syllabus coverage of the communications and navigation section of Module 13 (ATA 23/34). The book assumes a basic understanding of aircraft flight controls as well as an appreciation of electricity and electronics (broadly equivalent to Modules 3 and 4 of the EASA Part-66 syllabus).

It is important to be aware that this book is not designed to replace aircraft maintenance manuals. Nor does it attempt to provide the level of detail required by those engaged in the maintenance of specific aircraft types. Any maintenance statements made in this book are for training/educational purposes only. Always refer to the approved aircraft data and applicable safety instructions.

Chapter 1 sets the scene by providing an explanation of electromagnetic wave propagation and the radio frequency spectrum. The chapter also describes the various mechanisms by which radio waves propagate together with a detailed description of the behaviour of the ionosphere and its effect on radio signals.

Antennas are introduced in Chapter 2. This chapter explains the principles of isotropic and directional radiating elements and introduces several important concepts including radiation resistance, antenna impedance, radiated power, gain and efficiency. Several practical forms of antenna are described including dipoles, Yagi beam antennas, quarter wave (Marconi) antennas, corner reflectors, horn and parabolic dish radiators. Chapter 2 also provides an introduction to feeders (including coaxial cable and open-wire types), connectors and standing wave ratio (SWR). The chapter concludes with a brief introduction to waveguide systems.

Radio transmitters and receivers are the subject of Chapter 3. This chapter provides readers with an introduction to the operating principles of AM and FM transmitters as well as tuned radio frequency (TRF) and supersonic-heterodyne (superhet) receivers. Selectivity, image channel rejection and automatic gain control (AGC) are important requirements of a modern radio receiver, and these topics are introduced before moving on to describe more complex receiving equipment. Modern aircraft radio equipment is increasingly based on the use of digital frequency synthesis, and the basic principles of phase-locked loops and digital synthesisers are described and explained.

Very high frequency (VHF) radio has long been the primary means of communication between aircraft and the ground. Chapter 4 describes the principles of VHF communications (both voice and data). The chapter also provides an introduction to the Aircraft Communications Addressing and Reporting System (ACARS).

High frequency (HF) radio provides aircraft with an effective means of communicating over long distance oceanic and trans-polar routes. In addition, global data communication has recently been made possible using strategically located HF datalink (HFDL) ground stations. Chapter 5 describes the principles of HF radio communication as well as the equipment and technology used.

As well as communication with ground stations, modern passenger aircraft require facilities for local communication within the aircraft. Chapter 6 describes flight-deck audio systems including the interphone system and all-important cockpit voice recorder (CVR) which captures audio signals so that they can be later analysed in the event of a serious malfunction of the aircraft or of any of its systems.

The detection and location of the site of an air crash is vitally important to the search and rescue (SAR) teams and potential survivors. Chapter 7 describes the construction and operation of emergency locator transmitters (ELT) fitted to modern passenger aircraft. The chapter also provides a brief introduction to satellite-based location techniques.

Chapter 8 introduces the subject of aircraft navigation; this sets the scene for the remaining chapters of the book. Navigation is the science of conducting journeys over land and/or sea. This chapter reviews some basic features of the earth's geometry as it relates to navigation, and introduces some basic aircraft navigation terminology, e.g., latitude, longitude, dead reckoning, etc. The chapter concludes by reviewing a range of navigation systems used on modern transport and military aircraft. Many aircraft navigation systems utilise radio frequency methods to determine a position fix; this links very well into the previous chapters of the book describing fundamental principles of radio transmitters, receivers and antennas.

Radio waves have directional characteristics as described in the early chapters of the book. This is the basis of the automatic direction finder (ADF), one of earliest forms of radio navigation that is still in use today. ADF is a short/medium-range (200 nm) navigation system providing directional information.

Chapter 9 looks at the historical background to radio navigation, reviews some typical ADF hardware that is fitted to modern commercial transport aircraft, and concludes with some practical aspects associated with the operational use of ADF.

During the late 1940s, it was evident to the aviation world that an accurate and reliable short-range navigation system was needed. Since radio communication systems based on VHF were being successfully deployed, a decision was made to develop a radio navigation system based on VHF. This system became the VHF omni range (VOR) system and is described in Chapter 10. This system is in widespread use throughout the world today. VOR is the basis of the current network of 'airways' that are used in navigation charts.

Chapter 11 develops this theme with a system for measuring distance to a navigation aid. The advent of radar in the 1940s led to the development of a number of navigation aids including distance measuring equipment (DME). This is a short/medium-range navigation system, often used in conjunction with the VOR system to provide accurate navigation fixes. The system is based on secondary radar principles.

ADF, VOR and DME navigation aids are installed at airfields to assist with approaches to those airfields. These navigation aids cannot however be used for precision approaches and landings. The standard approach and landing system installed at airfields around the world is the instrument landing system (ILS). Chapter 12 describes how the ILS can be used for approach through to Autoland. The ILS uses a combination of VHF and UHF radio waves and has been in operation since 1946.

The advent of computers, in particular the increasing capabilities of integrated circuits using digital techniques, has led to several important advances in aircraft navigation. One example of this is the area navigation system (RNAV); this is described in Chapter 14. Area navigation is a means of combining, or filtering, inputs from one or more navigation sensors and defining positions that are not necessarily co-located with ground-based navigation aids.

A major advance in aircraft navigation came with the introduction of the inertial navigation system (INS); this is the subject of Chapter 15. The inertial navigation system is an autonomous dead reckoning system, i.e., it requires no external inputs or references from ground stations. The system was developed in the 1950s for use by the US military and subsequently the space programmes. Inertial navigation systems (INS) were introduced into commercial aircraft service during the early 1970s. The system is able to compute navigation data such as present position, distance to waypoint, heading, ground speed, wind speed, wind direction, etc. The system does not need radio navigation inputs, and it does not transmit radio frequencies. Being self-contained, the system can be used for long distance navigation over oceans and undeveloped areas of the globe.

Navigation by reference to the stars and planets has been employed since ancient times; aircraft navigators have utilised periscopes to take celestial fixes for long distance navigation. An artificial constellation of navigation aids was initiated in 1973 and referred to as Navstar (navigation system with timing and ranging). This global positioning system (GPS) was developed for use by the US military; it is now widely available for use in many applications including aircraft navigation. Chapter 16 looks at GPS and other global navigation satellite systems that are in use or planned for future deployment.

The term 'navigation' can be applied in both the lateral and vertical senses for aircraft applications. Vertical navigation is concerned with optimising the performance of the aircraft to reduce operating costs; this is the subject of Chapter 17. During the 1980s, lateral navigation and performance management functions were combined into a single system known as the flight management system (FMS). Various tasks previously routinely performed by the crew can now be automated with the intention of reducing crew workload.

Chapter 18 reviews how the planned journey from A to B could be affected by adverse weather conditions. Radar was introduced onto passenger aircraft during the 1950s to allow pilots to identify weather conditions and subsequently reroute around these conditions for the safety and comfort of passengers. A secondary use of weather radar is the terrain-mapping mode that allows the pilot to identify features of the ground, e.g. rivers, coastlines and mountains.

Increasing traffic density, particularly around airports, means that we need a method of air traffic control (ATC) to manage the flow of traffic and maintain safe separation of aircraft. The ATC system is based on secondary surveillance radar (SSR). Ground controllers use the system to address individual aircraft. An emerging ATC technology is ADS-B; this is also covered in Chapter 19.

With ever-increasing air traffic congestion, and the subsequent demands on ATC resources, the risk of a mid-air collision increases. The need for improved traffic flow led to the introduction of the traffic alert and collision avoidance system (TCAS); this is the subject of Chapter 20. TCAS is an automatic surveillance system that helps aircrews and ATC to maintain safe separation of aircraft. TCAS is an airborne system based on secondary radar that interrogates and replies directly with aircraft via a high-integrity datalink. The system is functionally independent of ground stations and alerts the crew if another aircraft comes within a predetermined time to a potential collision.

It is an unfortunate fact of life that the operation of virtually all items of electronic equipment can potentially disturb the operation of other nearby items of electronic equipment. This final chapter deals with electromagnetic compatibility (EMC) and begins by describing the electromagnetic environment within a modern aircraft and its potential for impacting on its increasingly complex avionic systems. The classification, potential sources and effects of electromagnetic interference (EMI) are discussed together with frequency spectral analysis of some typical sources of EMI. Various means of reducing EMI, such as wiring and cabling, grounding, bonding, shielding, and filtering are described. The chapter includes a case study of the risk to an aircraft's altimeter system from interference caused by radiation from 5G masts. This topic has generated considerable debate and controversy and further underlines the need to be fully aware of the electromagnetic environment in which all modern aircraft operate.

The book concludes with four useful appendices, including a comprehensive list of abbreviations and acronyms used with aircraft communications and navigation systems.

The review questions at the end of each chapter are typical of these used in CAA and other examinations. Further examination practice can be gained from the four revision papers given in

Appendix 2. Other features that will be particularly useful if you are an independent learner are the 'Key Points' and 'Test your understanding' questions interspersed throughout the text.

Preface to the second edition

When producing the second edition of the book the authors considered advances in technology as well as suggestions and feedback from readers (students and lecturers) and a review by the National Aerospace Library (NAL), one of the most prestigious aerospace and aeronautical library collections in the world.

New content developed specifically for the second edition included:

- Satellite communications (SATCOM). SATCOM is becoming increasingly important in its ability to facilitate global communication between aircraft and the ground for both crew and passenger use. Chapter 1 was updated accordingly.
- Space weather. The subject of space weather was not included in the first edition but has been given some attention in recent EASA and CAA publications. The UK CAA (CAP 1428) recommends the aviation industry should "initiate educational programmes that provide staff with a greater understanding of the impact of severe space weather events on their operations and to ensure that the risk of extreme space weather is captured in their Safety Management System (SMS)". Conditions on the Sun and in the solar wind, magnetosphere, ionosphere, and thermosphere can influence the performance and reliability of aircraft avionic systems. Space weather can impact on several of the systems included in this book, e.g., HF communications, GNSS and satellite communications. Introductory notes have therefore been included in Chapter 1.
- GPS. Although GPS operates in the same was as it always has, there are more aircraft flying now using GPS for approaches and landings. Chapter 16 was updated accordingly.

- Electronic instruments/displays and Integrated systems. (e.g., GA combined nav/com/GPS). Chapter 17 was updated to reflect these changes.
- Weather radar. Chapter 18 was updated to incorporate additional technologies used on General Aviation (GA) aircraft: lightning detection, and satellite datalink weather services.
- Air traffic control (ATC). ADS-B is currently being introduced in Europe, North America, and other areas worldwide. Chapter 21 has therefore been expanded to include Single European Sky and Future Navigation Systems (FANS).
- Traffic alert and collision avoidance systems. The low-cost FLARM® traffic awareness system for GA is becoming increasingly popular. A new section on FLARM® was added to Chapter 20.

In addition to this important new material the authors attempted to respond to requests and suggestions from readers. For example, a review via the National Aerospace Library noted that the low range radio altimeter (LRRA) was "buried" within Chapter 12 (Instrument landing systems). It was not the intention of the authors to diminish the purpose or value of an LRRA. Although the LRRA is an integral part of ILS/Autoland, the authors acknowledge that radio altimeters are often used as stand-alone systems, e.g., on rotorcraft for low-level operations. Accordingly, the section describing LRRA has been relocated to Chapter 8.

New EASA rules for aircraft maintenance for GA were introduced via EASA 208598 leaflet 03; Flying in the EU'. This leaflet described simpler Part-66 licences for GA aircraft mechanics (B2L and L licences). It also described CS-STAN; this has been introduced for standard changes and standard repairs. These can be incorporated on the aircraft by a mechanic immediately, i.e., there is no need to have it approved by the EASA or by a design organisation. Examples include installation of VHF voice communication systems and FLARM® equipment (as mentioned earlier, the latter has been added as a new section in Chapter 20.)

A new section was added into Chapter 8 for minimum navigation performance specifications

(MNPS) required for remote areas, e.g., in the North Atlantic. This new section also references reduced vertical separation minimums (RVSM); this allows aircraft to fly with a vertical separation of 1000 feet between FL290 and FL410 inclusive. Further information on RVSM is given in another title in this book series, AFIGS. Chapter 8 also includes a new section on electronic flight bags (EFB); these are replacing traditional paper-based documents, e.g., navigation charts.

Several high-profile accidents have occurred since the first edition was published, particularly flight AF447 (June 2009) and flight MH370 (March 2014). These accidents have highlighted shortcomings in the way that flights are currently tracked. The subsequent investigations into these two accidents is beyond the scope of this book, and indeed the book series; however, the subjects covered by the book series are very topical in the context of the accident investigations:

- Air Traffic Control (Chapter 19)
- Emergency Locator Transmitters (Chapter 7)
- Flight Data Recorders (covered in AEES)
- Satellite Communications (Chapter 1)
- Satellite Navigation (Chapter 16).

Aeroplanes flying over land with a high density of population are permanently tracked by air traffic control (ATC) systems, as described in Chapter 19. Aeroplanes flying over remote regions, e.g., oceans, polar routes, etc., will have limited, if any, ATC surveillance. The frequency of position reports by pilots or aircraft to ATC in remote and oceanic airspace is not systematic. Instead, it varies at certain intervals, depending on the density of the airspace and the procedures in place. In order to improve the positioning of aeroplanes in remote areas, the International Civil Aviation Organization (ICAO) has adopted international standards for the:

- location of aircraft and recorders using underwater beacons
- tracking of aircraft
- location system for aircraft in distress
- fast recovery of data from flight recorders.

One technology that has become both an opportunity and a threat is that of remotely piloted aircraft systems (RPAS), unmanned aerial vehicles (UAV), or 'drones'. A recent EASA report, 'Drone Collision' Task Force Final Report (04/10/16), illustrates the increasing number of RPAS occurrences per year between 2010 (virtually zero) and 2015 (circa 500). The report discusses airborne conflict (defined as a potential collision between a drone and an aircraft in the air) being the most common type of occurrence. There are increasing commercial and military requirements for using RPAS in controlled airspace. This subject is not specifically addressed in this new edition. However, some of the technology described in this book could be adapted in the future to account for RPAS. The reader is encouraged to follow industry media on this subject, e.g., mandates and rulemaking.

A recent article published by the RAeS (Aerospace, November 2016) discusses the subject of "remote control tower" technology, i.e., enabling air traffic controllers to manage flights from remote airports without the need for a manned control tower. The authors believe that this will not affect the on-board systems, e.g., VHF communications, transponders, etc., and so the technology is not described in any detail in this second edition. At the time of publishing this second edition, trials are still being conducted; the reader is encouraged to monitor developments via the industry press, e.g., mandates and rulemaking.

Unscheduled maintenance for any aircraft system is more expensive than scheduled maintenance. Furthermore, an avionic unit that is removed, but subsequently tested and found to be serviceable is deemed No Fault Found. The cost of NFFs, i.e., cost of time to remove, logistics costs for returning it to a repair shop, costly test equipment, higher inventories etc. can all be reduced when the maintenance technician has a thorough understanding of systems and equipment. NFFs may also cause expensive flight delays or cancellations. Although NFFs are common with avionic equipment, it is in everyone's interest to minimise occurrences.

Importantly, this second edition introduced "Key Points" covering very basic system testing to give additional understanding of systems. Troubleshooting starts (where possible) with a debriefing from the pilots and asking the right questions. The latter can be greatly enhanced by having a thorough understanding of how the system works, theory of operation, etc.

Finally, please note that, in order to be consistent with the EASA definitions, the book adopts the following terminology:

- 'Aeroplane' means an engine-driven, fixed-wing aircraft heavier than air that is supported in flight by the dynamic reaction of the air against its wings.
- 'Rotorcraft' means a heavier-than-air aircraft that depends principally for its support in flight on the lift generated by one or more rotors.
- 'Helicopter' means a rotorcraft that, for its horizontal motion, depends principally on its engine-driven rotors.
- 'Aircraft' means a machine that can derive support in the atmosphere from the reactions of the air other than the reactions of the air against the earth's surface. In this book, 'aircraft' applies to both aeroplanes and rotorcraft.

Preface to the third edition

With global availability of the internet it is hardly surprising that aircraft manufacturers and operators have looked at adapting the Aircraft Communications and Reporting System (ACARS) for use with Internet Protocol (IP) and, whilst ACARS is introduced in our companion title, Aircraft Digital Electronics and Computer Systems, we have now included an introduction in Chapter 1 together with some examples of ACARS message.

Vector network analysis (VNA) is increasingly used to analyse the performance of antennas and transmission lines as well as a wide variety of RF components. Chapter 2 now includes an introduction to VNA, including the use of Smith Charts for impedance measurement. The chapter also provides an introduction to the use of j-notation for expressing complex impedances. Several practical examples are included.

Software defined radio (SDR) technology is finding its way into the latest avionic equipment where signal processing (including filtering, modulation/demodulation and encoding/decoding) is performed by software rather than hardware. Accordingly, we have included an introduction to SDR in Chapter 3.

Microwave landing systems (MLS) are still in the EASA Part 66 syllabus, but the authors have decided to take it out of the 3rd edition because MLS is now superseded by satellite-based augmentation systems (SBAS). The MLS content of the previous edition of ACNS will be archived in the author's website http://www.key2study.com/66web.

Very low frequency (VLF) navigation is deleted from the EASA Part 66 syllabus and is now considered redundant for commercial aircraft; this chapter has been removed from the 3rd edition. The VLF navigation content of the previous edition of ACNS will be archived in the author's website.

Doppler navigation is deleted from the EASA Part 66 syllabus; however, the authors decided it should be kept in the 3rd edition because the system is still being used, albeit in specialised applications.

Radio altimeter systems still perform a critical function in an aircraft's operation, providing a direct measurement of the aircraft's clearance over the terrain. At the time of publishing the 3rd edition, radio altimeters were being affected by the fifth generation (5G) telecommunications standard for broadband cellular networks. (5G telecommunications began deploying worldwide in 2019.) Chapter 21 of this book gives a description of how 5G can cause interference with avionic systems in general. The radio altimeter is used as a sensor in many aircraft systems, including the instrument landing system (ILS), terrain awareness and collision avoidance (TCAS) and terrain awareness system (TAWS). The ILS and TCAS are described in more detail in this book; TAWS is described in another book title in the series, Aircraft electrical and electronic systems (AEES). The proposed solution for this problem is to incorporate filters in the radio altimeter equipment. Chapter 21 gives more details of 5G interference in the form of a case study. Readers are encouraged to monitor industry developments.

Precision area navigation (P-RNAV) is introduced into the area navigation chapter of the

3rd edition. P-RNAV is the European terminal airspace application evolved from B-RNAV. P-RNAV applications are intended primarily for terminal airspace where there is higher traffic density, terrain clearance considerations etc. These considerations require more accuracy than en-route B-RNAV navigation. The area navigation chapter has also been updated with a description of actual navigation performance (ANP) where the aircraft's navigation system monitors its actual navigation performance during flight.

The global navigation satellite system (GNSS) chapter is updated with a description of how loss of the GNSS sensor can be mitigated by using other navigation systems, e.g., distance measuring equipment (DME). This subject continues in the flight management system chapter with automatic DME tuning and scanning.

Four-dimensional (4D) navigation is introduced in the flight management system (FMS) chapter. This is a key air traffic control function performed by the flight management system. 4D navigation is a fundamental component of the air traffic control (ATC) evolution being undertaken via several programmes (SESAR in Europe and NextGen in the USA); these subjects are both discussed further in the ATC chapter of this book.

The weather radar chapter is updated with a description of solid-state weather radar antennas based on phased array, or electronically steered antenna (ESA) technology. Aside from no moving parts and reduced weight, phased array weather radar systems incorporate several advanced features, e.g., faster scanning, automatic threat analysis that adjusts the antenna's "scan and tilt" patterns to accurately profile weather cells, predictive windshear etc.

The air traffic control (ATC) chapter is updated with numerous subjects and new technology. Automatic dependent surveillance-contract (ADS-C) is a development of ADS-B, where the data exchanges are via an open contract between the air navigation service provider (ANSP) and a specific aircraft. Remote airfield control towers are now being deployed, enabling air traffic controllers to manage flights without the need for staff to be physically present in the airfield control tower. The chapter gives updates to the Single European Sky (SES) initiative; the

FAA's Next Generation Air Transportation System (NextGen); future air navigation system (FANS).

There is an increasing use of unmanned aerial vehicle/systems (UAV/UAS), also known as "drones" for applications in agriculture, surveying, delivery of food, medicines etc. Work continues with integrating drones into national airspace systems. Along with commercial aircraft, drones need to provide tracking and identification whilst in flight. Although drone collision avoidance technology is outside the scope of this book, a brief introduction is given to this subject.

With the increasing use of radio and wireless and networked devices electromagnetic compatibility is becoming increasingly important in the design, operation and maintenance of all items of avionic equipment that must safely co-exist in a modern aircraft environment. Chapter 21 provides an overview of the avionic electromagnetic environment before discussing the effect and sources of electromagnetic interference (EMI). Methods of reducing EMI are discussed and the chapter ends with a useful Case Study based on the introduction of C-band 5G networks and their potential for interference with aircraft radio altimeters.

Acknowledgements

Thanks to Chris Brady for access to his website www.b737.org.uk with its comprehensive content on aircraft systems, in particular the flight management system.

Online resources

Additional resources and interactive materials are available from the book's companion website at www.66web.co.uk.

Chapter 1 Introduction

Maxwell first suggested the existence of electromagnetic waves in 1864. Later, Heinrich Rudolf Hertz used an arrangement of rudimentary resonators to demonstrate the existence of electromagnetic waves. Hertz's apparatus was extremely simple and comprised two resonant loops, one for transmitting and the other for receiving. Each loop acted both as a tuned circuit and as a resonant antenna (or 'aerial').

Hertz's transmitting loop was excited by means of an induction coil and battery. Some of the energy radiated by the transmitting loop was intercepted by the receiving loop, and the received energy was conveyed to a spark gap where it could be released as an arc. The energy radiated by the transmitting loop was in the form of an **electromagnetic wave**—a wave that has both electric and magnetic field components and travels at the speed of light.

In 1894, Marconi demonstrated the commercial potential of the phenomenon that Maxwell predicted and Hertz actually used in his apparatus. It was also Marconi that made radio a reality by pioneering the development of telegraphy without wires (i.e. 'wireless'). Marconi was able to demonstrate very effectively that information could be exchanged between distant locations without the need for a 'landline'.

Marconi's system of **wireless telegraphy** proved to be invaluable for maritime communications (ship to ship and ship to shore) and was to be instrumental in saving many lives. The military applications of radio were first exploited during the First World War (1914 to 1918) and, during that period, radio was first used in aircraft.

This first chapter has been designed to set the scene and to provide you with an introduction to the principles of radio communication systems. The various topics are developed more fully in the later chapters, but the information provided here is designed to provide you with a starting point for the theory that follows.

1.1 The radio frequency spectrum

Radio frequency signals are generally understood to occupy a frequency range that extends from a few tens of kilohertz (kHz) to several hundred gigahertz (GHz). The lowest part of the radio frequency range that is of practical use (below 30 kHz) is only suitable for narrow-band communication. At this frequency, signals propagate as ground waves (following the curvature of the earth) over very long distances. At the other extreme, the highest frequency range that is of practical importance extends above 30 GHz. At these microwave frequencies, considerable bandwidths are available (sufficient to transmit many television channels using point-to-point links or to permit very high definition radar systems), and signals tend to propagate strictly along line-of-sight paths.

At other frequencies signals may propagate by various means including reflection from ionised layers in the ionosphere. At frequencies between 3 MHz and 30 MHz ionospheric propagation regularly permits intercontinental broadcasting and communications.

For convenience, the radio frequency spectrum is divided into a number of bands (see Table 1.1), each spanning a decade of frequency. The use to which each frequency range is put depends upon a number of factors, paramount among which is the propagation characteristics within the band concerned. In addition, various bands are set aside for radar use, and these are designated by letters (see Table 1.2). The use to which each band is put depends largely on how radio waves propagate within the band in question and also on the working range and target resolution that can be achieved.

Other factors that need to be taken into account include the efficiency of practical aerial systems in the range concerned and the bandwidth available. It is also worth noting that, although it may appear from Figure 1.1 that a

DOI: 10.1201/9781003411932-1

Table 1.1   Frequency bands

Frequency range	Wavelength	Designation
300 Hz to 3 kHz	1000 km to 100 km	Extremely low frequency (ELF)
3 kHz to 30 kHz	100 km to 10 km	Very low frequency (VLF)
30 kHz to 300 kHz	10 km to 1 km	Low frequency (LF)
300 kHz to 3 MHz	1 km to 100 m	Medium frequency (MF)
3 MHz to 30 MHz	100 m to 10 m	High frequency (HF)
30 MHz to 300 MHz	10 m to 1 m	Very high frequency (VHF)
300 MHz to 3 GHz	1 m to 10 cm	Ultra high frequency (UHF)
3 GHz to 30 GHz	10 cm to 1 cm	Super high frequency (SHF)

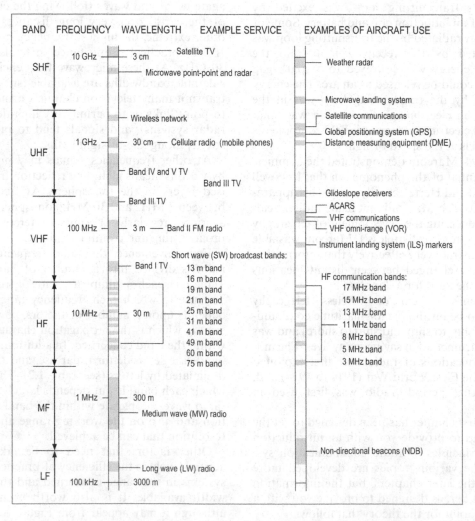

Figure 1.1   Some examples of frequency allocations within the radio frequency spectrum

Table 1.2 Radar bands

Band designation	Frequency	Wavelength	Typical aircraft radar application
L	1 GHz to 2 GHz	30 cm to 15 cm	Distance measuring equipment (DME)
S	2 GHz to 4 GHz	15 cm to 7.5 cm	Airport surveillance radar (ASR)
C	4 GHz to 8 GHz	7.5 cm to 3.75 cm	Microwave landing systems (MLS)
X	8 GHz to 12 GHz	3.75 cm to 2.5 cm	Weather radar (WXR)
K_u	12 GHz to 18 GHz	2.5 cm to 1.67 cm	
K	18 GHz to 26 GHz	1.67 cm to 1.11 cm	
K_a	26 GHz to 40 GHz	1.11 cm to 7.5 mm	Military combat aircraft

great deal of the radio frequency spectrum is not used, it should be stressed that competition for frequency space is fierce and there is, in fact, little vacant space! Frequency allocations are, therefore, ratified by international agreement and the various user services carefully safeguard their own areas of the spectrum.

1.2 Electromagnetic waves

As with light, radio waves propagate outwards from a source of energy (transmitter) and comprise electric (E) and magnetic (H) fields at right angles to one another. These two components, the **E-field** and the **H-field**, are inseparable. The resulting wave travels away from the source with the E and H lines mutually at right angles to the direction of **propagation**, as shown in Figure 1.2.

Radio waves are said to be **polarised** in the plane of the electric (E) field. Thus, if the E-field is vertical, the signal is said to be vertically polarised, whereas, if the E-field is horizontal, the signal is said to be horizontally polarised.

Figure 1.3 shows the electric E-field lines in the space between a transmitter and a receiver. The transmitter aerial (a simple dipole, see page 28) is supplied with a high frequency alternating current. This gives rise to an alternating electric field between the ends of the aerial and an alternating magnetic field around (and at right angles to) it.

The direction of the E-field lines is reversed on each cycle of the signal as the **wavefront** moves outwards from the source. The receiving aerial intercepts the moving field and voltage and current is induced in it as a consequence.

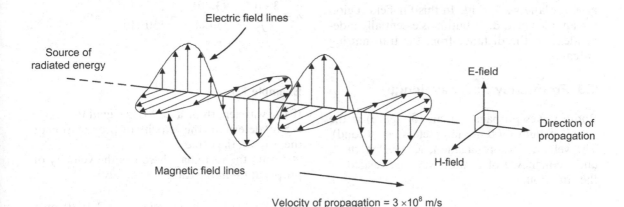

Velocity of propagation = 3×10^8 m/s

Figure 1.2 An electromagnetic wave

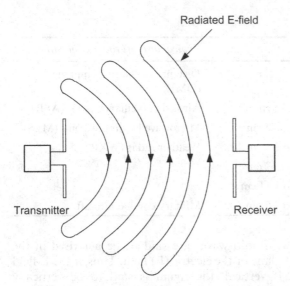

Figure 1.3 Electric field pattern in the near field region between a transmitter and a receiver (the magnetic field has not been shown but is perpendicular to the electric field)

This voltage and current is similar (but of smaller amplitude) to that produced by the transmitter.

Note that in Figure 1.3 (where the transmitter and receiver are close together) the field is shown spreading out in a spherical pattern (this is known more correctly as the **near field**). In practice there will be some considerable distance between the transmitter and the receiver, and so the wave that reaches the receiving antenna will have a plane wavefront. In this far field region the angular field distribution is essentially independent of the distance from the transmitting antenna.

1.3 Frequency and wavelength

Radio waves propagate in air (or space) at the speed of light (300 million meters per second). The velocity of propagation, v, wavelength, λ, and frequency, f of a radio wave are related by the equation:

$$v = f\lambda = 3 \times 10^8 \, \text{m} / \text{s}$$

This equation can be arranged to make f or λ the subject, as follows:

$$f = \frac{3 \times 10^8}{\lambda} \, \text{Hz and } \lambda = \frac{3 \times 10^8}{f} \, \text{m}$$

As an example, a signal at a frequency of 1 MHz will have a wavelength of 300 m, whereas a signal at a frequency of 10 MHz will have a wavelength of 30 m.

When a radio wave travels in a cable (rather than in air or 'free space') it usually travels at a speed that is between 60% and 80% of that of the speed of light.

Example 1.1

Determine the frequency of a radio signal that has a wavelength of 15 m.

Here we will use the formula $f = \dfrac{3 \times 10^8}{\lambda} \, \text{Hz}$

Putting $\lambda = 15$ m gives:

$$f = \frac{3 \times 10^8}{15} = \frac{300 \times 10^6}{15} = 20 \times 10^6 \, \text{Hz or } 20 \, \text{MHz}$$

Example 1.2

Determine the wavelength of a radio signal that has a frequency of 150 MHz.

In this case we will use $\lambda = \dfrac{3 \times 10^8}{f} \, \text{m}$

Putting $f = 150$ MHz gives:

$$\lambda = \frac{3 \times 10^8}{f} = \frac{3 \times 10^8}{150 \times 10^6} = \frac{300 \times 10^6}{150 \times 10^6} = 2 \, \text{m}$$

Example 1.3

If the wavelength of a 30 MHz signal in a cable is 8 m, determine the velocity of propagation of the wave in the cable.

Using the formula where v is the velocity of propagation in the cable, $v = f\lambda$, gives:

$$v = f\lambda = 30 \times 10^6 \times 8 \, \text{m} = 240 \times 10^6 = 2.4 \times 10^8 \, \text{m/s}$$

Test your understanding 1.1

An HF communications signal has a frequency of 25.674 MHz. Determine the wavelength of the signal.

Test your understanding 1.2

A VHF communications link operates at a wavelength of 1.2 m. Determine the frequency at which the link operates.

1.4 The atmosphere

The earth's atmosphere (see Figure 1.4) can be divided into five concentric regions having boundaries that are not clearly defined. These layers, starting with the layer nearest the earth's surface, are known as the troposphere, stratosphere, mesosphere, thermosphere and exosphere.

The boundary between the troposphere and the stratosphere is known as the **tropopause** and this region varies in height above the earth's surface from about 7.5 km at the poles to 18 km at the equator. An average value for the height of the tropopause is around 11 km or 36,000 feet (about the same as the cruising height for most international passenger aircraft).

The thermosphere and the upper parts of the mesosphere are often referred to as the **ionosphere** and it is this region that has a major role to play in the long distance propagation of radio waves, as we shall see later.

The lowest part of the earth's atmosphere is called the **troposphere** and it extends from the surface up to about 10 km (6 miles). The atmosphere above 10 km is called the stratosphere, followed by the mesosphere. It is in the stratosphere that incoming solar radiation creates the ozone layer.

1.5 Radio wave propagation

Depending on a number of complex factors, radio waves can propagate through the atmosphere in various ways, as shown in Figure 1.5. These include:

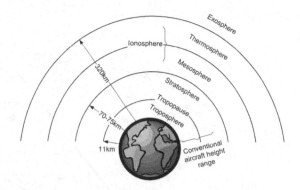

Figure 1.4 Zones of the atmosphere

- ground waves
- ionospheric waves
- space waves
- tropospheric waves.

As their name suggests, **ground waves** (or **surface waves**) travel close to the surface of the earth and propagate for relatively short distances at HF and VHF but for much greater distances at MF and LF. For example, at 100 kHz the range of a ground wave might be in excess of 500 km whilst at 1 MHz (using the same radiated power) the range might be no more than 150 km and at 10 MHz no more than about 15 km. Ground waves have two basic components; a **direct wave** and a **ground reflected wave** (as shown in Figure 1.6). The **direct path** is that which exists on a line-of-sight (**LOS**) basis between the transmitter and receiver. An example of the use of a direct path is that which is used by terrestrial microwave repeater stations which are typically spaced 20 to 30 km apart on a LOS basis. Another example of the direct path is that used for satellite TV reception. In order to receive signals from the satellite the receiving antenna must be able to 'see' the satellite. In this case, and since the wave travels largely undeviated through the atmosphere, the direct wave is often referred to as a **space wave**. Such waves travel over LOS paths at VHF, UHF and beyond.

As shown in Figure 1.6, signals can arrive at a receiving antenna by both the direct path and by means of reflection from the ground. **Ground reflection** depends very much on the quality of the ground, with sandy soils being a poor reflector of radio signals and flat marshy ground

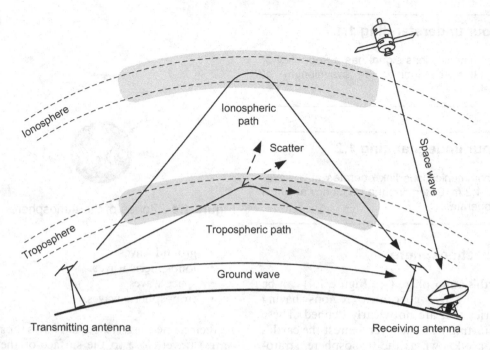

Figure 1.5 Radio wave propagation through the atmosphere

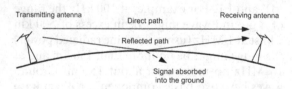

Figure 1.6 Constituents of a ground wave

being an excellent reflecting surface. Note that a proportion of the incident radio signal is absorbed into the ground and not all of it is usefully reflected. An example of the use of a mixture of direct path and ground (or building) reflected radio signals is the reception of FM broadcast signals in a car. It is also worth mentioning that, in many cases, the reflected signals can be stronger than the direct path (or the direct path may not exist at all if the car happens to be in a heavily built-up area).

Ionospheric waves (or **sky waves**) can travel for long distances at MF, HF and exceptionally also at VHF under certain conditions. Such waves are predominant at frequencies below VHF, and we shall examine this phenomenon in greater detail a little later, but before we do it is worth describing what can happen when waves meet

certain types of discontinuity in the atmosphere or when they encounter a physical obstruction. In both cases, the direction of travel can be significantly affected according to the nature and size of the obstruction or discontinuity. Four different effects can occur (see Figure 1.7) and they are known as:

- reflection
- refraction
- diffraction
- scattering.

Reflection occurs when a plane wave meets a plane object that is large relative to the wavelength of the signal. In such cases the wave is reflected back with minimal distortion and without any change in velocity. The effect is similar to the reflection of a beam of light when it arrives at a mirrored surface.

Refraction occurs when a wave moves from one medium into another in which it travels at a different speed. For example, when moving from a more dense to a less dense medium the wave is bent away from the normal (i.e. an imaginary line constructed at right angles to the boundary). Conversely, when moving from a less dense

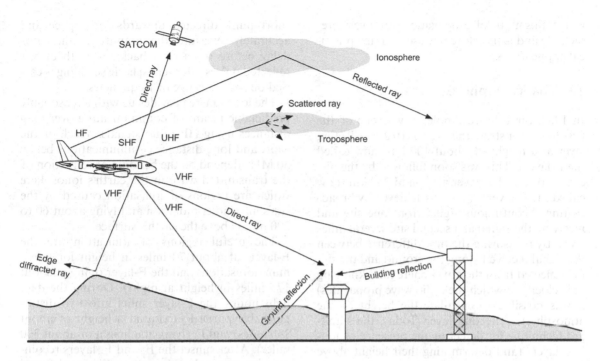

Figure 1.7 Various propagation effects

to a more dense medium, a wave will bend towards the normal. The effect is similar to that experienced by a beam of light when it encounters a glass prism.

Diffraction occurs when a wave meets an edge (i.e. a sudden impenetrable surface discontinuity) which has dimensions that are large relative to the wavelength of the signal. In such cases the wave is bent so that it follows the profile of the discontinuity. Diffraction occurs more readily at lower frequencies (typically VHF and below). An example of diffraction is the bending experienced by VHF broadcast signals when they encounter a sharply defined mountain ridge. Such signals can be received at some distance beyond the 'knife edge' even though they are well beyond the normal LOS range.

Scattering occurs when a wave encounters one or more objects in its path having a size that is a fraction of the wavelength of the signal. When a wave encounters an obstruction of this type it will become fragmented and re-radiated over a wide angle. Scattering occurs more readily at higher frequencies (typically VHF and above) and regularly occurs in the troposphere at UHF and EHF.

Radio signals can also be directed upwards (by suitable choice of antenna) so that signals enter the troposphere or ionosphere. In the former case, signals can be become scattered (i.e. partially dispersed) in the troposphere so that a small proportion arrives back at the ground. **Tropospheric scatter** requires high-power transmitting equipment and high-gain antennas but is regularly used for transmission beyond the horizon particularly where conditions in the troposphere (i.e. rapid changes of temperature and humidity with height) can support this mode of communication. Tropospheric scatter of radio waves is analogous to the scattering of a light beam (e.g. a torch or car headlights) when shone into a heavy fog or mist.

In addition to tropospheric scatter there is also **tropospheric ducting** (not shown in Figure 1.7) in which radio signals can become trapped as a result of the change of refractive index at a boundary between air masses having different temperature and humidity. Ducting usually occurs when a large mass of cold air is overrun by warm air (this is referred to as a temperature inversion). Although this condition may occur frequently in certain parts of the world, this mode of propagation is not very predictable and is therefore not used for any practical applications.

1.6 The ionosphere

In 1924, Sir Edward Appleton was one of the first to demonstrate the existence of a reflecting layer at a height of about 100 km (now called the E-layer). This was soon followed by the discovery of another layer at around 250 km (now called the F-layer). This was achieved by broadcasting a continuous signal from one site and receiving the signal at a second site several miles away. By measuring the time difference between the signal received along the ground and the signal reflected from the atmosphere (and knowing the velocity at which the radio wave propagates) it was possible to calculate the height of the atmospheric reflecting layer. Today, the standard technique for detecting the presence of ionised layers (and determining their height above the surface of the earth) is to transmit a very

short pulse directed upwards into space and accurately measure the amplitude and time delay before the arrival back on earth of the reflected pulses. This **ionospheric sounding** is carried out over a range of frequencies.

The ionosphere provides us with a reasonably predictable means of communicating over long distances using HF radio signals. Much of the short and long distance communications below 30 MHz depend on the bending or refraction of the transmitted wave in the earth's ionosphere which are regions of ionisation caused by the sun's ultraviolet radiation and lying about 60 to 200 miles above the earth's surface.

The useful regions of ionisation are the E-layer (at about 70 miles in height for maximum ionisation) and the F-layer (lying at about 175 miles in height at night). During the daylight hours, the F-layer splits into two distinguishable parts: F_1 (lying at a height of about 140 miles) and F_2 (lying at a height of about 200 miles). After sunset the F_1 and F_2 layers recombine into a single F-layer (see Figures 1.8 and

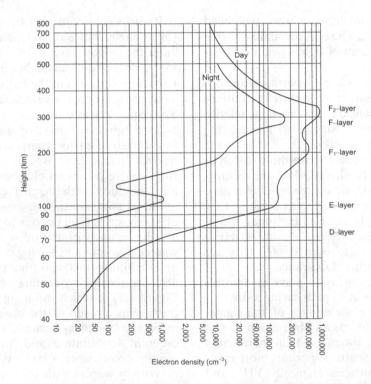

Figure 1.8 Typical variation of electron density versus height (note the use of logarithmic scales for both height and electron density)

1.10). During daylight, a lower layer of ionisation known as the D-layer exists in proportion
to the sun's height, peaking at local noon and
largely dissipating after sunset. This lower layer
primarily acts to absorb energy in the low end
of the high frequency (HF) band. The F-layer
ionisation regions are primarily responsible for
long distance communication using **sky waves** at
distances of up to several thousand km (greatly
in excess of those distances that can be achieved
using VHF **direct wave** communication, see
Figure 1.9). The characteristics of the ionised

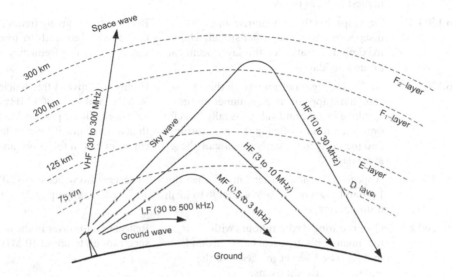

Figure 1.9   Effect of ionised layers on radio signals at various frequencies

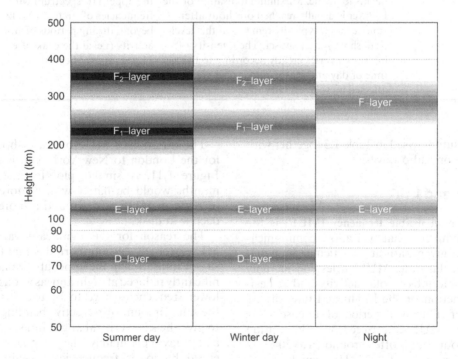

Figure 1.10   Position of ionised layers at day and night

Table 1.3 Ionospheric layers

Layer	Height (km)	Characteristics	Effect on radio waves
D	50 to 95 km	Develops shortly after sunrise and disappears shortly after sunset. Reaches maximum ionisation when the sun is at its highest point in the sky	Responsible for the absorption of radio waves at lower frequencies (e.g. below 4 MHz) during daylight hours
E	95 to 150 km	Develops shortly after sunrise and disappears a few hours after sunset. The maximum ionisation of this layer occurs at around midday	Reflects waves having frequencies less than 5 MHz but tends to absorb radio signals above this frequency
E_s	80 to 120 km	An intense region of ionisation that sometimes appears in the summer months (peaking in June and July). Usually lasts for only a few hours (often in the late morning and recurring in the early evening of the same day)	Highly reflective at frequencies above 30 MHz and up to 300 MHz on some occasions. Of no practical use other than as a means of long-distance VHF communication for radio amateurs
F	250 to 450 km	Appears a few hours after sunset, when the F_1- and F_2-layers (see below) merge to form a single layer	Reflects radio waves up to 20 MHz and occasionally up to 25 MHz
F_1	150 to 200 km	Occurs during daylight hours with maximum ionisation reached at around midday. The F_1-layer merges with the F_2-layer shortly after sunset	Reflects radio waves in the low HF spectrum up to about 10 MHz
F_2	250 to 450 km	Develops just before sunrise as the F-layer begins to divide. Maximum ionisation of the F_2-layer is usually reached one hour after sunrise and it typically remains at this level until shortly after sunset. The intensity of ionisation varies greatly according to the time of day and season and is also greatly affected by solar activity	Capable of reflecting radio waves in the upper HF spectrum with frequencies of up to 30 MHz and beyond during periods of intense solar activity (i.e. at the peak of each 11-year sunspot cycle)

layers are summarised in Table 1.3 together with their effect on radio waves.

1.7 MUF and LUF

The **maximum usable frequency (MUF)** is the highest frequency that will allow communication over a given path at a particular time and on a particular date. MUF varies considerably with the amount of solar activity and is basically a function of the height and intensity of the F-layer. During a period of intense solar activity the MUF can exceed 30 MHz during daylight hours but is often around 16 to 20 MHz by day and around 8 to 10 MHz by night.

The variation of MUF over a 24-hour period for the London to New York path is shown in Figure 1.11. A similar plot for the summer months would be flatter with a more gradual increase in MUF at dawn and a more gradual decline at dusk.

The reason for the significant variation of MUF over any 24-hour period is that the intensity of ionisation in the upper atmosphere is significantly reduced at night and, as a consequence, lower frequencies have to be used to produce the same amount of refractive bending and also to give the same critical angle and skip distance as by day. Fortunately, the attenuation experienced by lower frequencies travelling in the

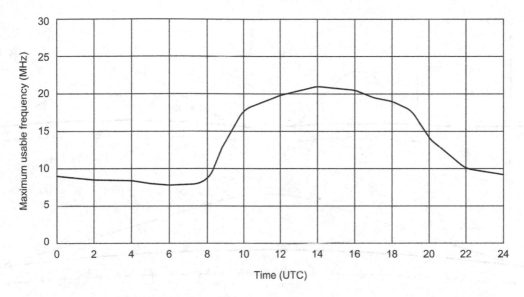

Figure 1.11   Variation of MUF with time for London–New York on 16th October 2006

ionosphere is much reduced at night and this makes it possible to use the lower frequencies required for effective communication. The important fact to remember from this is simply that, for a given path, the frequency used at night is about half that used for daytime communication.

The **lowest usable frequency (LUF)** is the lowest frequency that will support communication over a given path at a particular time and on a particular date. LUF is dependent on the amount of absorption experienced by a radio wave. This absorption is worse when the D-layer is most intense (i.e. during daylight). Hence, as with MUF, the LUF rises during the day and falls during the night. A typical value of LUF is 4 to 6 MHz during the day, falling rapidly at sunset to 2 MHz.

The frequency chosen for HF communication must therefore be somewhere above the LUF and below the MUF for a given path, day and time. A typical example might be a working frequency of 5 MHz at a time when the MUF is 10 MHz and the LUF is 2 MHz.

Figure 1.12 shows the typical MUF for various angles of attack together with the corresponding working ranges. This diagram assumes a **critical frequency** of 5 MHz. This is the lowest frequency that would be returned from the ionosphere using a path of vertical incidence (see ionospheric sounding on page 8).

The relationship between the critical frequency, $f_{crit.}$, and electron density, N, is given by:

$$f_{crit} = 9 \times 10^{-3} \times \sqrt{N}$$

where N is the electron density expressed in electrons per cm³.

The angle of attack, α, is the angle of the transmitted wave relative to the horizon.

The relationship between the MUF, $f_{m.u.f.}$, the critical frequency, $f_{crit.}$, and the angle of attack, α, is given by:

$$f_{m.u.f} = \frac{f_{crit}}{\sin \alpha}$$

Example 1.4

Given that the electron density in the ionosphere is 5×10^5 electrons per cm³, determine the critical frequency and the MUF for an angle of attack of 15°.

Now using the relationship $f_{crit} = 9 \times 10^{-3} \times \sqrt{N}$ gives:

$$f_{crit} = 9 \times 10^{-3} \times \sqrt{5 \times 10^5} = 6.34 \text{ MHz}$$

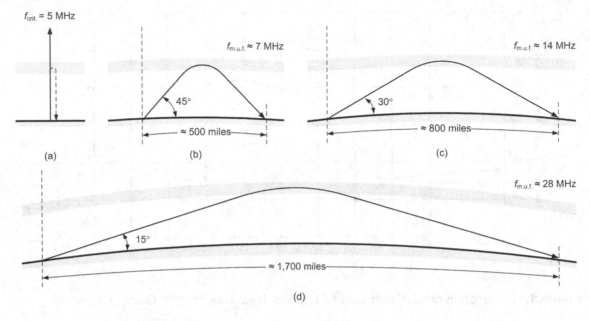

Figure 1.12   Effect of angle of attack on range and MUF

The MUF can now be calculated using:

$$f_{m.u.f.} = \frac{f_{crit}}{\sin \alpha} = \frac{6.364}{\sin 15°} = \frac{6.364}{0.259} = 24.57\,\text{MHz}$$

Test your understanding 1.3

Determine the electron density in the ionosphere when the MUF is 18 MHz for a critical angle of 20°.

1.8 Silent zone and skip distance

The **silent zone** is simply the region that exists between the extent of the coverage of the ground wave signal and the point at which the sky wave returns to earth (see Figure 1.13). Note also that, depending on local topography and soil characteristics, when a signal returns to earth from the ionosphere it is sometimes possible for it to experience a reflection from the ground, as shown in Figure 1.13. The onward reflected signal will suffer attenuation but in some circumstances may be sufficient to provide a further hop and an approximate doubling of the working range. The condition is known as **multi hop** propagation.

The **skip distance** is simply the distance between the point at which the sky wave is radiated and the point at which it returns to earth (see Figure 1.14).

Note that where signals are received simultaneously by ground wave and sky wave paths, the signals will combine both constructively and destructively due to the different path lengths and this, in turn, will produce an effect known as **fading**. This effect can often be heard during the early evening on medium wave radio signals as the D-layer weakens and sky waves first begin to appear in Table 1.4.

Test your understanding 1.4

Table 1.3 shows corresponding values of time and maximum usable frequency (MUF) for London to Lisbon on 28th August 2006. Plot a graph showing the variation of MUF with time and explain the shape of the graph.

1.9 Space weather

Conditions on the Sun and in the solar wind, magnetosphere, ionosphere and thermosphere

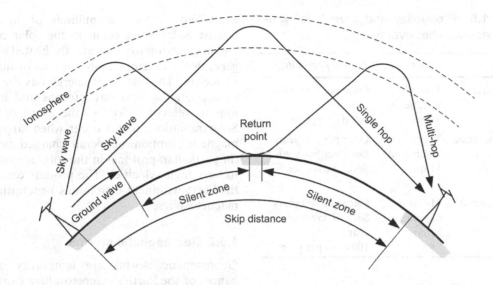

Figure 1.13 Silent zone and skip distance

Figure 1.14 Aircraft flying at high altitudes can be particularly susceptible to the effects of space weather

can have a major impact on the performance and reliability of aircraft avionic systems. Space weather can affect several of the systems described in this book, e.g. HF communications,

satellite navigation and satellite communications. Since modern aircraft fly at a considerable altitude (see Figure 1.14) they can be particularly susceptible to the effects of solar weather, particularly solar radiation.

Test your understanding 1.5

Explain the following terms in relation to HF radio propagation:

(a) silent zone
(b) skip distance
(c) multi-hop propagation.

Space weather is not a new phenomenon; however, the aviation industry continues to evolve, e.g. higher flight levels, reduced lateral/longitudinal spacing, increased use of satellite communications/datalinks, etc. Forecasts and warnings are already standard procedures for

Table 1.4   See Test your understanding 1.4

Time (UTC)	0	1	2	3	4	5	6	7	8	9	10	11	12
MUF (MHz)	9.7	9.0	8.4	7.9	7.5	9.6	12.3	14.1	15.4	16.4	17.1	17.6	17.9
Time (UTC)	13	14	15	16	17	18	19	20	21	22	23	24	
MUF (MHz)	18.1	18.1	17.9	17.7	17.2	16.5	15.6	14.2	12.7	11.5	10.6	9.7	

Table 1.5 Probability and severity of various space weather events

Event	Effect	Typical probability
Solar flares	Major	Once per 100 years
	Significant	Once per year
	Minor	100 days per year
Solar radiation	Major	Once per 100 years
	Significant	Once per five years
	Minor	Several days per year
Geomagnetic	Major	Once per 100 years
	Significant	Several days per year
	Minor	100 days per year

weather events that occur in the troposphere. Pilots also need to be aware of space weather events so that warnings can be given when threshold levels of radiation, communication and navigation disturbance are exceeded. The probability and severity of various space weather events is shown in Table 1.5.

1.9.1 Solar flares

Solar flares (see Figure 1.15) are violent explosions in the Sun's atmosphere with an energy

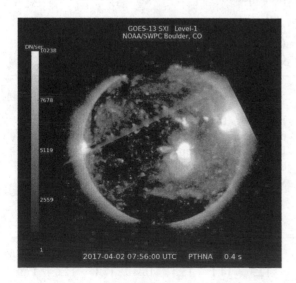

Figure 1.15 M-class solar flare observed on 2nd April 2017

equivalent to tens of millions of hydrogen bombs. Solar flares occur in the solar corona and in the chromosphere, and the heated plasma generated is at temperatures of tens of millions of Kelvin. The release of energy has the effect of accelerating electrons, protons and heavier ions to velocities close to the speed of light. Solar radiation storms occur when large-scale magnetic eruptions accelerate charged particles (in particular, protons) in the solar atmosphere to very high velocities. The protons can reach the Earth within 30–60 minutes, penetrating the magnetosphere.

1.9.2 Geomagnetic storms

Geomagnetic storms are temporary disturbances of the Earth's magnetosphere caused by a solar wind and/or cloud of magnetic field interacting with the Earth's magnetic field. The effect on HF communications, satellite communications and satellite navigation ranges from major disruption through to no significant effect, as illustrated in Table 1.5.

1.9.3 Classification of solar flares

Solar flares are capable of producing electromagnetic radiation across the entire electromagnetic spectrum from long-wave radio to the shortest wavelength Gamma rays. Most flares occur around visible sunspots, where intense magnetic fields emerge from the Sun's surface into the corona. The energy efficiency associated with solar flares may take several hours or even days to build up, but most flares take only a matter of minutes to release their energy.

Depending on the peak flux emitted, solar flares are classified as A, B, C, M or X (see Table 1.6). The peak X-ray flux (in Watts per square metre, W/m^2) is currently monitored by the Geostationary Operational Environmental Satellite (GOES) system operated by the U.S. National Oceanic and Atmospheric Administration (NOAA). Figure 1.16 shows a typical GOES X-ray chart for the period in which the solar flare of Figure 1.15 occurred. Figure 1.17 shows the position and extent of the HF radio blackout produced by the solar flare. Current X-ray charts are freely available from the NOAA/GOES Space Weather Prediction

Table 1.6   Classification of solar flares

Physical measure	Average frequency (1 cycle = 11 years)	Severity	Effect
X20 (2×10^{-3})	Less than 1 per cycle	Extreme	**HF Radio**: Complete HF radio blackout on the entire sunlit side of the Earth lasting for many hours. No HF communication is possible **Navigation**: LF navigation systems experience outages on the sunlit side of the Earth causing loss of positioning information. Increased satellite navigation errors for several hours which may spread into the night side of the Earth
X10 (1×10^{-3})	8 per cycle (8 days per cycle)	Severe	**HF Radio**: HF radio blackout on most of the sunlit side of the Earth lasting for several hours. HF communication can be lost for some time **Navigation**: Outages of LF navigation signals resulting in positioning errors lasting for several hours. Minor disruption of satellite navigation on the sunlit side of the Earth
X1 (1×10^{-4})	175 per cycle (140 days per cycle)	Strong	**HF Radio**: Wide-area blackout of HF radio communication. Loss of radio contact for about one hour on the sunlit side of the Earth **Navigation**: LF navigation signals degraded for about an hour. Negligible effect on satellite systems
M5 (5×10^{-5})	350 per cycle (300 days per cycle)	Moderate	**HF Radio**: Limited blackout of HF radio communication on the sunlit side of the Earth. Loss of radio contact lasting for tens of minutes **Navigation**: Degradation of LF navigation signals lasting for tens of minutes. No impact on satellite systems
M1 (1×10^{-5})	2000 per cycle (950 days per cycle)	Minor	**HF Radio**: Weak or minor degradation of HF radio communication on the sunlit side of the Earth. Occasional loss of radio contact for brief periods **Navigation**: Occasional degradation of LF navigation signals. No impact on satellite systems

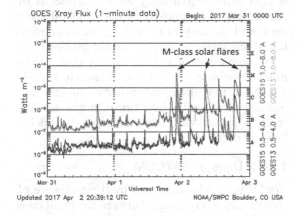

Figure 1.16 GOES X-ray chart showing three M-class flares in a 24-hour period

website at http://www.swpc.noaa.gov/products/goes-x-ray-flux.

When examining the X-ray chart shown in Figure 1.16 you should note that the vertical scale is logarithmic and each class has a peak flux ten times greater than the preceding one, with X-class flares having a peak flux of order 10^{-4} W/m^2. Within a class there is a linear scale from 1 to 9, so an X2 flare is twice as powerful as an X1 flare and is four times more powerful than an M5 flare. The chart in Figure 1.15 shows three M-class solar flares occurring in a 24-hour period. Each of these flares produced an HF radio blackout extending for up to several hours (see Figure 1.17).

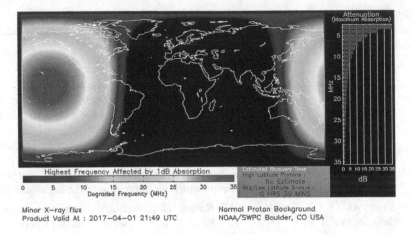

Highest Frequency Affected by 1dB Absorption

0   5   10   15   20   25   30   35
Degraded Frequency (MHz)

Minor X—ray flux
Product Valid At : 2017—04—01 21:49 UTC

Normal Proton Background
NOAA/SWPC Boulder, CO USA

Figure 1.17   Radio blackout resulting from an M-class solar flare (see Figure 1.16)

1.9.4 Coronal mass ejection

A coronal mass ejection (CME) is an ejection of material from the solar corona. The ejected material consists of plasma consisting primarily of electrons and protons (in addition to small quantities of heavier elements such as helium, oxygen and iron), plus the entrained coronal magnetic field. When the solar cloud reaches the Earth as an ICME (Interplanetary CME), it may disrupt the Earth's magnetosphere, compressing it on the dayside and extending the night-side tail. When the magnetosphere reconnects on the night-side, it creates energy that's directed back towards the Earth's upper atmosphere. This process can cause particularly strong aurora, known as the Northern Lights (in the Northern Hemisphere) and the Southern Lights (in the Southern Hemisphere).

CME events, along with solar flares, can disrupt radio transmissions, cause power outages (blackouts) and cause damage to satellites and electrical transmission lines. Fortunately, a major CME directed towards the Earth is a relatively rare occurrence. However, a large CME directed directly towards the earth has the capability of causing major power and telecommunications outages as well as the potential damage and disruption to any satellites and avionics directly in its path.

1.10 Satellite communications (SATCOM)

SATCOM voice and data services facilitate aircraft to ground communication in locations and in situations where conventional HF and VHF radio communication is not possible. Apart from the potential for income generation and improving airline competitiveness, the principal benefits of SATCOM are improved operational efficiency and better traffic management, both of which can be instrumental in achieving considerable cost savings. SATCOM is now being regularly used for airline operational control (AOC) and air traffic control (ATC). SATCOM is also being increasingly used as a means of connecting to ground-based networks allowing crew and passengers to make calls to conventional telephone equipment on the ground.

A comparison of the two largest global communication satellite providers is shown in Table 1.6. Note that these and other companies are making considerable investment in new satellite technology so this information will quickly become dated.

1.10.1 Uses of SATCOM

Traditionally characterized by the exchange of short plain text messages between the flight crew and the ground-based company executives, the

AOC use of SATCOM has increased markedly due to a significant increase in bandwidth and data rate supported by SATCOM systems. This has made it possible to monitor aircraft health and performance in real-time, ensuring that maintenance is provided whenever and wherever it is needed.

The use of SATCOM for ATC leads to improved flight planning, more efficient rerouting procedures and optimised arrivals, all of which have made it safer to have more aircraft present in the same airspace with a level of separation that would have been impossible to safely achieve several decades ago.

Intranet-type data communication is also possible using a SATCOM-based network. This can provide passengers with telephone, Internet and e-mail access. A typical satellite communication system based on Iridium is shown in Figure 1.18. This system uses three principal components:

- satellite network
- ground network (based on gateways)

- subscriber products (including phones and data modems).

1.10.2 The Iridium SATCOM network

The Iridium network allows voice and data messages to be routed anywhere in the world (see Figure 1.18). The system operates in L-band (see Table 1.7). Voice and data messages are relayed from one satellite to another until they reach the receiving terminal; the signal is then relayed back to the gateway. When an Iridium customer places a call from a handset or terminal, it connects to the nearest satellite and is relayed among satellites around the globe to whatever satellite is above the appropriate gateway; this downlinks the call and transfers it to the global public voice network or Internet so that it reaches the recipient. Users can access the network via aircraft earth stations (AES) or Iridium subscriber units (ISU).

The Iridium network comprises 66 active satellites in near polar orbits at an altitude of

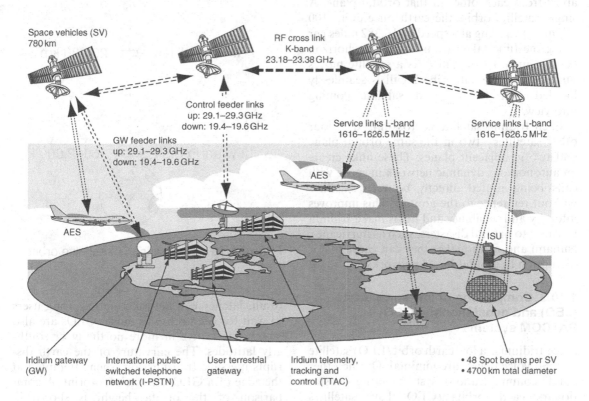

Figure 1.18   Iridium satellite communication system

Table 1.7 Main global communication satellite system providers

Specification	Iridium	Inmarsat
Number of satellites	66 plus multiple in-orbit backup satellites	I3: 5 satellites I4: 4 satellites
Type of orbit	Low-earth (LEO)	Geostationary (GEO)
Orbital height	485 miles	22,000 miles
Orbital period	1 hour, 40 minutes	Stationary
Satellite weight and size	1500 pounds, 14 feet long, 3.5 feet high	13,200 pounds, 23 feet long, 10 feet high (I4)
Spot beams	48 per satellite (30 miles in diameter per beam)	I3: 1 global, 7 spot beams I4: 1 global, 19 regional, 228 narrow
Frequencies	1616–1626.5 MHz (downlink) 1616–1626.5 MHz (uplink)	1525.0–1559.0 MHz 1626.5–1660.5 MHz
Data transmission rate	2.4–10 kbps	Up to 432 kbps+
Satellite life	7 to 9 years	10 years

485 miles (780 km). The satellites fly in formation in six orbital planes, evenly spaced around the planet, each with 11 satellites equally spaced apart from each other in that orbital plane. A single satellite orbits the earth once every 100 minutes, travelling at a speed of 16,382 miles per hour; the time taken from horizon to horizon being a mere ten minutes. As a satellite moves out of view, the subscriber's call is seamlessly handed over to the next satellite coming into view.

Each Iridium satellite is cross-linked to four other satellites—two in the same orbital plane and two in adjacent planes. These links create an autonomous dynamic network in space with calls being routed directly between satellites without reference to the ground. This improves integrity and reliability and helps make Iridium resistant to natural disasters, such as hurricanes, tsunami and earthquakes that can cause major disruption to ground-based networks.

1.10.3 Comparison of low earth orbit (LEO) and geostationary (GEO) SATCOM systems

Since Iridium is a low earth orbit (LEO) satellite system, voice delays are minimal. On the other hand, communications systems using geostationary earth orbits (GEO) have satellites located 22,300 miles above the equator. As a

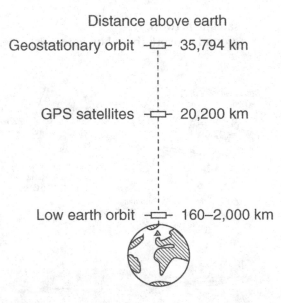

Figure 1.19 Comparison of satellite orbits

result, latency can be quite high, causing users to wait for each other to finish. GEOs are also largely ineffective in more northerly or southerly latitudes. The curvature of the earth disrupts message transmission when attempted at the edge of a GEO satellite's footprint. A comparison of the orbital height is shown in Figure 1.19.

1.10.4 Iridium ground network

The Iridium **ground network** comprises the system control segment and telephony gateways used to connect with the terrestrial telephone system. With centralized management of the Iridium network, the system control segment supplies global operational support and control services for the satellite constellation, delivers satellite tracking data to the gateways and controls the termination of Iridium messaging services. The system control segment consists of three primary components:

- four telemetry tracking and command/control (TTAC) stations
- the operational support network
- the satellite network operation centre (SNOC).

Ku-band (12 to 18 GHz) feeder links and cross-links throughout the satellite constellation supply the connections among the system control segment, the satellites and the gateways. Telephone gateways are the ground-based antennas and electronics that provide voice and data services, messaging and prepaid and post-paid billing services, as well as other customer services. The gateways are responsible for the support and management of mobile subscribers and the interconnection of the Iridium network to the terrestrial phone system. Gateways also provide management functions for their own networks and links.

1.10.5 Aircraft equipment for SATCOM

A typical aircraft SATCOM installation can support datalink channels for packet data services as well as voice channels. Aircraft on-board equipment for SATCOM usually includes a satellite data unit (SDU) together with a power amplifier and an antenna with a steerable beam. The SDU supplies all primary services necessary to provide air/ground communications through satellites (connection protocols and signal conversion). The SATCOM avionics converts voice/data signals into L-band microwave signals (and vice-versa). A diplexer/low noise amplifier (D/LNA) separates transmitted and received signals and amplifies the received signal. A Beam Steering Unit (BSU) controls the pointing of the air-borne antenna.

Operational experience suggests that SATCOM is capable of providing reliable data-link coverage in oceanic and remote continental environments. As a result, it seems likely that SATCOM will make an increasingly significant contribution to air traffic management (ATM) replacing, to a large extent, dependency on VHF and HF radio communication. The Asia/Pacific region, in particular, has been a focus for SATCOM development. However, since SATCOM functionality depends largely on the availability of geostationary satellites, coverage could remain poor in polar regions in which case HFDL (HF data link) might continue to be used for some years to come (see Chapter 5).

1.11 Communication systems integration and management

In recent years, considerable effort has been directed towards the inter-operation and seamless integration of the multiple communication systems present in an aircraft. Figure 1.20 shows the various different communication systems that can be present in a typical modern passenger aircraft. The systems present include HF and VHF radio transceivers, SATCOM and cabin and flight interphone as well as networks that provide crew and passenger access to Internet, 'phone and e-mail services.

The systems shown in Figure 1.20 occupy different locations within an aircraft and thus require a significant amount of cabling. For example, the VHF transceivers that are usually mounted in the avionics bay of a large aircraft will all require externally mounted antennas as well as a means of control from the flight deck. It would make little sense to have multiple headsets, loudspeakers and microphones so a means of switching, routing and controlling the level of audio signals is essential.

The crew can communicate by using HF and VHF radio communication systems as well as the SATCOM system in Figure 1.20. Air-to-ground communication is available in voice and data mode whilst air-to-air communication is available in voice mode only. HF radio provides longdistance coverage while VHF is used for

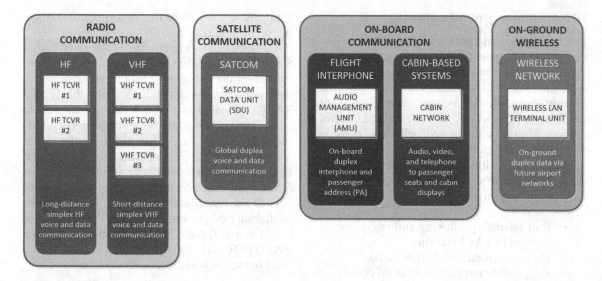

Figure 1.20   Potential communication systems present in modern transport aircraft

relatively short distances. Operation is simplex (i.e. one way at a time) and further details can be found in Chapters 4 and 5 respectively. Note that several HF and VHF transceivers (TCVR) are usually fitted in a large aircraft.

Global voice and data communication is supported by SATCOM (see section 1.10) while on-board communication is facilitated by means of the flight/cabin interphone systems (see Chapter 6) as well as a cabin network that supports data communication for passenger use.

In modern aircraft, high-speed duplex (two-way) communication can be supported by means of a wireless airport communication system (WACS). This, currently 'optional equipment', can provide effective communication when an aircraft is at the gate.

1.11.1 Signal routing and control

Audio signals need to be routed to the different systems capable of voice communications. To communicate outside the aircraft, the operators can use radio communications such as HF and VHF systems or the satellite communication (SATCOM) system. The flight and cabin interphone systems facilitate on-board communications.

Figure 1.21 shows how the audio and control signals are routed for HF and VHF radio communication. This arrangement involves several principal components:

- audio control panel (ACP)
- audio management unit (AMU)
- HF and VHF transceivers
- HF and VHF antennas
- radio management panel (RMP).

1.11.2 Audio management unit (AMU)

The AMU acts as an interface between users and the multiple communications and navigation systems present in the aircraft. The AMU provides the following functions:

- radio transmission
- radio and navigation reception
- visual and aural warnings of calls and alerts from the ground crew and cabin attendants
- flight interphone
- interface with the cockpit voice recorder (CVR)
- SELCAL calls
- emergency function for the Captain and the First Officer.

For transmission, the AMU routes the microphone inputs from the various acoustic devices (hand microphone, boom microphone, etc.) and directs the signals to the radio communication transceivers selected on the audio control panels (ACP).

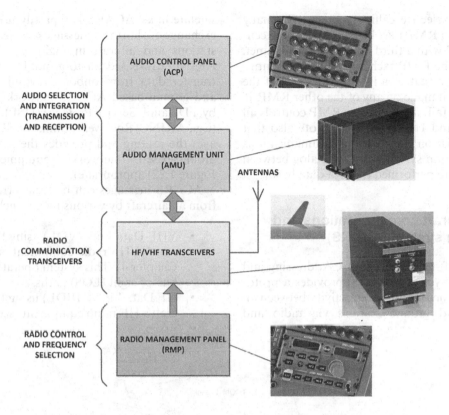

Figure 1.21 Management, control and integration of HF/VHF radio communication

For reception, the AMU collects the audio outputs from the various communications and navigation systems and directs them to the various crew stations and acoustic equipment, as dictated by the current ACP settings.

The flight interphone provides a voice link via telephone handsets between the various crew stations and, when required, between the flight deck and the ground mechanic via an externally accessible power control panel. For further details see Chapter 6.

1.11.3 Audio control panels (ACP)

On large aircraft, audio control panels are provided on the flight deck for the Captain and First Officer as well as a third occupant of the flight deck. Other ACPs may also be installed (for example, one in the avionics bay to facilitate ground service). The ACP are connected to the audio management unit (AMU) via a data bus (e.g. ARINC 429). Each ACP enables:

- the use of various radio communication and radio navigation facilities installed in the aircraft for transmission and reception of the audio signals
- the display of various calls including selective calling (SELCAL), ground crew calls and calls from the Cabin Attendants
- the use of flight, cabin and service interphone systems (see Chapter 6).

1.11.4 Radio management panels (RMP)

Large passenger aircraft usually have several radio management panels (RMP). To transmit, the flight crew uses the audio control panel (ACP) to select a VHF or HF system. The ACP works through the audio management unit (AMU). Each system is connected to the RMPs, for frequency selection, and to the AMU for connection to the audio integrating and

SELCAL (selective calling) systems. On many aircraft, two RMP may be mounted on the centre pedestal with a third on the overhead panel. Note that the RMP exchange data on a continuous basis so that each RMP is informed of the last selection made on any of the other RMP. If two RMPs fail, the remaining RMP controls all the VHF and HF transceivers. Note also that the transmission of data to the communications and navigation systems and the dialog between the RMP are performed through data buses.

1.12 Aircraft communications and reporting system (ACARS)

The Aircraft Communications, Addressing and Reporting System (ACARS) provides a protocol for character-based connectivity between an aircraft and ground stations via radio and satellite links. ACARS is typically used for the exchange of short text messages between ground stations and aircraft in flight. ACARS uses ARINC 618-based air-to-ground protocols to transfer data from onboard avionics systems and ground-based ACARS networks operated by a Datalink Service Provider (DSP). The DSP (e.g., ARINC, Rockwell Collins, or SITA) manages the routing and provides the satellite and ground-based network equipment (see Figure 1.22) appropriate).

ACARS messages can be transferred to and from an aircraft by various means including:

- VHF Data Link (VDL) using the aircraft's VHF radio equipment (see Chapter 4). This system operates over line-of-sight (LOS) paths.
- HF Data Link (HFDL) using the aircraft's HF radio equipment (see

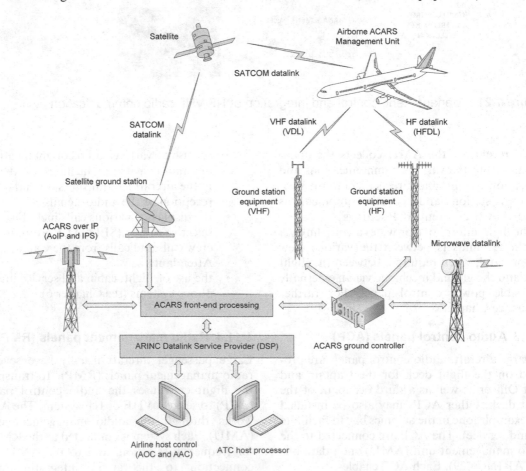

Figure 1.22 ACARS with LOS and BLOS connectivity

Chapter 5). This system supports an extended range, including oceanic and polar regions.

- Satellite Communication (SATCOM) using airborne satellite terminals. Satcom ACARS operates over most of the Earth's surface with the exception of extreme polar regions.

Three basic types of ACARS message are supported:

- Air Traffic Control (ATC) including requests for, and issue of, clearances and other instructions to aircraft.
- Aeronautical Operational Control (AOC) used for operational communications between an aircraft and an airline operator. Typical messages include flight planning, position reporting, and weather updates.
- Airline Administrative Control (AAC). Typical administrative messages include aircraft status, passenger, and seat information.

A typical partially decoded ACARS message is shown in Figure 1.23. The information in the message shows that it was sent on 21/01/2023 at 12:41:34 from G-YMMN (a British Airways Boeing 777-236) with flight number BA0118. The aircraft departed from Kempegowda International Airport, India (VOBL) and its destination is Heathrow Airport, London (EGLL). The message contains the current position of the aircraft (latitude and longitude). Note that although there are many common features, ACARS coding of AOC and AAC data tends to vary with different airlines and operators.

Airborne ACARS equipment comprises a Management Unit (MU) or a Communication Management Unit (CMU) providing the same functionality. This unit supervises the transmission and reception of ACARS messages using the three air-ground sub-networks: HF, VHF, or SATCOM. The MU/CMU determines which of the three sub-networks should be used for optimal routing. The MU/CMU provides the interface with the aircraft's other avionic systems including the flight-deck printer, Flight Management System (FMS), Aircraft Personality Module (APM), etc.

With global availability of the internet, it is hardly surprising that aircraft manufacturers and operators have looked at adapting ACARS for use with Internet Protocol (IP). ACARS over Internet Protocol (AoIP) exploits the availability of broadband cellular networks as well as IP satellite connectivity. AoIP can help reduce operational costs and improve access to ACARS in remote areas. By offering multiple routing options AoIP can improve robustness and fault tolerance. In the future it is expected that AoIP will further transition to Internet Protocol Suite (PS). This can be thought of as a network that combines line-of-sight (LOS) with beyond-line-of-sight (BLOS) sub-networks in a protected spectrum, as shown previously in Figure 1.22.

Key point

In partnership with a data service provider (DSP), ACARS provides character-based data connectivity via HF, VHF and SATCOM datalinks.

Key point

ACARS over IP (AoIP) enables an airline to use text messaging via global networks that were previously unavailable.

```
[#1 (F:0.132 L: -5 E:0) 21/01/2023 12:41:34 -----------------------------
Mode : 2 Label : 15 Id : 3 Ack : X
Aircraft reg: .G-YMMN Flight id: BA0118
No: M07A
FST01VOBLEGLLN515902W0004590070  79 754  M8C
1006126526324911600012501241
############################
Destination Airport : EGLL
Departure Airport : VOBL
```

Figure 1.23 A partially decoded ACARS message

Key point

Figure 1.24 shows four partially decoded ACARS message.

1. What are the registrations and flight numbers of the four aircraft?
2. Over what time interval have the messages been sent?
3. Which of the aircraft is sending a maintenance report?
4. Which of the aircraft is sending its current position?

Key point

It is important to be aware of the characteristics and limitations of radio communication at HF, VHF and by SATCOM. HF radio communication is possible over long distances but can be very dependent on ionospheric conditions at the time. VHF radio communication is generally line-of-sight with limited range determined by aircraft altitude. SATCOM systems are less prone to disturbances and can cater for both voice and data but rely on the availability of one or more satellites that are 'visible' from the aircraft.

```
[#1 (F:0.132 L: -2 E:0) 21/01/2023 12:44:13 ------------------
Mode : 2 Label : B0ld : 2 Ack : X
Aircraft reg: .G-VGBR Flight id: VS0131
No: J89A
/EGTT.AFN/FMHVIR131,.G-VGBR,4067D6,124358/FPON50592W001208,1/
FCOADS,01/FCOATC,010F98

[#1 (F:0.132 L: -6 E:0) 21/01/2023 12:44:12 ------------------
Mode : 2 Label : H1ld : 1 Ack : X
Aircraft reg: .N774AN Flight id: AA0156
No: O03B
ETB

[#1 (F:0.132 L: +0 E:0) 21/01/2023 12:44:10 ------------------
Mode : 2 Label : H1ld : 3 Ack : X
Aircraft reg: .G-VIIT Flight id: BA2205
No: O08K
ETB

[#1 (F:0.132 L: -4 E:0) 21/01/2023 12:44:06 ------------------
Mode : 2 Label : H1ld : 5 Ack : X
Aircraft reg: .N787AL Flight id: AA0173
No: C17F
#CFB0804-02-001
SN 31348
LRU I/O CARD (2) IN P210
LPN 0804-02-001
SN 22815
LRU I/O CARD (3) IN P210
```

Figure 1.24   See Test your understanding 1.6

1.13 Multiple choice questions

1. A transmitted radio wave will have a plane wavefront:
   (a) in the near field
   (b) in the far field
   (c) close to the antenna.

2. The lowest layer in the earth's atmosphere is:
   (a) the ionosphere
   (b) the stratosphere
   (c) the troposphere.

3. A radio wave at 115 kHz is most likely to propagate as:
   (a) a ground wave
   (b) a sky wave
   (c) a space wave.

4. The height of the E-layer is approximately:
   (a) 100 km
   (b) 200 km
   (c) 400 km.

5. When a large mass of cold air is overrun by warm air the temperature inversion Produced will often result in:
   (a) ionospheric reflection
   (b) stratospheric refraction
   (c) tropospheric ducting.

6. Ionospheric sounding is used to determine:
   (a) the maximum distance that a ground wave will travel
   (b) the presence of temperature inversions in the upper atmosphere
   (c) the critical angle and maximum usable frequency for a given path.

7. The critical frequency is directly proportional to:
   (a) the electron density
   (b) the square of the electron density
   (c) the square root of the electron density.

8. The MF range extends from:
   (a) 300 kHz to 3 MHz
   (b) 3 MHz to 30 MHz
   (c) 30 MHz to 300 MHz.

9. A radio wave has a frequency of 15 MHz. Which one of the following gives the wavelength of the wave?
   (a) 2 m
   (b) 15 m
   (c) 20 m.

10. Which one of the following gives the velocity at which a radio wave propagates?
    (a) 300 m/s
    (b) 3 × 108 m/s
    (c) 3 million m/s.

11. The main cause of ionisation in the upper atmosphere is:
    (a) solar radiation
    (b) ozone
    (c) currents of warm air.

12. The F_2-layer is:
    (a) higher at the equator than at the poles
    (b) lower at the equator than at the poles
    (c) the same height at the equator as at the poles.

13. The free-space path loss experienced by a radio wave:
    (a) increases with frequency but decreases with distance
    (b) decreases with frequency but increases with distance
    (c) increases with both frequency and distance.

14. For a given HF radio path, the MUF changes most rapidly at:
    (a) mid-day
    (b) mid-night
    (c) dawn and dusk.

15. Radio waves tend to propagate mainly as line of sight signals in the:
    (a) MF band
    (b) HF band
    (c) VHF band.

16. In the HF band radio waves tend to propagate over long distances as:
    (a) ground waves
    (b) space waves
    (c) ionospheric waves.

17. The maximum distance that can be achieved from a single-hop reflection from the F-layer is in the region:
    (a) 500 to 2000 km
    (b) 2000 to 3500 km
    (c) 3500 to 5000 km.

18. The F_1- and F_2-layers combine:
    (a) only at about mid-day
    (b) during the day
    (c) during the night.

19. The path of a VHF or UHF radio wave can be bent by a sharply defined obstruction such as a building or a mountain top. This phenomenon is known as:
    (a) ducting
    (b) reflection
    (c) diffraction.

20. Radio waves at HF can be subject to reflections in ionised regions of the upper atmosphere. This phenomenon is known as:
    (a) ionospheric reflection
    (b) tropospheric scatter
    (c) atmospheric ducting.

21. Radio waves at UHF can sometimes be subject to dispersion over a wide angle in regions of humid air in the atmosphere. This phenomenon is known as:
    (a) ionospheric reflection
    (b) tropospheric scatter
    (c) atmospheric ducting.

22. Radio waves at VHF and UHF can sometimes propagate for long distances in the lower atmosphere due to the presence of a temperature inversion. This phenomenon is known as:
    (a) ionospheric reflection
    (b) tropospheric scatter
    (c) atmospheric ducting.

23. The layer in the atmosphere that is mainly responsible for the absorption of MF radio waves during the day is:
    (a) the D-layer
    (b) the E-layer
    (c) the F-layer.

24. The layer in the atmosphere that is mainly
    responsible for the reflection of HF radio
    waves during the day is:
    (a) the D-layer
    (b) the E-layer
    (c) the F-layer.

25. On the sunlit side of the Earth, an M-class
    solar flare is likely to cause:
    (a) negligible impact on HF radio
        communication
    (b) minor or moderate degradation of HF
        radio communication
    (c) complete blackout of HF radio
        communication over a wide area.

26. Low earth orbiting (LEO) satellites operate
    at an altitude of approximately:
    (a) 485 miles
    (b) 1800 miles
    (c) 22,000 miles.

27. ACARS messages can be transferred to
    and from an aircraft:
    (a) only using VHF radio equipment
    (b) only using HF and VHF radio
        equipment
    (c) using HF and VHF radio and
        SATCOM datalinks.

Chapter 2   Antennas

It may not be apparent from an inspection of the external profile of an aircraft that most large aircraft carry several dozen antennas of different types. To illustrate this point, Figure 2.1 shows just a few of the antennas carried by a Boeing 757. What should be apparent from this is that many of the antennas are of the low-profile variety which is essential to reduce drag.

Antennas are used both for transmission and reception. A transmitting antenna converts the high frequency electrical energy supplied to it into electromagnetic energy which is launched or radiated into the space surrounding the antenna. A receiving antenna captures the electromagnetic energy in the surrounding space and converts this into high frequency electrical energy which is then passed on to the receiving system. The **law of reciprocity** indicates that an antenna will have the same gain and directional properties when used for transmission as it does when used for reception.

2.1 The isotropic radiator

The most fundamental form of antenna (which cannot be realised in practice) is the isotropic radiator. This theoretical type of antenna is often used for comparison purposes and as a reference when calculating the gain and directional characteristics of a real antenna.

Isotropic antennas radiate uniformly in all directions. In other words, when placed at the centre of a sphere such an antenna would illuminate the internal surface of the sphere uniformly, as shown in Figure 2.2(a). All practical antennas have directional characteristics as illustrated in Figure 2.2(b). Furthermore, such characteristics may be more or less pronounced according to the antenna's application. We shall look at antenna gain and directivity in more detail later on, but before we do that we shall introduce you to some common types of antenna.

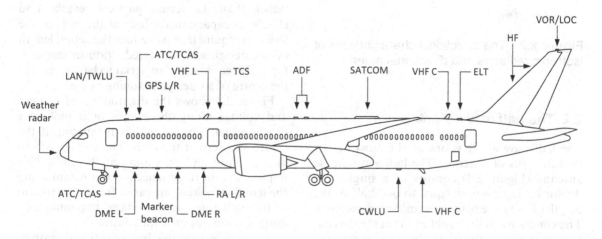

Figure 2.1   Typical antenna locations. 1, VOR; 2, HF comms.; 3, 5 and 6, VHF comms.; 4, ADF; 7, TCAS (upper); 8, weather radar

DOI: 10.1201/9781003411932-2

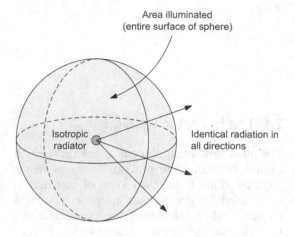

(a) Isotropic radiator

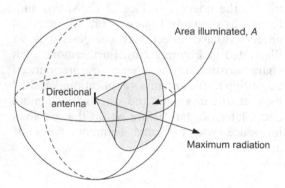

(b) Directional antenna

Figure 2.2 The directional characteristics of isotropic radiators and directional antennas

2.2 The half-wave dipole

The **half-wave dipole** is one of the most fundamental types of antenna. The half-wave dipole antenna (Figure 2.3) consists of a single conductor having a length equal to one-half of the length of the wave being transmitted or received. The conductor is then split in the centre to enable connection to the feeder. In practice, because of the capacitance effects between the ends of the an tenna and ground, the antenna is cut a little shorter than a half wavelength.

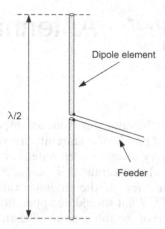

Figure 2.3 A half-wave dipole antenna

The length of the antenna (from end to end) is equal to one half wavelength, hence:

$$l = \frac{\lambda}{2}$$

Now since $v = f \times \lambda$ we can conclude that, for a half-wave dipole:

$$l = \frac{v}{2f}$$

Note that l is the **electrical length** of the antenna rather than its actual **physical length**. End effects, or capacitance effects at the ends of the antenna require that we reduce the actual length of the aerial, and a 5% reduction in length is typically required for an aerial to be resonant at the centre of its designed tuning range.

Figure 2.4 shows the distribution of current and voltage along the length of a half-wave dipole aerial. The current is maximum at the centre and zero at the ends. The voltage is zero at the centre and maximum at the ends. This implies that the impedance is not constant along the length of aerial but varies from a maximum at the ends (maximum voltage, minimum current) to a minimum at the centre.

The dipole antenna has directional properties, illustrated in Figures 2.5 to 2.7. Figure 2.5 shows the radiation pattern of the antenna in the plane of the antenna's electric field (i.e. the

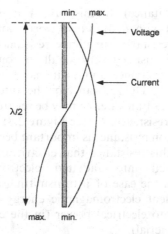

Figure 2.4 Voltage and current distribution in the half-wave dipole antenna

E-field plane) whilst Figure 2.6 shows the radiation pattern in the plane of the antenna's magnetic field (i.e. the H-field plane).

The 3D plot shown in Figure 2.7 combines these two plots into a single 'doughnut' shape. Things to note from these three diagrams are that:

- in the case of Figure 2.5, minimum radiation occurs along the axis of the antenna whilst the two zones of maximum radiation are at 90° (i.e. are 'normal to') the dipole elements

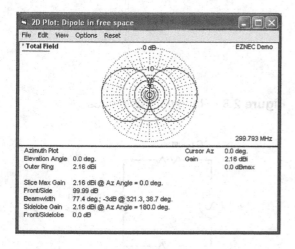

Figure 2.5 E-field polar radiation pattern for a half-wave dipole

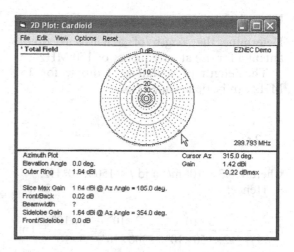

Figure 2.6 H-field polar radiation pattern for a half-wave dipole

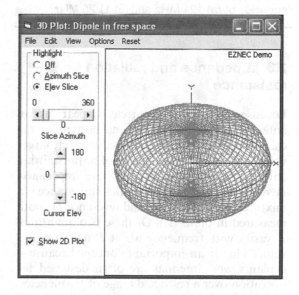

Figure 2.7 3D polar radiation pattern for a half-wave dipole (note the 'doughnut' shape)

- in the case of Figure 2.6, the antenna radiates uniformly in all directions.

Hence, a vertical dipole will have **omnidirectional** characteristics whilst a horizontal dipole will have a **bi-directional** radiation pattern. This is an important point as we shall see later.

Example 2.1

Determine the length of a half-wave dipole antenna for use at a frequency of 150 MHz.

The length of a half-wave dipole for 150 MHz can be determined from:

$$l = \frac{v}{2f}$$

where $v = 3 \times 10^8$ m/s and $f = 150 \times 10^6$ Hz.

Hence:

$$l = \frac{v}{2f} = \frac{3 \times 10^8}{2 \times 150 \times 10^6} = \frac{3 \times 10^8}{300 \times 10^6} = \frac{3 \times 10^6}{3 \times 10^6} = 1\,\text{m}$$

Test your understanding 2.1

Determine the length of a half-wave dipole for frequencies of (a) 121 MHz and (b) 11.25 MHz.

2.3 Impedance and radiation resistance

Because voltage and current appear in an antenna (a minute voltage and current in the case of a receiving antenna and a much larger voltage and current in the case of a transmitting antenna) an aerial is said to have **impedance**. Here it's worth remembering that impedance is a mixture of resistance, R, and reactance, X, both measured in ohms (Ω). Of these two quantities, X varies with frequency whilst R remains constant. This is an important concept because it explains why antennas are often designed for operation over a restricted range of frequencies.

The impedance, Z, of an aerial is the ratio of the voltage, E, across its terminals to the current, I, flowing in it. Hence:

$$Z = \frac{E}{I}\,\Omega$$

You might infer from Figure 2.7 that the impedance at the centre of the half-wave dipole should be zero. In practice the impedance is usually between 70 Ω and 75 Ω. Furthermore, at resonance the impedance is purely resistive and contains no reactive component (i.e. inductance

and capacitance). In this case X is negligible compared with R. It is also worth noting that the DC resistance (or **ohmic resistance**) of an antenna is usually very small in comparison with its impedance and so it may be ignored. Ignoring the DC resistance of the antenna, the impedance of an antenna may be regarded as its **radiation resistance**, R_r (see Figure 2.8).

Radiation resistance is important because it is through this resistance that electrical power is transformed into radiated electromagnetic energy (in the case of a transmitting antenna) and incident electromagnetic energy is transformed into electrical power (in the case of a receiving aerial).

The equivalent circuit of an antenna is shown in Figure 2.9. The three series-connected components that make up the antenna's impedance are:

- the DC resistance, $R_{d.c.}$
- the radiation resistance, R_r
- the 'off-tune' reactance, X.

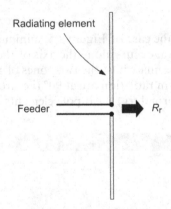

Figure 2.8   Radiation resistance

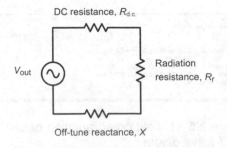

Figure 2.9   Equivalent circuit of an antenna

Note that when the antenna is operated at a frequency that lies in the centre of its pass-band (i.e. when it is *on-tune*) the off-tune reactance is zero. It is also worth bearing in mind that the radiation resistance of a half-wave dipole varies according to its height above ground. The 70 Ω to 75 Ω impedance normally associated with a half-wave dipole is only realised when the antenna is mounted at an elevation of 0.2 wavelengths, or more.

Test your understanding 2.2

A half-wave dipole is operated at its centre frequency (zero off-tune reactance). If the antenna has a total DC loss resistance of 2.5 Ω and is supplied with a current of 2 A and a voltage of 25 V, determine:

(a) the radiation resistance of the antenna
(b) the power loss in the antenna.

2.4 Radiated power and efficiency

In the case of a transmitting antenna, the radiated power, P_r, produced by the antenna is given by:

$$P_r = I_a^2 \times R_r \text{ W}$$

where I_a is the antenna current, in amperes, and R_r is the radiation resistance in ohms. In most practical applications it is important to ensure that P_r is maximised, and this is achieved by ensuring that R_r is much larger than the DC resistance of the antenna elements.

The efficiency of an antenna is given by the relationship:

$$\text{Radiation efficiency} = \frac{P_r}{P_r} + P_{\text{loss}} \times 100\%$$

Where P_{loss} is the power dissipated in the DC resistance present. At this point it is worth stating that, whilst efficiency is vitally important in the case of a transmitting antenna, it is generally unimportant in the case of a receiving antenna. This explains why a random length of wire can make a good receiving aerial but not a good transmitting antenna!

Example 2.2

An HF transmitting antenna has a radiation resistance of 12 Ω. If a current of 0.5 A is supplied to it, determine the radiated power.

Now:

$$P_r = I_a^2 \times R_r = (0.5)^2 \times 12 = 0.25 \times 12 = 3 \text{ W}$$

Example 2.3

If the aerial in Example 2.2 has a DC resistance of 2 Ω, determine the power loss and the radiation efficiency of the antenna.

From the equivalent circuit shown in Figure 2.9, the same current flows in the DC resistance, $R_{\text{d.c.}}$, as flows in the antenna's radiation resistance, R_r.

Hence $I_a = 0.5$ A and $R_{\text{d.c.}} = 2$ Ω
Since $P_{\text{loss}} = I_a^2 \times R_{\text{d.c.}}$

$$P_{\text{loss}} = (0.5)^2 \times 2 = 0.25 \times 2 = 0.5 \text{ W}$$

The radiation efficiency is given by:

$$\text{Radiation efficiency} = \frac{P_r}{P_r} + P_{\text{loss}} \times 100\%$$

$$= \frac{4}{4+0.5} \times 100\% = \frac{4}{4.5} \times 100\% = 89\%$$

In this example, more than 10% of the power output is actually wasted! It is also worth noting that in order to ensure a high value of radiation efficiency, the loss resistance must be kept very low in comparison with the radiation resistance.

2.5 Antenna gain

The field strength produced by an antenna is proportional to the amount of current flowing in it. However, since different types of antenna produce different values of field strength for the same applied RF power level, we attribute a power gain to the antenna. This power gain is specified in relation to a **reference antenna** (often either a half-wave dipole or a theoretical isotropic radiator), and it is usually specified in decibels (dB)—see Appendix 4.

In order to distinguish between the two types of reference antenna we use subscripts **i** and **d** to denote isotropic and half-wave dipole reference antennas respectively. As an example, an aerial having a gain of 10 dBi produces ten times power gain when compared with a theoretical isotropic radiator. Similarly, an antenna having a gain of 13 dBd produces twenty times power gain when compared with a half-wave dipole. Putting this another way, to maintain the same field strength at a given point, you would have to apply 20 W to a half-wave dipole or just 1 W to the antenna in question! Some comparative values of antenna gain are shown on page 40.

2.6 The Yagi beam antenna

Originally invented by two Japanese engineers, Yagi and Uda, the Yagi antenna has remained extremely popular in a wide variety of applications and, in particular, for fixed domestic FM radio and TV receiving aerials. In order to explain in simple terms how the Yagi antenna works we shall use a simple light analogy.

An ordinary filament lamp radiates light in all directions. Just like an antenna, the lamp converts electrical energy into electromagnetic energy. The only real difference is that we can *see* the energy that it produces!

The action of the filament lamp is comparable with our fundamental dipole antenna. In the case of the dipole, electromagnetic radiation will occur all around the dipole elements (in three dimensions the radiation pattern will take on a doughnut shape). In the plane that we have shown in Figure 2.10(c), the directional pattern will be a figure-of-eight that has two lobes of equal size. In order to concentrate the radiation into just one of the radiation lobes we could simply place a reflecting mirror on one side of the filament lamp as shown in Figure 2.11. The radiation will be reflected (during which the reflected light will undergo a 180° phase change), and this will reinforce the light on one side of the filament lamp. In order to achieve the same effect in our antenna system we need to place a conducting element about one quarter of a wavelength behind the dipole element. This element is referred to as a **reflector** and it is said to be 'parasitic' (i.e. it is not actually connected to the

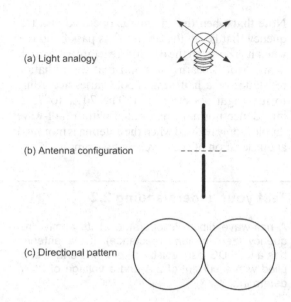

(a) Light analogy

(b) Antenna configuration

(c) Directional pattern

Figure 2.10   Dipole antenna light analogy

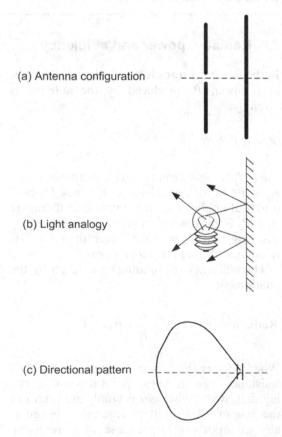

(a) Antenna configuration

(b) Light analogy

(c) Directional pattern

Figure 2.11   Light analogy for a dipole and reflector

Antennas 33

feeder). The reflector needs to be cut slightly longer than the driven dipole element. The resulting directional pattern will now only have one **major lobe** because the energy radiated will be concentrated into just one-half of the figure-of-eight pattern that we started with).

Continuing with our optical analogy, in order to further concentrate the light energy into a narrow beam we can add a lens in front of the lamp. This will have the effect of bending the light emerging from the lamp towards the normal line (see Figure 2.12). In order to achieve the same effect in our antenna system we need to place a conducting element, known as a **director**, on the other side of the dipole and about one-quarter of a wavelength from it. Once again, this element is parasitic, but in this case it needs to be cut slightly shorter than the driven dipole element. The resulting directional pattern will now have a narrower major lobe as the energy becomes concentrated in the normal direction (i.e. at right angles to the dipole elements). The resulting antenna is known as a three-element Yagi aerial, see Figure 2.13.

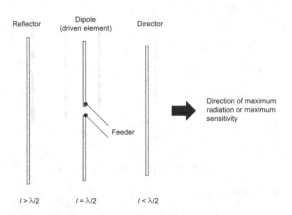

Figure 2.13 A three-element Yagi antenna

If desired, additional directors can be added to further increase the gain and reduce the **beamwidth** (i.e. the angle between the halfpower or –3 dB power points on the polar characteristic) of Yagi aerials (see Figures 2.14 and 2.15). Some comparative gain and beamwidth figures are shown on page 40.

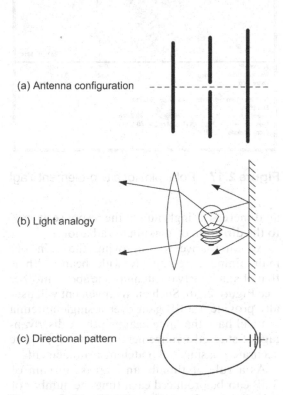

Figure 2.12 Light analogy for a dipole, reflector and director

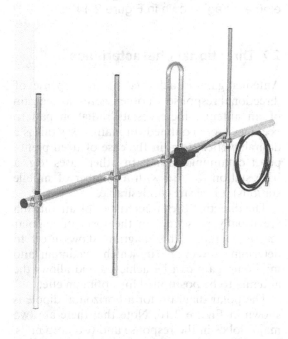

Figure 2.14 A four-element Yagi antenna (note how the dipole element has been 'folded' in order to increase its impedance and provide a better match to the 50 Ω feeder system)

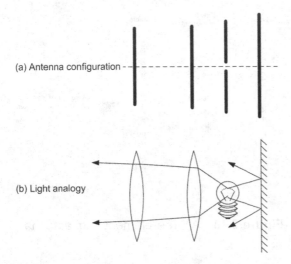

(a) Antenna configuration

(b) Light analogy

(c) Directional pattern

Figure 2.15 Light analogy for the four-element Yagi shown in Figure 2.14

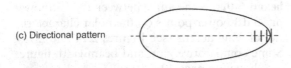

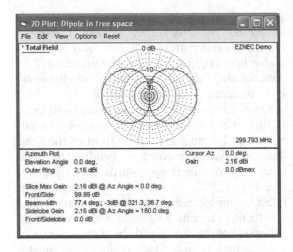

Figure 2.16 Polar plot for a horizontal dipole

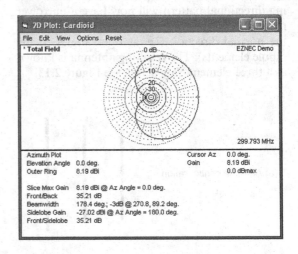

Figure 2.17 Polar plot for a two-element Yagi

2.7 Directional characteristics

Antenna gain is achieved at the expense of directional response. In other words, as the gain of an antenna increases its radiation pattern becomes more confined. In many cases this is a desirable effect (e.g. in the case of fixed point–point communications). In other cases (e.g. a base station for use with a number of mobile stations) it is clearly undesirable.

The directional characteristics of an antenna are usually presented in the form of a polar response graph. This diagram allows users to determine directions in which maximum and minimum gain can be achieved and allows the antenna to be positioned for optimum effect.

The polar diagram for a horizontal dipole is shown in Figure 2.16. Note that there are two major lobes in the response and two deep nulls. The antenna is thus said to be bi-directional.

Figure 2.17 shows the polar diagram for a dipole plus reflector. The radiation from this antenna is concentrated into a single major lobe,

an d there is a single null in the response at 180° to the direction of maximum radiation.

An alternative to improving the gain but maintaining a reasonably wide beamwidth is that of stacking two antennas one above another (see Figure 2.20). Such an arrangement will usually provide a 3 dB gain over a single antenna but will have the same beamwidth. A disadvantage of stacked arrangements is that they require accurate phasing and matching arrangements.

As a rule of thumb, an increase in gain of 3 dB can be produced each time the number of elements is doubled. Thus a two-element antenna will offer a gain of about 3 dBd, a four-element antenna will produce 6 dBd, an eight-element Yagi will realise 9 dBd, and so on.

Test your understanding 2.3

Identify the antenna shown in Figure 2.19. Sketch a typical horizontal radiation pattern for this antenna.

Test your understanding 2.4

Identify the antenna shown in Figure 2.20. Sketch a typical horizontal radiation pattern for this antenna.

Test your understanding 2.5

Figure 2.18 shows the polar response of a Yagi beam antenna (the gain has been specified relative to a standard reference dipole). Use the polar plot to determine:

(a) the gain of the antenna
(b) the beamwidth of the antenna
(c) the size and position of any 'side lobes'
(d) the 'front-to-back' ratio (i.e. the size of the major lobe in comparison to the response of the antenna at 180° to it).

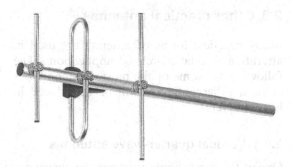

Figure 2.19   See Test your understanding 2.3

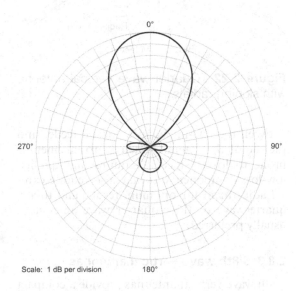

Figure 2.18   Polar diagram for a Yagi beam antenna (see Test your understanding 2.5)

Figure 2.20   See Test your understanding 2.4

2.8 Other practical antennas

Many practical forms of antenna are used in aircraft and aviation-related applications. The following are some of the most common types (several other antennas will be introduced in later chapters).

2.8.1 Vertical quarter-wave antennas

One of the most simple antennas to construct is the quarter-wave antenna (also known as a **Marconi antenna**). Such antennas produce an omnidirectional radiation pattern in the horizontal plane and radiate vertically polarised signals. Practical quarter-wave antennas can be produced for the high-HF and VHF bands, but their length is prohibitive for use on the low-HF and LF bands.

In order to produce a reasonably flat radiation pattern (and prevent maximum radiation being directed upwards into space) it is essential to incorporate an effective ground plane. At VHF, this can be achieved using just four quarter-wave radial elements at 90° to the vertical radiating element (see Figure 2.21). All four radials are grounded at the feed-point to the outer screen of the coaxial feeder cable.

A slight improvement on the arrangement in Figure 2.21 can be achieved by sloping the radial elements at about 45° (see Figure 2.22). This arrangement produces a flatter radiation pattern.

At HF rather than VHF, the ground plane can be the earth itself. However, to reduce the earth resistance and increase the efficiency of the antenna, it is usually necessary to incorporate some buried earth radials (see Figure 2.23). These radial wires simply consist of quarter-wave lengths of insulated stranded copper wire grounded to the outer screen of the coaxial feeder at the antenna feed point.

2.8.2 Vertical half-wave antennas

An alternative to the use of a quarter-wave radiating element is that of a half-wave element. This type of antenna must be *voltage fed* (rather than *current fed* as is the case with the quarter-wave antenna). A voltage-fed antenna requires the use of a resonant transformer connected

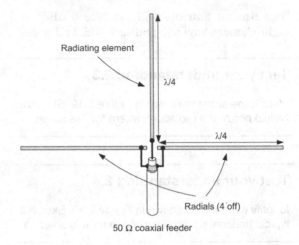

Figure 2.21   Quarter-wave vertical antenna

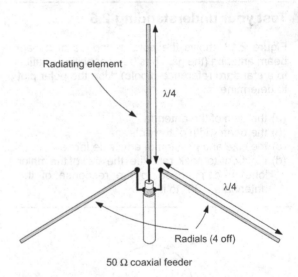

Figure 2.22   Quarter-wave vertical antenna with sloping radials

between the low-impedance coaxial feeder and the end of the antenna. Such an arrangement is prone to losses since it requires high-quality, low-loss components. It may also require careful adjustment for optimum results and thus a quarterwave or three-quarter wave antenna is usually preferred.

2.8.3 5/8th wave vertical antennas

5/8th wave vertical antennas provide a compact solution to the need for an omnidirectional

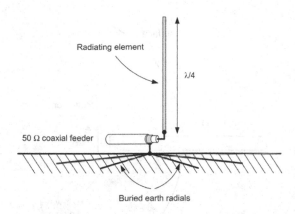

Figure 2.23 Quarter wave vertical antenna with buried earth radials

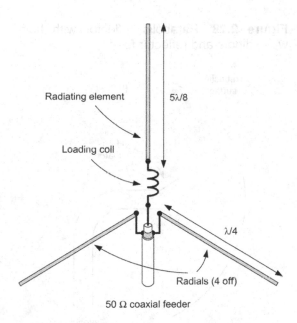

Figure 2.24 5/8th wave vertical antenna with sloping radials

VHF/UHF antenna offering some gain over a basic quarter-wave antenna. In fact, a 5/8th wave vertical antenna behaves electrically as a three-quarter wave antenna (i.e. it is current fed from the bottom and there is a voltage maximum at the top). In order to match the antenna, an inductive loading coil is incorporated at the feed-point. A typical 5/8th wave vertical antenna with sloping ground plane is shown in Figure 2.24.

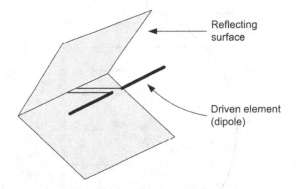

Figure 2.25 High-gain corner reflector antenna with dipole feed

2.8.4 Corner reflectors

An alternative to the Yagi antenna (described earlier) is that of a corner reflecting arrangement like that shown in Figure 2.25. The two reflecting surfaces (which may be solid or perforated to reduce wind resistance) are inclined at an angle of about 90°. This type of aerial is compact in comparison with a Yagi and also relatively unobtrusive.

2.8.5 Parabolic reflectors

The need for very high gain coupled with directional response at UHF or microwave frequencies is often satisfied by the use of a parabolic reflector in conjunction with a radiating element positioned at the feed-point of the dish (see Figure 2.26). In order to be efficient, the diameter of a parabolic reflecting surface must be large in comparison with the wavelength of the

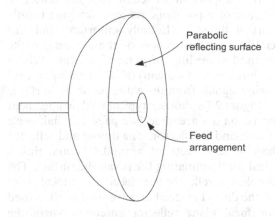

Figure 2.26 Parabolic reflector antenna

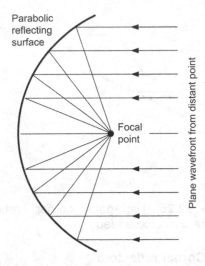

Figure 2.27   Principle of the parabolic reflector

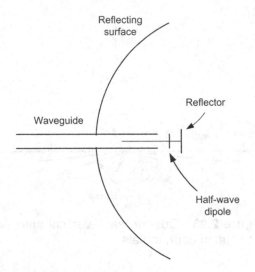

Figure 2.28   Parabolic reflector with half-wave dipole and reflector feed

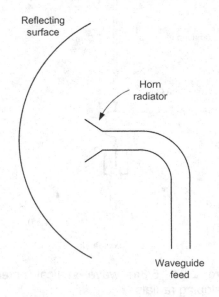

Figure 2.29   Parabolic reflector with horn and waveguide

signal. The gain of such an antenna depends on various factors but is directly proportional to the ratio of diameter to wavelength.

The principle of the parabolic reflector antenna is shown in Figure 2.27. Signals arriving from a distant transmitter will be reflected so that they pass through the focal point of the parabolic surface (as shown). With a conventional parabolic surface, the focal point lies directly on the axis directly in front of the reflecting surface. Placing a radiating element (together with its supporting structure) at the focal point may thus have the undesirable effect of partially obscuring the parabolic surface! In order to overcome this problem the surface may be modified so that the focus is offset from the central axis.

It is important to realise that the reflecting surface of a parabolic reflector antenna is only part of the story. Equally important (and crucial to the effectiveness of the antenna) is the method of feeding the parabolic surface. What's required here is a means of illuminating or capturing signals from the entire parabolic surface.

Figure 2.28 shows a typical feed arrangement based on a waveguide (see page 50), half-wave dipole and a reflector. The dipole and reflector have a beamwidth of around 90°, and this is ideal for illuminating the parabolic surface. The dipole and reflector are placed at the focal point of the dish. This feed arrangement is often used for **focal plane reflector** antennas where the outer edge of the dish is in the same plane as the halfwave dipole plus reflector feed.

An alternative arrangement using a waveguide and small horn radiator (see page 39) is shown in Figure 2.29. The horn aerial offers some modest gain (usually 6 to 10 dB, or so), and this can be instrumental in increasing the overall gain of the arrangement. Such antennas are generally not focal plane types, and the horn feed will usually require support above the parabolic surface. Some typical parabolic reflector antennas are shown in Figures 2.30 and 2.31.

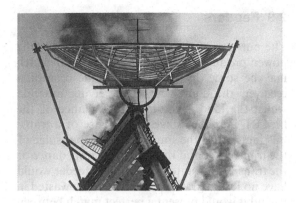

Figure 2.30 Parabolic reflector antenna with dipole and reflector feed

Figure 2.31 High-gain earth station antenna with parabolic reflector and horn feed

---

Test your understanding 2.6

Identify an antenna type suitable for use in the following applications. Give reasons for your answers:

(a) an SHF satellite earth station
(b) a low frequency non-directional beacon
(c) an airfield communication system
(d) a long-range HF communication system
(e) a microwave link between two fixed points.

2.8.6 Horn antennas

Like parabolic reflector antennas, horn antennas (see Figure 2.32) are commonly used at microwave frequencies. Horn aerials may be used alone or as a means of illuminating a parabolic (or other) reflecting surface. Horn antennas are ideal for use with waveguide feeds; the transition from waveguide (see page 50) to the free space aperture being accomplished over several wavelengths as the waveguide is gradually flared out in both planes. During the transition from waveguide to free space, the impedance changes gradually. The gain of a horn aerial is directly related to the ratio of its aperture (i.e. the size of the horn's opening) and the wavelength. However, as the gain increases, the beamwidth becomes reduced.

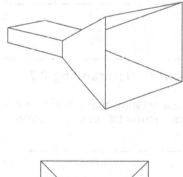

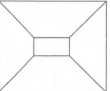

Front view

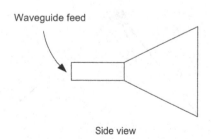

Waveguide feed

Side view

Figure 2.32 A horn antenna

Table 2.1 Typical characteristics of some common antennas

Application	Gain (dB$_d$)	Beamwidth (degrees)
Vertical half-wave dipole	0	360
Vertical quarter-wave with ground plane	0	360
Four-element Yagi	6	43
UHF corner reflector	9	27
Two stacked vertical halfwave dipoles	3	360
5/8th wave vertical with ground plane	2	360
Small horn antenna for use at 10 GHz	10	20
3-m diameter parabolic antenna for tracking space vehicles at UHF	40	4

Test your understanding 2.7

Identify the antenna shown in Figure 2.33. Sketch a typical horizontal radiation pattern for this antenna.

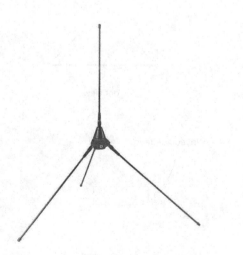

Figure 2.33 See Test your understanding 2.7

2.9 Feeders

The purpose of the feeder line is to convey the power produced by a source to a load which may be some distance away. In the case of a receiver, the source is the receiving antenna whilst the load is the input impedance of the first RF amplifier stage. In the case of a transmitting system, the source is the output stage of the transmitter and the load is the impedance of the transmitting antenna. Ideally, a feeder would have no losses (i.e. no power would be wasted in it) and it would present a perfect match between the impedance of the source to that of the load. In practice, this is seldom the case. This section explains the basic principles and describes the construction of most common types of feeder.

2.9.1 Characteristic impedance

The impedance of a feeder (known as its **characteristic impedance**) is the impedance that would be seen looking into an infinite length of the feeder at the working frequency. The characteristic impedance, Z_0, is a function of the inductance, L, and capacitance, C, of the feeder and may be approximately represented by:

$$Z_0 = \sqrt{\frac{L}{C}}\ \Omega$$

L and C are referred to as the **primary constants** of a feeder. In this respect, L is the loop inductance per unit length whilst C is the shunt capacitance per unit length (see Figure 2.34).

In practice, a small amount of DC resistance will be present in the feeder, but this is usually negligible. For the twin open wire shown in Figure 2.35(a), the inductance, L, and capacitance, C, of the line depend on the spacing between the wires and the diameter of the two conductors. For the coaxial cable shown in Figure 2.35(b) the characteristic impedance depends upon the ratio of the diameters of the inner and outer conductors.

Example 2.4

A cable has a loop inductance of 20 nH and a capacitance of 100 pF. Determine the characteristic impedance of the cable.

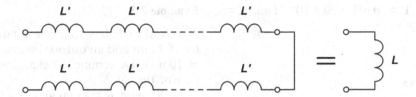

(a) Loop inductance (line short-circuit at the far end)

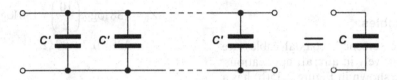

(b) Loop capacitance (line open-circuit at the far end)

Figure 2.34 Loop inductance and loop capacitance

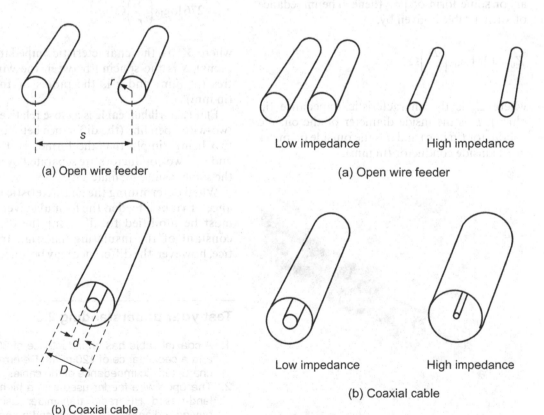

(a) Open wire feeder

(b) Coaxial cable

Figure 2.35 Dimensions of flat twin feeder and coaxial cables (see also Figure 2.36)

Low impedance   High impedance

(a) Open wire feeder

Low impedance   High impedance

(b) Coaxial cable

Figure 2.36 Effect of dimensions on the characteristic impedance of open wire feeder and coaxial cable

In this case, L = 20 nH = 20×10^{-9} H and C = 100 pF = 100×10^{-12} F.

Using

$$Z_0 = \sqrt{\frac{L}{C}}\,\Omega \text{ gives :}$$

$$Z_0 = \sqrt{\frac{180 \times 10^{-9}}{100 \times 10^{-12}}} = \sqrt{1.8 \times 10^3} = \sqrt{1800} = 42\,\Omega$$

2.9.2 Coaxial cables

Because they are screened, coaxial cables are used almost exclusively in aircraft applications. The coaxial cable shown in Figure 2.35(b) has a centre conductor (either solid or stranded wire) and an outer conductor that completely shields the inner conductor as depicted in Figure 2.37. The two conductors are concentric and separated by an insulating dielectric that is usually air or some form of polythene. The impedance of such a cable is given by:

$$Z_0 = 138 \log_{10}\left(\frac{D}{d}\right)\Omega$$

where Z_0 is the characteristic impedance (in ohms), D is the inside diameter of the outside conductor (in mm) and d is the outside diameter of the inside conductor (in mm).

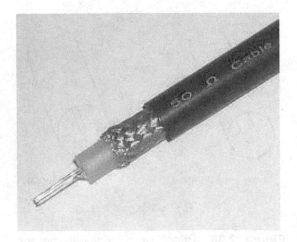

Figure 2.37 Construction of a high-quality coaxial cable (50 Ω impedance)

Example 2.5

A coaxial cable has an inside conductor diameter of 2 mm and an outside conductor diameter of 10 mm. Determine the characteristic impedance of the cable.

In this case, $d = 2$ mm and $D = 10$ mm.

Using $Z_0 = 138 \log_{10}\left(\dfrac{D}{d}\right)\Omega$ gives:

$$Z_0 = 138 \log_{10}\left(\frac{10}{2}\right) = 138 \log_{10}(5)$$

$$= 138 \times 0.7 = 97\,\Omega$$

2.9.3 Two-wire open feeder

The characteristic impedance of the two-wire open feeder shown in Figure 2.35(a) is given by:

$$Z_0 = 276 \log_{10}\left(\frac{S}{r}\right)\Omega$$

where Z_0 is the characteristic impedance (in ohms), s is the spacing between the wire centres (in mm) and r is the radius of the wire (in mm).

Flat twin **ribbon cable** is a close relative of the two-wire open line (the difference between these two being simply that the former is insulated and the two conductors are separated by a rib of the same insulating material).

When determining the characteristic impedance of ribbon feeder, the formula given above must be modified to allow for the dielectric constant of the insulating material. In practice, however, the difference may be quite small.

Test your understanding 2.8

1. A coaxial cable has an inductance of 30 nH/m and a capacitance of 120 pF/m. Determine the characteristic impedance of the cable.
2. The open wire feeder used with a high-power land-based HF radio transmitter uses wire having a diameter of 2.5 mm and a spacing of 15 mm. Determine the characteristic impedance of the feeder.

2.9.4 Attenuation

The attenuation of a feeder is directly proportional to the DC resistance of the feeder and inversely proportional to the impedance of the line. Obviously, the lower the resistance of the feeder, the smaller will be the power losses. The attenuation is given by:

$$A = 0.143 \frac{R}{Z_0} \, \text{dB}$$

where A is the attenuation in dB (per meter), R is the resistance in ohms (per meter) and Z is the characteristic impedance (in ohms).

Whilst the attenuation of a feeder remains reasonably constant throughout its specified frequency range, it is usually subject to a progressive increase beyond the upper frequency limit (see Figure 2.38). It is important when choosing a feeder or cable for a particular application to ensure that the operating frequency is within that specified by the manufacturer. As an example, RG178B/U coaxial cable has a loss that increases with frequency from 0.18 dB at 10 MHz, to 0.44 dB at 100 MHz, to 0.95 dB at 400 MHz and to 1.4 dB at 1 GHz.

Note that, in a 'loss-free' feeder, R and G are both very small and can be ignored (i.e. $R = 0$

and $G = 0$) but with a real feeder both R and G are present.

2.9.5 Primary constants

The two types of feeder that we have already described differ in that one type (the coaxial feeder) is unbalanced whilst the other (the two-wire **transmission line**) is balanced. In order to fully understand the behaviour of a feeder, whether balanced or unbalanced, it is necessary to consider its equivalent circuit in terms of four conventional component values: resistance, inductance, capacitance and conductance, as shown in Figure 2.39. These four parameters are known as primary constants, and they are summarised in Table 2.2.

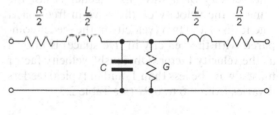

(a) Unbalanced feeder

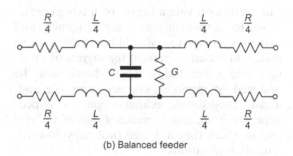

(b) Balanced feeder

Figure 2.39 Equivalent circuit of balanced and unbalanced feeders

Table 2.2 The primary constants of a feeder

Constant	Symbol	Units
Resistance	R	Ohms (Ω)
Inductance	L	Henries (H)
Capacitance	C	Farads (F)
Conductance	G	Siemens (S)

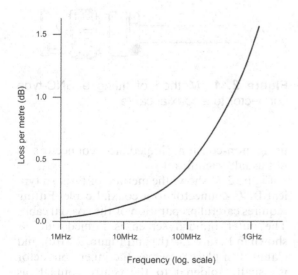

Figure 2.38 Attenuation of a typical coaxial cable feeder

Table 2.3 Velocity factor for various types of feeder

Type of feeder	Velocity factor
Two-wire open line (wire with air dielectric)	0.975
Parallel tubing (air dielectric)	0.95
Coaxial line (air dielectric)	0.85
Coaxial line (solid plastic dielectric)	0.66
Two-wire line (wire with plastic dielectric)	0.68 to 0.82
Twisted-pair line (rubber dielectric)	0.56 to 0.65

2.9.6 Velocity factor

The velocity of a wave in a feeder is not the same as the velocity of the wave in free space. The ratio of the two (velocity in the feeder compared with the velocity in free space) is known as the **velocity factor**. Obviously, velocity factor must always be less than 1, and in typical feeders it varies from 0.6 to 0.97 (see Table 2.3).

2.10 Connectors

Connectors provide a means of linking coaxial cables to transmitting/receiving equipment and antennas. Connectors should be reliable, easy to mate and sealed to prevent the ingress of moisture and other fluids. They should also be designed to minimise contact resistance and, ideally, they should exhibit a constant impedance which accurately matches that of the system in which they are used (normally 50 Ω for aircraft applications).

Coaxial connectors are available in various formats (see Figure 2.40). Of these, the BNC- and N-type connectors are low-loss constant impedance types.

The need for constant impedance connectors (e.g. BNC and N-type connectors) rather than cheaper non-constant impedance connectors (e.g. PL-259) becomes increasingly critical as the frequency increases. As a general rule, constant impedance connectors should be used for applications at frequencies of above 200 MHz. Below this frequency, the loss associated with

Figure 2.40 Coaxial connectors (from left to right: PL-259, BNC and N-type)

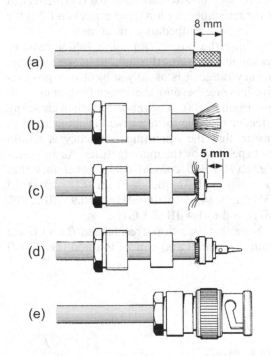

Figure 2.41 Method of fitting a BNC-type connector to a coaxial cable

using non-constant impedance connectors is not usually significant.

Figure 2.41 shows the method of fitting a typical BNC connector to a coaxial cable. Fitting requires careful preparation of the coaxial cable. The outer braided screen is fanned out, as shown in Figure 2.41(b) and Figure 2.41(c), and clamped in place, whereas the inner conductor is usually soldered to the centre contact, as shown in Figure 2.41(d).

2.11 Standing wave ratio

Matching a source (such as a transmitter) to a load (such as an aerial) is an important consideration because it allows the maximum transfer of power from one to the other. Ideally, a feeder should present a perfect match between the impedance of the source and the impedance of the load. Unfortunately this is seldom the case and all too often there is some degree of mismatch present. This section explains the consequences of mismatching a source to a load and describes how the effect of a mismatch can be quantified in terms of standing wave ratio (SWR).

Where the impedance of the transmission line or feeder perfectly matches that of the aerial, all of the energy delivered by the line will be transferred to the load (i.e. the aerial). Under these conditions, no energy will be reflected back to the source.

If the match between source and load is imperfect, a proportion of the energy arriving at the load will be reflected back to the source. The result of this is that a standing wave pattern of voltage and current will appear along the feeder (see Figure 2.42).

The standing wave shown in Figure 2.42 occurs when the wave travelling from the source to the load (i.e. the **forward wave**) interacts with the wave travelling from the load to the source (the **reflected wave**). It is important to note that both the forward and reflected waves are moving but in opposite directions. The standing wave, on the other hand, is stationary.

As indicated in Figure 2.42, when a standing wave is present, at certain points along the feeder the voltage will be a maximum whilst at others it will take a minimum value. The current distribution along the feeder will have a similar pattern (note, however, that the voltage maxima will coincide with the current minima, and vice versa).

Four possible scenarios are shown in Figure 2.43. In Figure 2.43(a) the feeder is perfectly matched to the load. Only the forward wave is present, and there is no standing wave. This is the ideal case in which all of the energy generated by the source is absorbed by the load.

In Figure 2.43(b) the load is short-circuit. This represents one of the two worst-case scenarios as the voltage varies from zero to a very high positive value. In this condition, all of the generated power is reflected back to the source.

In Figure 2.43(c) the load is open-circuit. This represents the other worst-case scenario. Here again, the voltage varies from zero to a high positive value and, once again, all of the generated power is reflected back to the source.

In Figure 2.43(d) the feeder is terminated by an impedance that is different from the feeder's characteristic impedance but is neither a short-circuit nor an open-circuit. This condition lies somewhere between the extreme and perfectly matched cases.

The **standing wave ratio** (SWR) of a feeder or transmission line is an indicator of the effectiveness of the impedance match between the transmission line and the antenna. The SWR is the ratio of the maximum to the minimum current along the length of the transmission line, or the ratio of the maximum to the minimum voltage. When the line is absolutely matched the SWR is unity. In other words, we get unity SWR when there is no variation in voltage or current along the transmission line. The greater the number representing SWR, the larger the mismatch. Also, I^2R losses increase with increasing SWR.

For a *purely resistive* load:

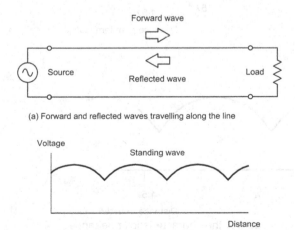

(a) Forward and reflected waves travelling along the line

(b) Voltage standing wave produced

Figure 2.42 Forward and reflected waves when a load is mismatched

$$\mathrm{SWR} = \frac{Z_r}{Z_0}\left(\text{when } Z_r > Z_0\right)$$

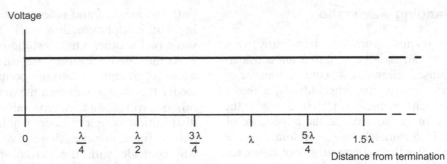

(a) Correctly terminated feeder

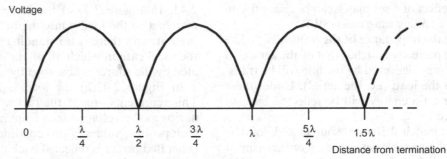

(b) Feeder terminated by a short-circuit

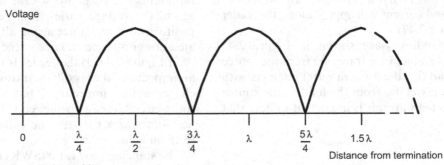

(c) Feeder terminated by an open-circuit

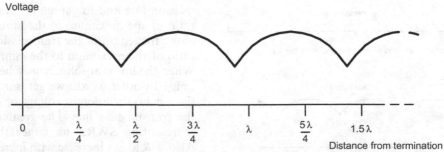

(d) Feeder terminated by an impedance that is not equal to the characteristic impedance

Figure 2.43   Effect of different types of mismatch

and

$$SWR = \frac{Z_0}{Z_r} \left(\text{when } Z_r < Z_0 \right)$$

where Z_0 is the characteristic impedance of the transmission line (in ohms) and Z_r is the impedance of the load (also in ohms). Note that, since SWR is a ratio, it has no units.

SWR is optimum (i.e. unity) when Z_r is equal to Z_0. It is unimportant as to which of these terms is in the numerator. Since SWR cannot be less than 1, it makes sense to put whichever is the larger of the two numbers in the numerator.

The average values of RF current and voltage become larger as the SWR increases. This, in turn, results in increased power loss in the series loss resistance and leakage conductance respectively.

For values of SWR between 1 and 2 this additional feeder loss is not usually significant and is typically less than 0.5 dB. However, when the SWR exceeds 2.5 or 3, the additional loss becomes increasingly significant and steps should be taken to reduce it to a more acceptable value.

2.11.1 SWR measurement

Standing wave ratio is easily measured using an instrument known as an SWR bridge, SWR meter or a combined SWR/power meter (see Figure 2.44). Despite the different forms of this instrument the operating principle involves sensing the forward and reflected power and displaying the difference between them.

Figure 2.44 A combined RF power and SWR meter

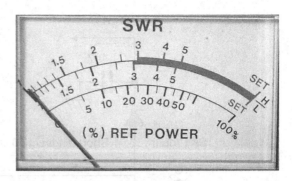

Figure 2.45 A typical SWR meter scale (showing SWR and % reflected power)

Figure 2.45 shows the scale calibration for the SWR meter circuit shown in Figure 2.46. The instrument comprises a short length of transmission line with two inductively and capacitively coupled secondary lines. Each of these secondary lines is terminated with a matched resistive load and each has a signal detector. Secondary line L1 (and associated components D1 and R1) is arranged so that it senses the forward wave, whilst secondary line L2 (and associated components D2 and R2) is connected so that it senses the reflected wave.

In use, RF power is applied to the system, the meter is switched to indicate the forward power and VR1 is adjusted for full-scale deflection. Next, the meter is switched to indicate the reflected power and the SWR is read directly from the meter scale. More complex instruments use cross-point meter movements where the two pointers simultaneously indicate forward and reflected power, and the point at which they intersect (read from a third scale) gives the value of SWR present.

The point at which the SWR in a system is measured is important. To obtain the most meaningful indication of the SWR of an aerial the SWR should ideally be measured at the far end of the feeder (i.e. at the point at which the feeder is connected to the aerial). The measured SWR will actually be *lower* at the other end of the feeder (i.e. at the point at which the feeder is connected to the transmitter). The reason for this apparent anomaly is simply that the loss present in the feeder serves to improve the apparent SWR seen by the transmitter. The more lossy the feeder the better the SWR!

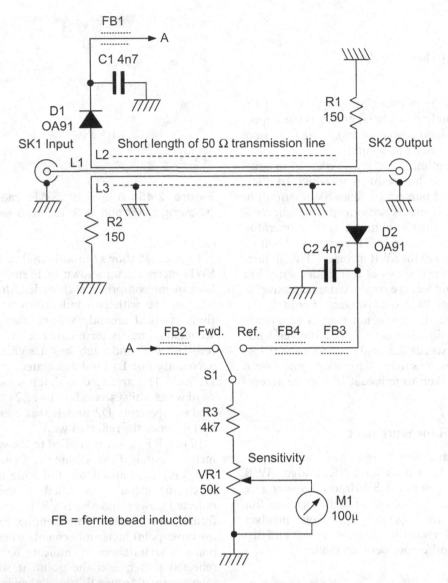

Figure 2.46 A typical SWR meter Figure 2.47 Variation of resistance and reactance for the 250 MHz half-wave dipole antenna

Sudden deterioration of antenna performance and an equally sudden increase in SWR usually points either to mechanical failure of the elements or to electrical failure of the feed-point connection, feeder or RF connectors. Gradual deterioration, on the other hand, is usually associated with corrosion or ingress of fluids into the antenna structure, feeder or antenna termination.

The SWR of virtually all practical aerial/feeder arrangements is liable to some

considerable variation with frequency. For this reason, it is advisable to make measurements at the extreme limits of the frequency range as well as at the centre frequency. In the case of a typical transmitting aerial, the SWR can vary from 2:1 at the band edges to 1.2:1 at the centre. Wideband aerials, particularly those designed primarily for receiving applications, often exhibit significantly higher values of SWR. This makes them unsuited to transmitting applications.

2.11.2 A design example

In order to pursue this a little further it's worth taking an example with some measured values to confirm that the SWR of a half-wave dipole (see page 28) really does change in the way that we have predicted. This example further underlines the importance of SWR and the need to have an accurate means of measuring it.

Assume that we are dealing with a simple half-wave dipole aerial that is designed with the following parameters:

Centre frequency:	250 MHz
Feed-line impedance:	75 ohm
Dipole length:	0.564 meters
Element diameter:	5 mm
Bandwidth:	51 MHz
Q-factor::	4.9

The calculated resistance of this aerial varies from about 52 Ω at 235 MHz to 72 Ω at 265 MHz. Over the same frequency range its reactance varies from about −37 Ω (a capacitive reactance) to +38 Ω (an inductive reactance). As predicted, zero reactance at the feed point occurs at a frequency of 250 MHz for the dipole length in question. This relationship is shown in Figure 2.47.

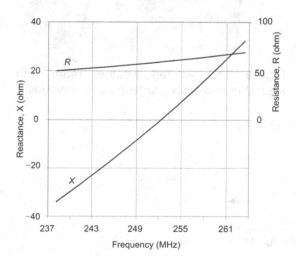

Figure 2.47 Variation of resistance and reactance for the 250 MHz half-wave dipole antenna

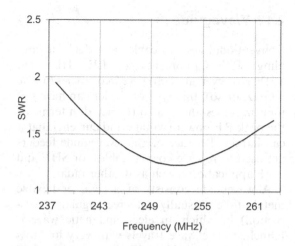

Figure 2.48 Variation of SWR for the 250 MHz half-wave dipole antenna

Measurements of SWR show a minimum value of about 1.23 occurring at about 251 MHz and an expected gradual rise either side of this value (see Figure 2.48). This graph shows that the transmitting bandwidth is actually around 33 MHz (extending from 237 MHz to around 270 MHz) for an SWR of 2:1 instead of the intended 51 MHz. Clearly this could be a problem in an application where a transmitter is to be operated with a maximum SWR of 2:1

The band wid th limitation of a system (comprising transmitter, feeder and aerial) is usually attributable to the inability of a transmitter to operate into a load that has any appreciable amo unt of reactance present rather tha n to an inability of the aerial to radiate effectively. Most aerials will radiate happily at frequencies that are some distance away from their resonant frequency—the problem is more one of actually getting the power that the transmitter is capable of delivering into them!

Key point

SWR tests can provide a rapid check of antennas and matching arrangements. VSWR will vary with frequency, but for most purposes a VSWR of 1.5:1 (or even 2:1) will be indicative of satisfactory antenna/matching performance. Always carry out VSWR checks on installed/modified/disturbed antenna systems using an approved test set and keep well away from a radiating antenna.

2.12 Waveguide

Conventional coaxial cables are ideal for coupling RF equipment at LF, HF and VHF. However, at microwave frequencies (above 3 GHz, or so), this type of feeder can have significant losses and is also restricted in terms of the peak RF power (voltage and current) that it can handle. Because of this, waveguide feeders are used to replace coaxial cables for SHF and EHF applications, such as weather radar.

A waveguide consists of a rigid or flexible metal tube (usually of rectangular cross-section) in which an electromagnetic wave is launched. The wave travels with very low loss inside the waveguide with its magnetic field component (the H-field) aligning with the broad dimension of the waveguide and the electric field component (the E-field) aligning with the narrow dimension of the waveguide (see Figure 2.49).

A simple waveguide system is shown in Figure 2.50. The SHF signal is applied to a quarter wavelength coaxial probe. The wave launched in the guide is reflected from the plane blanked-off end of the waveguide and travels through sections of waveguide to the load (in this case a horn antenna, see page 39). An example of the use of a waveguide is shown in Figure 2.51. In this application a flexible waveguide is used to feed the weather radar antenna mounted in the nose of a large passenger aircraft. The antenna comprises a flat steerable plate with a large number of radiating slots (each equivalent to a half-wave dipole fed in phase).

Figure 2.51 Aircraft weather radar with steerable microwave antenna and waveguide

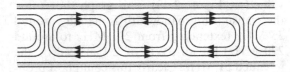

(a) Waveguide (broad dimension) showing H-field

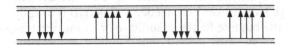

(b) Waveguide (narrow dimension) showing E-field

Figure 2.49 E- and H-fields in a rectangular waveguide

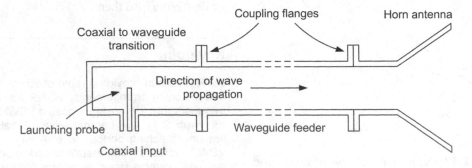

Figure 2.50 A simple waveguide system comprising launcher, waveguide and horn antenna

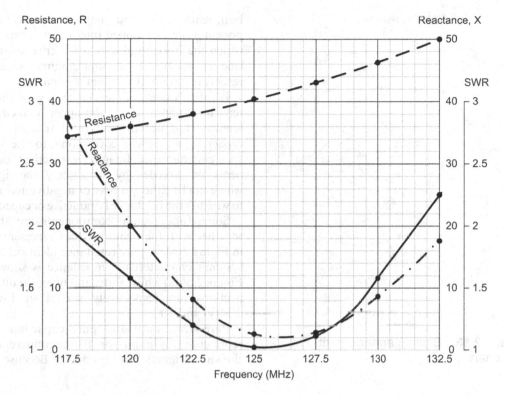

Figure 2.52   See Test your understanding 2.9

Test your understanding 2.9

Figure 2.52 shows the frequency response of a vertical quarter-wave antenna used for local VHF communications. Use the graph to determine the following:

(a) The frequency at which the SWR is minimum
(b) The 2:1 SWR bandwidth of the antenna
(c) The reactance of the antenna at 120 MHz
(d) The resistance of the antenna at 120 MHz
(e) The frequency at which the reactance of the antenna is a minimum
(f) The frequency at which the resistance of the antenna is 50 Ω.

Test your understanding 2.10

Explain what is meant by standing wave ratio (SWR) and why this is important in determining the performance of an antenna/feeder combin ation.

2.13 Vector network analysis

Vector Network Analysis (VNA) provides a powerful method for investigating antennas and matching arrangements as well as RF networks generally. To interpret the results of VNA analysis it is important to be familiar with a method of graphical representation known as a Smith Chart. We will start this section by explaining how the Smith Chart works before moving on to using it to perform vector network analysis of antennas.

2.13.1 The Smith chart

Used in conjunction with VNA, a Smith Chart is a valuable tool that helps you visualise the behaviour of an antenna and the arrangements for feeding and matching it to an avionic system. Figure 2.53 shows the basic arrangement of a Smith Chart. Although this may look somewhat intimidating, the basic concept is simply that a complex impedance (comprising

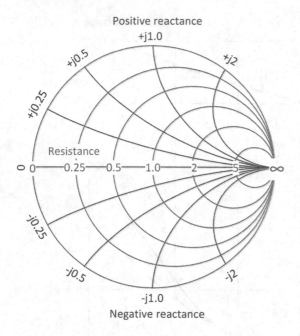

Figure 2.53 Basic arrangement of a Smith chart

both real/resistive and imaginary/reactive components) can be mapped into a circular space in which the horizontal axis represents resistance (the real part) and the outer circumference represents reactance (the imaginary part).

Positive values of reactance (i.e., inductive reactance) appear along the upper semi-circumference while negative values of reactance (i.e., capacitive reactance) correspond to the lower semi-circumference. Imaginary (reactive) components are preceded by the letter, j, the sign of which can be either positive or negative according to whether the reactance is inductive or capacitive.

Some fundamental components are shown together with their Smith Chart representation in Figure 2.54. A pure resistance identical to the system's characteristic impedance is shown in Figure 2.56(a) whilst short and open circuit connections are respectively shown in Figures 2.54(b) and (c).

Pure inductance and pure capacitance are represented by points on the circumference as shown in Figures 2.54(d) and (e). Because their

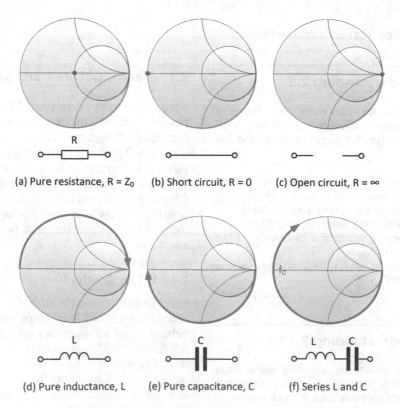

(a) Pure resistance, R = Z₀   (b) Short circuit, R = 0   (c) Open circuit, R = ∞

(d) Pure inductance, L   (e) Pure capacitance, C   (f) Series L and C

Figure 2.54 Fundamental components plotted on the Smith Chart

reactance change with frequency we have shown these as lines rather than dots with the arrow on the line showing the direction of increasing frequency.

Finally, Figure 2.54(f) shows a series combination of pure inductance and pure capacitance. Note that the impedance of this combination will become zero at the frequency of resonance, f_0. This is the frequency at which the inductive and capacitive reactance will become equal but of opposite.

2.13.2 *j*-notation

The operator, *j*, is used to indicate the reactive part of an impedance. When the sign of the *j*-term is positive the reactance will be inductive. Conversely, when the sign of the *j*-term is negative the component will be capacitive. All practical RF components and antennas exhibit complex impedances where both resistance and reactance are present. Furthermore, the complex impedance will vary with frequency due to the changing value of reactance.

Consider the case of a component that can be represented by a resistance of 50 Ω connected in series with an inductance of 100 nH. At 100 MHz the reactance of the inductor will be approximately 63 Ω and the component will have an impedance that can be expressed as (50 + *j*63) Ω. At 140 MHz the component's inductive reactance will increase to 88 Ω and so the impedance will then be (50+*j*63) Ω. Corresponding impedance values for a resistance of 50 Ω connected in series with a capacitor of 20 pF will be (50 − *j*80) Ω at 100 MHz and (50 − *j*57) Ω at 140 MHz.

2.13.3 Normalising

To cater for any given value of characteristic impedance (see page 40) values plotted on a Smith Chart must first be normalized by simply dividing the real/resistive and imaginary/reactive parts by the value of characteristic impedance. So, for example, an impedance of (50 + *j*25) Ω will be plotted as (1 + *j*0.5) Ω. Conversely, an impedance of (50 − *j*25) Ω will be normalised to (1 + *j*0.5) Ω. These conjugate points are shown plotted in Figure 2.55.

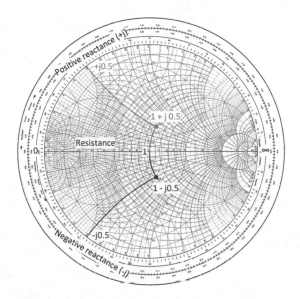

Figure 2.55   Plotting normalized impedances

2.13.4 Standing wave ratio

Standing wave ratio can be easily shown on a Smith Chart by superimposing circles of constant SWR on the chart, as shown in Figure 2.56. A unity value of SWR corresponds to a circle of zero radius (i.e., a point at the exact centre of the Smith Chart) while infinite SWR corresponds to the outer perimeter of the chart.

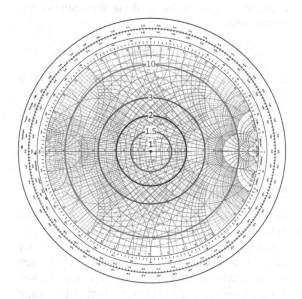

Figure 2.56   Circles of constant SWR

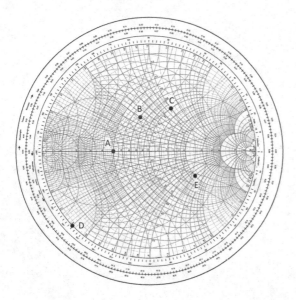

Figure 2.57 See Test your understanding 2.11

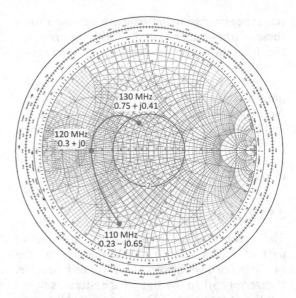

Figure 2.58 An impedance-frequency plot

When used as a tool to investigate matching, the RF designer will usually strives to ensure that load impedances are as close to the centre of the Smith Chart as possible.

Test your understanding 2.11

Figure 2.57 shows a Smith Chart with five different impedances plotted on it. If the impedances are normalised to 50 Ω:

1. Determine the normalised impedance of each of these values.
2. Write down an expression for each of these five impedances expressed in de-normalised form.
3. Which of the five points is purely resistive and which is purely reactive? Explain how this can be inferred from the diagram.

2.13.5 Impedance-frequency plots

Smith Charts are often used to visualise the behaviour of antennas and RF components over a range of frequencies. Figure 2.58 shows an example of the analysis of a component in which the normalised impedance varies from $(0.23 - j0.65)$ Ω to $(0.75 - j4.1)$ Ω over a frequency range extending from 110 MHz to 135 MHz. Note how the antenna becomes purely resistive at 120

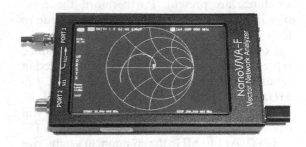

Figure 2.59 Typical low-cost VNA

MHz and how the SWR of the antenna has fallen below 2 above 130 MHz.

2.13.6 Antenna analysis

A vector network analyser (see Figure 2.59) can provide a quick check on the characteristics of an antenna and its feeder. Figure 2.60 shows some typical results of the analysis of a VHF antenna in a 50 Ω system. The analyser's frequency sweep has been set from 120 MHz to 135 MHz with markers displayed at 120 MHz, 125 MHz and 130 MHz. From Figure 2.60 we can quickly determine that:

- The antenna's SWR is less than 2 over the entire scan range

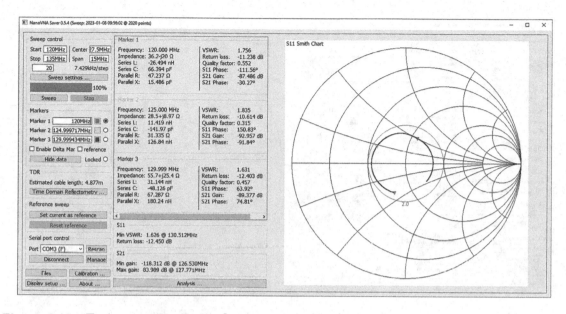

Figure 2.60 Typical VNA analysis of a VHF antenna

- The antenna is purely resistive at 123 MHz and 133 MHz
- The lowest SWR (1.6 approx.) is achieved at 130.5 MHz and the best match to a 50 Ω system is obtained at this frequency.

Test your understanding 2.12

Figure 2.61 shows the results of a VNA analysis of a VHF antenna.

1. Is the antenna inductive or capacitive at 130 MHz?
2. What is the SWR at 115 MHz?
3. What is the impedance (expressed in complex form) of the antenna at 123.5 MHz?
4. At what two frequencies is the antenna purely resistive?
5. What is the minimum value of SWR and at what frequency is it obtained?

2.14 Multiple choice questions

1. An isotropic radiator will radiate:
   (a) only in one direction
   (b) in two main directions
   (c) uniformly in all directions.

2. Another name for a quarter-wave vertical antenna is:
   (a) a Yagi antenna
   (b) a dipole antenna
   (c) a Marconi antenna.

3. A full-wave dipole fed at the centre must be:
   (a) current fed
   (b) voltage fed
   (c) impedance fed.

4. The radiation efficiency of an antenna:
   (a) increases with antenna loss resistance
   (b) decreases with antenna loss resistance
   (c) is unaffected by antenna loss resistance.

5. A vertical quarter-wave antenna will have a polar diagram in the horizontal plane which is:
   (a) unidirectional
   (b) omnidirectional
   (c) bi-directional.

6. Which one of the following gives the approximate length of a half-wave dipole for use at 300 MHz?
   (a) 50 cm
   (b) 1 m
   (c) 2 m.

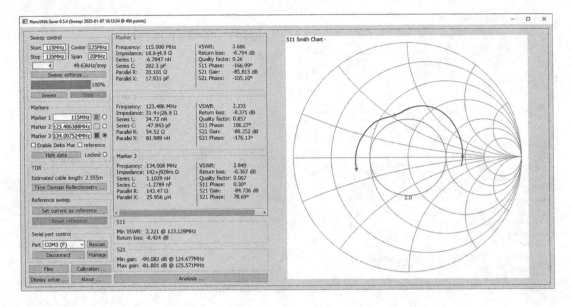

Figure 2.61 See Test your understanding 2.12

7. A standing wave ratio of 1:1 indicates:
    (a) that there will be no reflected power
    (b) that the reflected power will be the
        same as the forward power
    (c) that only half of the transmitted power
        will be radiated.

8. Which one of the following gives the 2:1
    SWR bandwidth of the antenna whose
    frequency response is shown in
    Figure 2.62?
    (a) 270 kHz
    (b) 520 kHz
    (c) 11.1 MHz.

9. Which one of the following antenna types
    would be most suitable for a fixed long
    distance HF communications link?
    (a) a corner reflector
    (b) two stacked vertical dipoles
    (c) a three-element horizontal Yagi.

10. What type of antenna is shown in
    Figure 2.63?
    (a) a folded dipole
    (b) a Yagi
    (c) a corner reflector.

11. When two antennas are vertically stacked
    the combination will have:
    (a) increased gain and decreased
        beamwidth
    (b) decreased gain and increased
        beamwidth
    (c) increased gain and unchanged
        beamwidth.

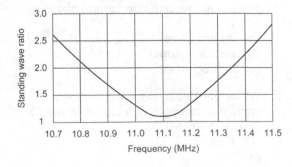

Figure 2.62 See Question 8

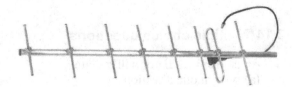

Figure 2.63 See Question 10

12. The characteristic impedance of a coaxial
cable depends on:
(a) the ratio of inductance to capacitance
(b) the ratio of resistance to inductance
(c) the product of the resistance and
reactance (either capacitive or inductive).

13. The attenuation of an RF signal in a
coaxial cable:
(a) increases with frequency
(b) decreases with frequency
(c) stays the same regardless of frequency.

14. If a transmission line is perfectly matched
to an aerial load it will:
(a) have no impedance
(b) be able to carry an infinite current
(c) appear to be infinitely long.

15. The characteristic impedance of an RF
coaxial cable is:
(a) usually between 50 and 75 Ω
(b) either 300 or 600 Ω
(c) greater than 600 Ω.

16. The beamwidth of an antenna is measured:
(a) between the 50% power points
(b) between the 70% power points
(c) between the 90% power points.

17. Circles superimposed on a Smith Chart
represent:
(a) zero SWR
(b) unity SWR
(c) constant SWR.

Chapter 3
Transmitters and receivers

Transmitters and receivers are used extensively in aircraft communications and navigation systems. In conjunction with one ore more antennas, they are responsible for implementing the vital link between the aircraft, ground stations, other aircraft and satellites. This chapter provides a general introduction to the basic principles and operation of transmitters and receivers. These themes are further developed in Chapters 4 and 5.

3.1 A simple radio system

Figure 3.1 shows a simple radio communication system comprising a **transmitter** and **receiver** for use with **continuous wave** (CW) signals. Communication is achieved by simply switching (or 'keying') the radio frequency signal on and off. Keying can be achieved by interrupting the supply to the power amplifier stage or even the oscillator stage; however, it is normally applied within the driver stage that operates at a more modest power level. Keying the oscillator stage usually results in impaired frequency stability. On the other hand, attempting to interrupt the appreciable currents and/or voltages that appear in the power amplifier stage can also prove to be somewhat problematic.

The simplest form of CW receiver consists of nothing more than a radio frequency amplifier (which provides gain and selectivity) followed by a detector and an audio amplifier. The **detector** stage mixes a locally generated radio frequency signal produced by the **beat frequency oscillator** (BFO) with the incoming signal to produce a signal that lies within the audio frequency range (typically between 300 Hz and 3.4 kHz).

As an example, assume that the incoming signal is at a frequency of 100 kHz and that the BFO is producing a signal at 99 kHz. A signal at the difference between these two frequencies (1 kHz) will appear at the output of the detector stage. This will then be amplified within the audio stage before being fed to the loudspeaker.

Example 3.1

A radio wave has a frequency of 162.5 kHz. If a beat frequency of 1.25 kHz is to be obtained, determine the two possible BFO frequencies.

The BFO can be above or below the incoming signal frequency by an amount that is equal to

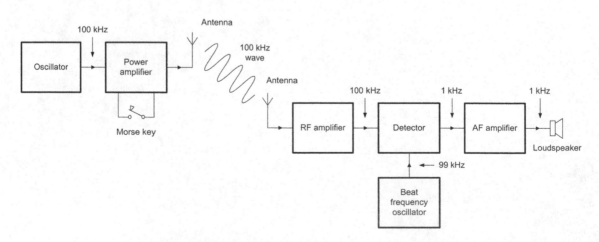

Figure 3.1 A simple radio communication system.

DOI: 10.1201/9781003411932-3

the beat frequency (i.e. the audible signal that results from the 'beating' of the two frequencies and which appears at the output of the detector stage).

Hence, $f_{BFO} = f_{RF} \pm f_{AF}$
from which:

$f_{BFO} = (162.5 \pm 1.25)$ kHz $= 160.25$ or 163.75 kHz

Test your understanding 3.1

An audio frequency signal of 850 Hz is produced when a BFO is set to 455.5 kHz. What is the input signal frequency to the detector?

3.1.1 Morse code

Transmitters and receivers for CW operation are extremely simple, but nevertheless they can be extremely efficient. This makes them particularly useful for disaster and emergency communication or for any situation that requires optimum use of low power equipment. Signals are transmitted using the code invented by Samuel Morse (see Figures 3.2 and 3.3).

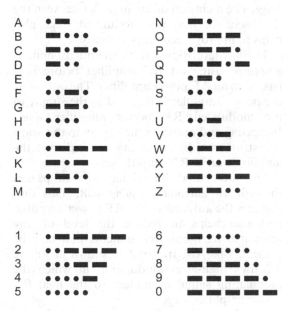

Figure 3.2   Morse code.

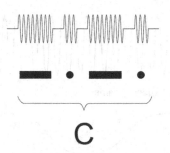

Figure 3.3   Morse code signal for the letter C

3.2 Modulation and demodulation

In order to convey information using a radio frequency carrier, the signal information must be superimposed or 'modulated' onto the carrier. **Modulation** is the name given to the process of changing a particular property of the carrier wave in sympathy with the instantaneous voltage (or current) signal.

The most commonly used methods of modulation are **amplitude modulation** (AM) and **frequency modulation** (FM). In the former case, the carrier amplitude (its peak voltage) varies according to the voltage, at any instant, of the modulating signal. In the latter case, the carrier frequency is varied in accordance with the voltage, at any instant, of the modulating signal.

Figure 3.4 shows the effect of amplitude and frequency modulating a sinusoidal carrier (note that the modulating signal is in this case also sinusoidal). In practice, many more cycles of the RF carrier would occur in the time-span of one cycle of the modulating signal. The process of modulating a carrier is undertaken by a **modulator** circuit, as shown in Figure 3.5. The input and output waveforms for amplitude and frequency modulator circuits are shown in Figure 3.6.

Demodulation is the reverse of modulation and is the means by which the signal information is recovered from the modulated carrier. Demodulation is achieved by means of a **demodulator** (sometimes also called a **detector**). The output of a demodulator consists of a reconstructed version of the original signal information present at the input of the modulator stage within the transmitter. The input and output waveforms for amplitude and frequency

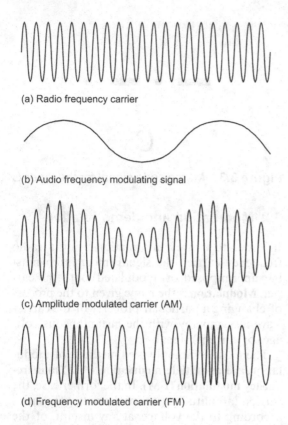

(a) Radio frequency carrier

(b) Audio frequency modulating signal

(c) Amplitude modulated carrier (AM)

(d) Frequency modulated carrier (FM)

Figure 3.4   Modulated waveforms

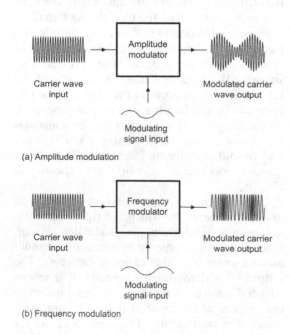

(a) Amplitude modulation

(b) Frequency modulation

Figure 3.5   Action of a modulator

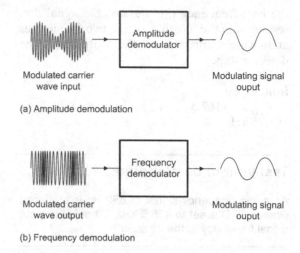

(a) Amplitude demodulation

(b) Frequency demodulation

Figure 3.6   Action of a demodulator

modulator circuits are shown in Figure 3.6. We shall see how this works a little later.

3.3 AM transmitters

Figure 3.7 shows the block schematic of a simple AM transmitter. An accurate and stable RF oscillator generates the radio frequency **carrier** signal. The output of this stage is then amplified and passed to a modulated RF power amplifier stage. The inclusion of an amplifier between the RF oscillator and the modulated stage also helps to improve frequency stability.

The low-level signal from the microphone is amplified using an AF amplifier before it is passed to an AF power amplifier. The output of the power amplifier is then fed as the supply to the modulated RF power amplifier stage. Increasing and reducing the supply to this stage is instrumental in increasing and reducing the amplitude of its RF output signal.

The modulated RF signal is then passed through an **antenna coupling unit**. This unit matches the antenna to the RF power amplifier and also helps to reduce the level of any unwanted harmonic components that may be present. The AM transmitter shown in Figure 3.7 uses high-level modulation in which the modulating signal is applied to the final RF power amplifier stage.

An alternative to high-level modulation of the carrier wave is shown in Figure 3.8. In this

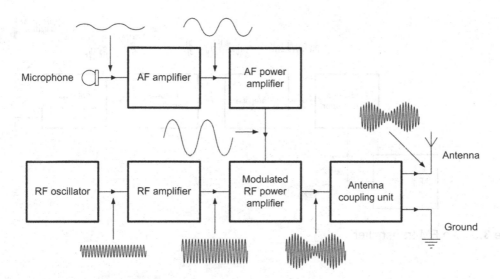

Figure 3.7 An AM transmitter using high-level modulation

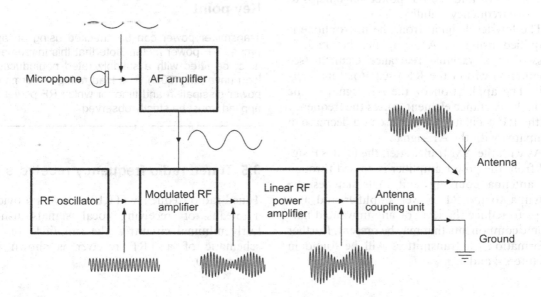

Figure 3.8 An AM transmitter using low-level modulation

arrangement the modulation is applied to a low-power RF amplifier stage, and the amplitude-modulated signal is then further amplified by the final RF power amplifier stage. In order to prevent distortion of the modulated waveform this final stage *must* operate in linear mode (the output waveform must be a faithful replica of the input waveform). Low-level modulation avoids the need for an AF power amplifier.

3.4 FM transmitters

Figure 3.9 shows the block schematic of a simple FM transmitter. Once again, an accurate and stable RF oscillator generates the radio frequency carrier signal. As with the AM transmitter, the output of this stage is amplified and passed to an RF power amplifier stage. Here again, the inclusion of an amplifier between the

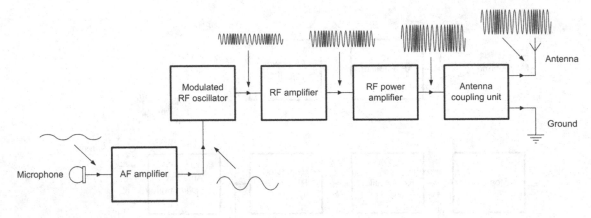

Figure 3.9   An FM transmitter

RF oscillator and the RF power stage helps to improve frequency stability.

The low-level signal from the microphone is amplified using an AF amplifier before it is passed to a **variable reactance** element (see Chapter 4) within the RF oscillator tuned circuit. The application of the AF signal to the variable reactance element causes the frequency of the RF oscillator to increase and decrease in sympathy with the AF signal.

As with the AM transmitter, the final RF signal from the power amplifier is passed through an antenna coupling unit that matches the antenna to the RF power amplifier and also helps to reduce the level of any unwanted harmonic components that may be present. Further information on transmitters will be found in Chapters 4 and 5.

Key point

Transmitter power can be checked using an approved RF power meter. Note that this instrument must be fitted with a suitably rated noninductive load (invariably 50Ω). The load rating, in terms of power dissipation and time for which RF power is applied, must be strictly observed.

3.5 Tuned radio frequency receivers

Tuned radio frequency (TRF) receivers provide a means of receiving local signals using fairly minimal circuitry. The simplified block schematic of a TRF receiver is shown in Figure 3.10.

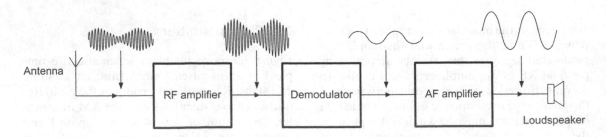

Figure 3.10   A TRF receiver

The signal from the antenna is applied to an RF amplifier stage. This stage provides a moderate amount of gain at the signal frequency. It also provides **selectivity** by incorporating one or more tuned circuits at the signal frequency. This helps the receiver to reject signals that may be present on adjacent channels.

The output of the RF amplifier stage is applied to the demodulator. This stage recovers the audio frequency signal from the modulated RF signal. The demodulator stage may also incorporate a tuned circuit to further improve the selectivity of the receiver.

The output of the demodulator stage is fed to the input of the AF amplifier stage. This stage increases the level of the audio signal from the demodulator so that it is sufficient to drive a loudspeaker.

TRF receivers have a number of limitations with regard to sensitivity and selectivity, and this makes this type of radio receiver generally unsuitable for use in commercial radio equipment.

3.6 Superhet receivers

Superhet receivers provide both improved **sensitivity** (the ability to receive weak signals) and improved **selectivity** (the ability to discriminate signals on adjacent channels) when compared with TRF receivers. Superhet receivers are based on the **supersonic-heterodyne** principle where the wanted input signal is converted to a fixed **intermediate frequency** (IF) at which the majority of the gain and selectivity is applied. The intermediate frequency chosen is generally 455 kHz or 1.6 MHz for AM receivers and 10.7 MHz for communications and FM receivers. The simplified block schematic of a simple superhet receiver is shown in Figure 3.11.

The signal from the antenna is applied to an **RF amplifier** stage. As with the TRF receiver, this stage provides a moderate amount of gain at the signal frequency. The stage also provides selectivity by incorporating one or more tuned circuits at the signal frequency.

The output of the RF amplifier stage is applied to the **mixer** stage. This stage combines the RF signal with the signal derived from the **local oscillator** (LO) stage in order to produce a signal at the **intermediate frequency** (IF). It is

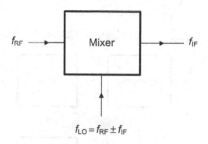

$f_{LO} = f_{RF} \pm f_{IF}$

Figure 3.11 A superhet receiver

worth noting that the output signal produced by the mixer actually contains a number of signal components, including the sum and difference of the signal and local oscillator frequencies as well as the original signals plus harmonic components. The wanted signal (i.e. that which corresponds to the IF) is passed (usually by some form of filter—see page 65) to the IF amplifier stage. This stage provides amplification as well as a high degree of selectivity.

The output of the IF amplifier stage is fed to the demodulator stage. As with the TRF receiver, this stage is used to recover the audio frequency signal from the modulated RF signal.

Finally, the AF signal from the demodulator stage is fed to the AF amplifier. As before, this stage increases the level of the audio signal from the demodulator so that it is sufficient to drive a loudspeaker.

In order to cope with a wide variation in signal amplitude, superhet receivers invariably incorporate some form of **automatic gain control** (AGC). In most circuits the DC level from the AM demodulator (see page 59 is used to control the gain of the IF and RF amplifier stages. As the signal level increases, the DC level from the demodulator stage increases, and this is used to reduce the gain of both the RF and IF amplifiers.

The superhet receiver's intermediate frequency f_{IF} is the difference between the signal frequency, f_{RF}, and the local oscillator frequency, f_{LO}. The desired local oscillator frequency can be calculated from the relationship:

$$f_{LO} = f_{RF} \pm f_{IF}$$

Note that in most cases (and in order to simplify tuning arrangements) the local oscillator

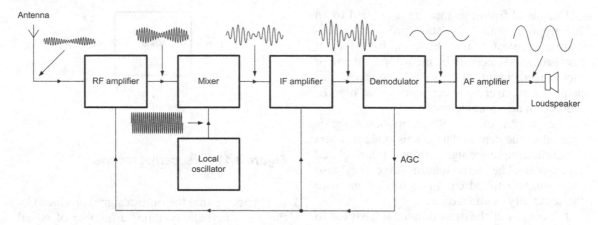

Figure 3.12   Action of a mixer stage in a superhet receiver

operates above the signal frequency, ie fLO = fRF + fIF So, for example, a superhet receiver with a 16 MHz IF tuned to receive a signal at 55 MHz will operate with an LO at (55 + 16) = 71 MHz

Figure 3.12 shows the relationship between the frequencies entering and leaving a mixer stage.

Example 3.2

A VHF Band II FM receiver with a 10.7 MHz IF covers the signal frequency range 88 MHz to 108 MHz. Over what frequency range should the local oscillator be tuned?

Using fLO = fRF + fIF when fRF = 88 MHz gives fLO = 88 MHz + 10.7 MHz = 98.7 MHz

Using fLO = fRF + fIF when fRF = 108 MHz gives fLO = 108 MHz + 10.7 MHz = 118.7 MHz.

3.7 Selectivity

Radio receivers use tuned circuits in order to discriminate between incoming signals at different frequencies. Figure 3.13 shows two basic configurations for a tuned circuit; series and parallel. The impedance-frequency characteristics of these circuits are shown in Figure 3.14. It is important to note that the impedance of the series tuned circuit falls to a very low value at the resonant frequency whilst that for a parallel tuned circuit increases to a very high value at resonance. For this reason, series tuned circuits are sometimes known as **acceptor circuits**.

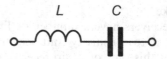

(a) Series tuned circuit

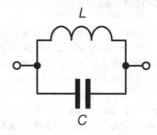

(b) Parallel tuned circuit

Figure 3.13 Series and parallel tuned circuits

Parallel tuned circuits, on the other hand, are sometimes referred to as **rejector circuits**.

The frequency response (voltage plotted against frequency) of a parallel tuned circuit is shown in Figure 3.15. This characteristic shows how the signal developed across the circuit reaches a maximum at the resonant frequency (f_o). The range of frequencies accepted by the circuit is normally defined in relation to the half-power (−3dB power) points. These points correspond to 70.7% of the maximum voltage,

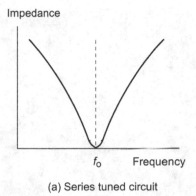

(a) Series tuned circuit

(b) Parallel tuned circuit

Figure 3.14 Frequency response of the tuned circuits shown in Figure 3.13

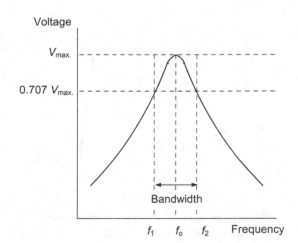

Figure 3.15 Frequency response for a parallel tuned circuit

and the frequency range between these points is referred to as the **bandwidth** of the tuned circuit.

A perennial problem with the design of the TRF receivers that we met earlier is the lack of selectivity due to the relatively wide bandwidth of the RF tuned circuits. An RF tuned circuit will normally exhibit a quality factor (**Q-factor**) of about 100. The relationship between bandwidth, Δf, Q-factor, Q and resonant frequency, f_o, for a tuned circuit is given by:

$$\Delta f = \frac{f_0}{Q}$$

As an example, consider a tuned circuit which has a resonant frequency of 10 MHz and a Q-factor of 100. Its bandwidth will be:

$$\Delta f = \frac{f_0}{Q} = \frac{10\,\text{MHz}}{100} = 100\,\text{kHz}$$

Clearly many strong signals will appear within this range, and a significant number of them may be stronger than the wanted signal. With only a single tuned circuit at the signal frequency, the receiver will simply be unable to differentiate between the wanted and unwanted signals.

Selectivity can be improved by adding additional tuned circuits at the signal frequency. Unfortunately, the use of multiple tuned circuits brings with it the problem of maintaining accurate tuning of each circuit throughout the tuning range of the receiver. Multiple 'ganged' variable capacitors (or accurately matched variable capacitance diodes) are required.

A **band-pass filter** can be constructed using two parallel tuned circuits coupled inductively (or capacitively), as shown in Figure 3.16. The

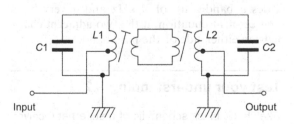

Figure 3.16 A typical band-pass filter

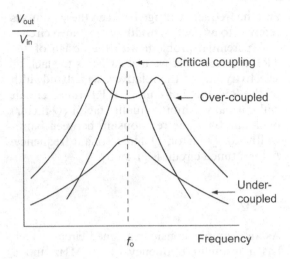

Figure 3.17 Response of coupled tuned circuits

Figure 3.18 Band-pass coupled tuned circuits in the RF stages of a VHF receiver

frequency response of this type of filter depends upon the degree of coupling between the two tuned circuits. Optimum results are obtained with a critical value of coupling (see Figure 3.17). Too great a degree of coupling results in a 'doublehumped' response whilst too little coupling results in a single peak in the response curve accompanied by a significant loss in signal. Critical coupling produces a relatively 'flat' pass-band characteristic accompanied by a reasonably steep fall-off either side of the pass-band.

Band-pass filters (see Figure 3.18) are often found in the IF stages of superhet receivers where they are used to define and improve the receiver's selectivity. Where necessary, a higher degree of selectivity and **adjacent channel rejection** can be achieved by using a multi-element ceramic, mechanical or crystal filter. A typical 455 kHz crystal filter (for use with an HF receiver) is shown in Figure 3.19. This filter provides a bandwidth of 9 kHz and a very high degree of attenuation at the two **adjacent channels** on either side of the pass-band.

Test your understanding 3.2

Sketch the block schematic of a superhet receiver and state the function of each of the blocks.

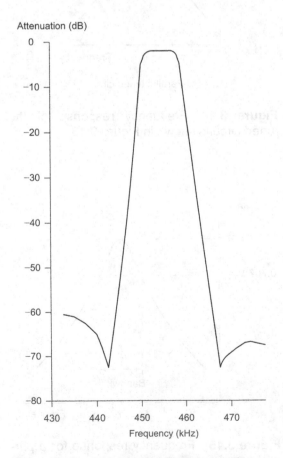

Figure 3.19 Mechanical IF filter response

Test your understanding 3.3

An HF communications receiver has an intermediate frequency of 455 kHz. What frequency must the local oscillator operate at when the receiver is tuned to 5.675 MHz?

Test your understanding 3.4

A tuned circuit IF filter is to operate with a centre frequency of 10.7 MHz and a bandwidth of 150 kHz. What Q-factor is required?

Test your understanding 3.5

The ability of a receiver to reject signals on adjacent channels is determined by the selectivity of its IF stages. Explain why this is.

Test your understanding 3.6

Sketch the frequency response of two coupled tuned circuits. In relation to your answer, explain what is meant by:

(a) overcoupling
(b) undercoupling
(c) critical coupling

3.8 Image channel rejection

Earlier we showed that a superhet receiver's intermediate frequency, f_{IF}, is the difference between the signal frequency, f_{RF}, and the local oscillator frequency, f_{LO}. We also derived the following formula for determining the frequency of the local oscillator signal:

$$f_{LO} = f_{RF} \pm f_{IF}$$

The formula can be rearranged to make f_{RF} the subject, as follows:

$$f_{RF} = f_{LO} \pm f_{IF}$$

In other words, there are two potential radio frequency signals that can mix with the local oscillator signal in order to provide the required IF. One of these is the wanted signal (i.e. the signal present on the channel to which the receiver is tuned) whilst the other is referred to as the image channel.

Being able to reject any signals that may just happen to be present on the image channel of a superhet receiver is an important requirement of any superhet receiver. This can be achieved by making the RF tuned circuits as selective as possible (so that the image channel lies well outside their pass-band). The problem of rejecting the image channel is, however, made easier by selecting a relatively high value of intermediate frequency (note that, in terms of frequency, the image channel is spaced at *twice* the IF away from the wanted signal).

Figure 3.20 shows the relationship that exists between the wanted signal, the local oscillator signal and the image channel for receivers with (a) a455 *kHz* IF and (b) a1.6 MHz IF. A typical response curve for the RF tuned circuits of the receiver (assuming a typical Q-factor) has been superimposed onto both of the graphs (the same response curve has been used in both cases but the frequency scale has been changed for the two different intermediate frequencies). From Figure 3.20 it should be clear that, whilst the image channel for the 455 kHz IF falls inside the RF tuned circuit response, that for the 1.6 MHz IF falls well outside the curve.

Test your understanding 3.7

An FM receiver tuned to 118.6 MHz has an IF of 10.7 MHz. Determine the frequency of the image channel given that the local oscillator operates above the signal frequency.

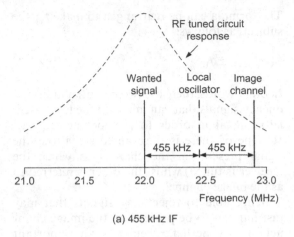

(a) 455 kHz IF

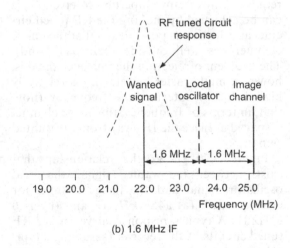

(b) 1.6 MHz IF

Figure 3.20   Image channel rejection

3.9 Automatic gain control

The signal levels derived from the antennas fitted to an aircraft can vary from as little as 1 μV to more than 1,000 μV. Unfortunately, this presents us with a problem when signals are to be amplified. The low-level signals benefit from the maximum amount of gain present in a system whilst the larger signals require correspondingly less gain in order to avoid non-linearity and consequent distortion of the signals and modulation. AM, CW and SSB receivers therefore usually incorporate some means of **automatic gain control** (AGC) that progressively reduces the signal gain as the amplitude of the input signal increases (see Figure 3.21).

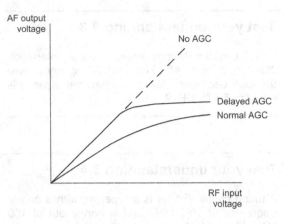

Figure 3.21   AGC action

In simple receivers, the AGC voltage (a DC voltage dependent on signal amplitude) is derived directly from the signal detector and is fed directly to the bias circuitry of the IF stages (see Figure 3.11). In more sophisticated equipment, the AGC voltage is amplified before being applied to the IF and RF stages. There is, in fact, no need to reduce the signal gain for small RF signals. Hence, in more sophisticated equipment, the AGC circuits may be designed to provide a 'delay' so that there is no gain reduction until a predetermined threshold voltage is exceeded. In receivers that feature **delayed AGC** there is no gain reduction until a certain threshold voltage is achieved. Beyond this, there is a progressive reduction in gain (see Figure 3.21).

3.10 Double superhet receivers

The basic superhet receiver shown in Figure 3.12 has an intermediate frequency (IF) of usually either 455 kHz, 1.6 MHz or 10.7 MHz. In order to achieve an acceptable degree of image channel rejection (recall that the image channel is spaced by twice the IF away from the wanted frequency) a 455 kHz IF will generally be satisfactory for the reception of frequencies up to about 5 MHz whilst an IF of 1.6 MHz (or greater) is often used at frequencies above this. At VHF, intermediate frequencies of 10.7 MHz (or higher) are often used.

Unfortunately, the disadvantage of using a high IF (1.6 MHz or 10.7 MHz) is simply that the bandwidth of conventional tuned circuits is too wide to provide a satisfactory degree of selectivity, and thus elaborate (and expensive) IF filters are required.

To avoid this problem and enjoy the best of both worlds, many high-performance receivers make use of *two* separate intermediate frequencies; the first IF provides a high degree of image channel rejection whilst the second IF provides a high degree of selectivity. Such receivers are said to use **dual conversion**.

A typical double superhet receiver is shown in Figure 3.22. The incoming signal frequency (26 MHz in the example) is converted to a first IF at 10.695 MHz by mixing the RF signal with a first local oscillator signal at 36.695 MHz (note that 36.695 MHz − 26 MHz = 10.695 MHz). The first IF signal is then filtered and amplified before it is passed to the second mixer stage.

The input of the second mixer (10.695 MHz) is then mixed with the second local oscillator signal at 10.240 MHz. This produces the second IF at 455 kHz (note that 10.695 MHz − 10.240 MHz = 455 kHz). The second IF signal is then filtered and amplified. It is worth noting that the

bulk of the gain is usually achieved in the second IF stages and there will normally be several stages of amplification at this frequency.

In order to tune the receiver, the first local oscillator is either made variable (using conventional tuned circuits) or is synthesised using digital phase-locked loop techniques (see page 70). The second local oscillator is almost invariably crystal controlled in order to ensure good stability and an accurate relationship between the two intermediate frequencies. Typical IF bandwidths in the receiver shown in Figure 3.22 are 75 kHz at the first IF and a mere 6 kHz in the second IF.

The first IF filter (not shown in Figure 3.22) is connected in the signal path between the first and second mixer. Where a stage of amplification is provided at the first IF, the filter precedes the amplifier stage. The requirements of the filter are not stringent since the ultimate selectivity of the receiver is defined by the second IF filter which operates at the much lower frequency of 455 kHz.

There are, however, some good reasons for using a filter which offers a high degree of rejection of the unwanted second mixer image

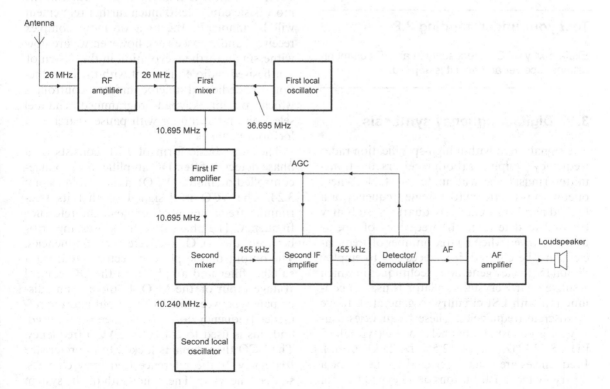

Figure 3.22 A double conversion superhet receiver

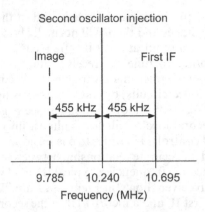

Figure 3.23 Second oscillator signal

response which occurs at 9.785 MHz. If this image is present at the input of the second mixer, it will mix with the second mixer injection at 10.240 MHz to produce a second IF component of 455 kHz, as shown in Figure 3.23. The function of the first IF filter is thus best described as **roofing**; bandwidth is a less important consideration.

Test your understanding 3.8

Explain why AGC is necessary in an HF communications receiver and how it is applied.

3.11 Digital frequency synthesis

The signals used within high-specification radio frequency equipment (both receivers and transmitters) must be both accurate and stable. Where operation is restricted to a single frequency or a limited number of channels, quartz crystals may be used to determine the frequency of operation. However, when a large number of frequencies must be covered, it is necessary to employ digital frequency generating techniques in which a single quartz crystal oscillator is used in conjunction with LSI circuitry to generate a range of discrete frequencies. These frequencies usually have a constant channel spacing (typically 3 kHz, 8.33 kHz, 9 kHz, 12.5 kHz, 25 kHz, etc.). Frequencies are usually selected by means of a rotary switch, push-buttons or a keypad but can also be stored in semiconductor memories.

Digital **phase-locked loop** (PLL) circuitry was first used in military communications equipment in the mid-1960s and resulted from the need to generate a very large number of highly accurate and stable frequencies in a multichannel frequency synthesiser. In this particular application cost was not a primary consideration and highly complex circuit arrangements could be employed involving large numbers of discrete components and integrated circuits.

Phase-locked loop techniques did not arrive in mass-produced equipment until the early 1970s. By comparison with today's equipment such arrangements were crude, employing as many as nine or ten integrated circuit devices. Complex as they were, these PLL circuits were more cost-effective than their comparable multi-crystal mixing synthesiser counterparts. With the advent of large-scale integration in the late 1970s, the frequency generating unit in most radio equipment could be reduced to one, or perhaps two, LSI devices together with a handful of additional discrete components. The cost-effectiveness of this approach is now beyond question, and it is unlikely that, at least in the most basic equipment, much further refinement will be made. In the area of more complex receivers and transceivers, however, we are now witnessing a further revolution in the design of synthesised radio equipment with the introduction of dedicated microcomputer controllers which permit keypad-programmed channel selection and scanning with pause, search and lock-out facilities.

The most basic form of PLL consists of a phase detector, filter, DC amplifier and voltage-controlled oscillator (VCO), as shown in Figure 3.24. The VCO is designed so that its free-running frequency is at, or near, the reference frequency. The phase detector senses any error between the VCO and reference frequencies. The output of the phase detector is fed, via a suitable filter and amplifier, to the DC control voltage input of the VCO. If there is any discrepancy between the VCO output and the reference frequency, an error voltage is produced, and this is used to correct the VCO frequency. The VCO thus remains locked to the reference frequency. If the reference frequency changes, so does the VCO. The bandwidth of the system is determined by the time constants of the loop

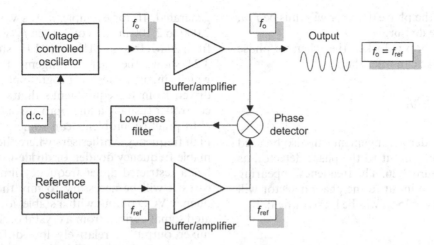

Figure 3.24   A simple phase-locked loop

filter. In practice, if the VCO and reference frequencies are very far apart, the PLL may be unable to lock.

The frequency range over which the circuit can achieve lock is known as the **capture range**. It should be noted that a PLL takes a finite time to achieve a locked condition and that the VCO locks to the mean value of the reference frequency.

The basic form of PLL shown in Figure 3.24 is limited in that the reference frequency is the same as that of the VCO and no provision is incorporated for changing it, other than by varying the frequency of the reference oscillator itself. In practice, it is normal for the phase detector to operate at a much lower frequency than that of the VCO output, and thus a frequency divider is incorporated in the VCO feedback path (see Figure 3.25). The frequency

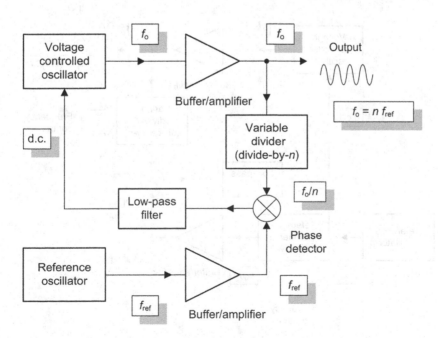

Figure 3.25   A phase-locked loop with frequency divider

presented to the phase detector will thus be f_o/n, where n is the **divisor**.

When the loop is locked (i.e. when no phase error exists) we can infer that:

$$f_{\text{ref}} = \frac{f_0}{n} \text{ or } f_o = nf_{\text{ref}}$$

A similar divider arrangement can also be used at the reference input to the phase detector, as shown in Figure 3.26. The frequency appearing at the reference input to the phase detector will be f_{ref}/m, and the loop will be locked when:

$$\frac{f_{\text{ref}}}{m} = \frac{f_0}{n} \text{ or } f_0 = \frac{n}{m} f_{\text{ref}}$$

Thus if f_{ref}, n and m were respectively 100 kHz, 2,000 and 10, the output frequency, f_{out}, would be:

$$(2,000 / 10) \times 100 \text{ kHz} = 20 \text{ MHz}$$

If the value of n can be made to change by replacing the fixed divider with a programmable divider, different output frequencies can be generated. If, for example, n was variable from 2,000 to 2,100 in steps of 1, then f_{out} would range from 20 MHz to 21 MHz in 10 kHz steps. Figure 3.24 shows the basic arrangement of a PLL which incorporates a programmable divider driven from the equipment's digital frequency controller (usually a microprocessor).

In practice, problems can sometimes arise in high frequency synthesisers where the programmable frequency divider, or **divide-by-n counter**, has a restricted upper frequency limit. In such cases it will be necessary to mix the high frequency VCO output with a stable, locally generated signal derived from a crystal oscillator. The mixer output (a relatively low difference frequency) will then be within the range of the programmable divider.

3.12 A design example

We shall bring this chapter to a conclusion by providing a design example of a complete HF communications receiver. This receiver was developed by the author for monitoring trans-Atlantic HF communications in the 5.5 MHz aircraft band. The circuit caters for the

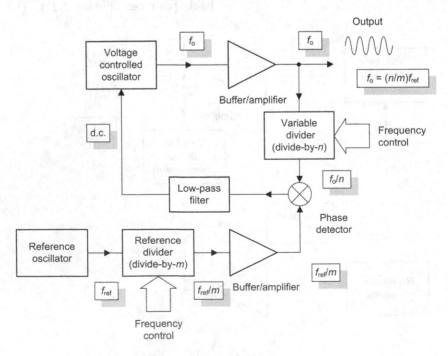

Figure 3.26   A complete digital frequency synthesiser

reception of AM, CW (Morse code) and SSB signals (see Chapter 5). To aid stability, the CIO/BFO frequency is controlled by means of a ceramic resonator. The RF performance is greatly enhanced by the use of dual-gate MOSFET devices in the RF amplifier, mixer and product detector stages and junction gate FETs in the local oscillator stage. These devices offer high gain with excellent strong-signal handling capability. They also permit simple and effective coupling between stages without the need for complex impedance matching.

The receiver is tunable over the frequency range 5.0 MHz to 6.0 MHz. Used in conjunction with a simple antenna, it offers reception of aircraft signals at distances in excess of 1000 km. The receiver is based on the single superhet principle operating with an intermediate frequency of 455 kHz. This frequency is low enough to ensure reasonable selectivity with just two stages of IF amplification and with the aid of a low-cost 455 kHz filter. Adequate image rejection is provided by two high-Q ganged RF tuned circuits.

The design uses conventional discrete component circuitry in all stages with the exception of the audio amplifier/output stage and voltage regulator. This approach ensures that the receiver is simple and straightforward to align and does not suffer from the limitations associated with several of the popular integrated circuit IF stages.

The block diagram of the receiver is shown in Figure 3.27 together with the circuits of its individual stages in Figures 2.28 to 2.34. The vast majority of the receiver's gain and selectivity is associated with the two IF stages, TR3 and TR4. These two stages provide over 40 dB of voltage gain, and the three IF tuned circuits and filter are instrumental in reducing the IF bandwidth to about 3.4 kHz for SSB reception. The RF stage (TR1) provides a modest amount of RF gain (about 20 dB at the maximum RF gain setting) together with a significant amount of image channel rejection.

The local oscillator stage (TR7) provides the necessary local oscillator signal which tunes from 5.455 MHz to 6.455 MHz. The local oscillator signal is isolated from the mixer stage and the LO output by means of the buffer stage, TR8.

The receiver incorporates two detector stages, one for AM and one for CW and SSB. The AM detector makes use of a simple diode envelope detector (D3) whilst the CW/SSB detector is based on a product detector (TR5). This stage offers excellent performance with both weak and strong CW and SSB signals. The 455 kHz carrier insertion is provided by means of the BFO/CIO stage (TR11). Amplified AGC is provided by means of TR9 and TR10.

A conventional integrated circuit audio amplifier stage (IC1) provides the audio gain necessary to drive a small loudspeaker. A 5-V regulator (IC2) is used to provide a stabilised low-voltage DC rail for the local oscillator and buffer stages. Diode switching is used to provide automatic changeover for the external DC or AC supplies. This circuitry also provides charging current for the internal nickel-metal hydride (NiMH) battery pack.

3.13 Software defined radio

In the radio architecture that we met previously (as well as that which we will be meeting in the next chapter) the entire signal path is implemented using application-specific hardware. In modern radio equipment there is an increasing use of software for control of the hardware and also for signal processing. This has led to the advent of two important technologies, software-controlled radio (SCR) and software defined radio (SDR). SCR is now extensively used in current equipment so that operating parameters can be configured using semiconductor electrically erasable read-only memories and digital controls such as selectors and rotary encoders. In software-controlled radio the signal path is implemented by hardware with some functionality controlled by software. Parameters that are often controlled by software include tuning, more selection, gain control, transmit power, etc.

Software defined radio takes this one step further with additionally much of the signal processing (including filtering, modulation/demodulation and encoding/decoding) being performed using software rather than hardware. Note that an SDR may still use conventional radio architecture in the front-end RF and mixer stages. SDR is an emerging technology showing considerable promise with state-of-the art implementation in commercial as well as military radio equipment.

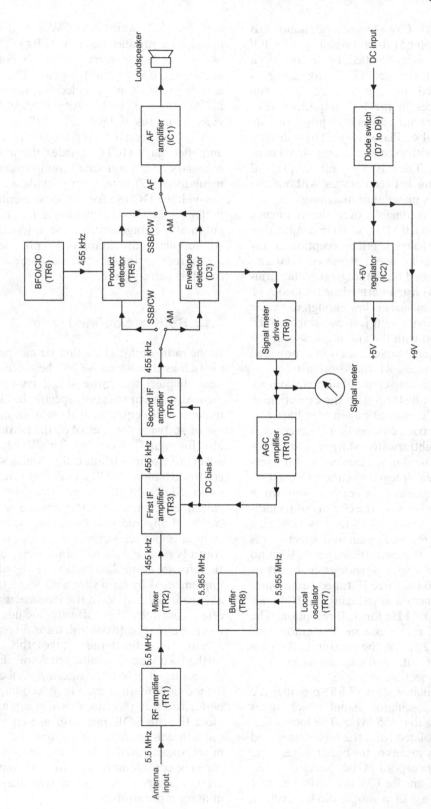

Figure 3.27 Superhet receiver design example

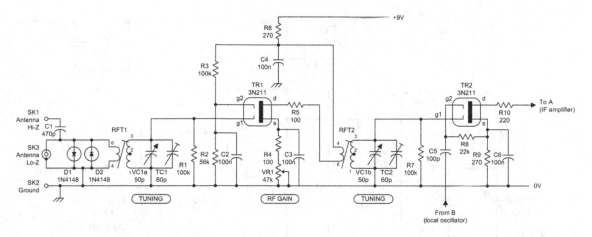

Figure 3.28 RF stages of the superhet receiver

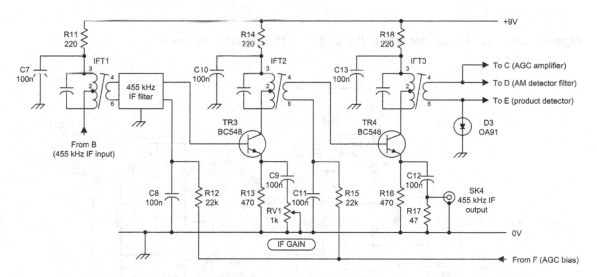

Figure 3.29 IF stages of the superhet receiver

With SDR technology the signal path can be easily reconfigured without the need for changes to the equipment hardware. Most of the digital signal processing within an SDR is conventionally implemented using one more field programmable logic arrays (FPGA) or an equivalent embedded processing device.

Key point

SDR use conventional analogue hardware architecture in their front-end RF circuitry with digital signal processing in the back end.

Key point

The signal path in SDR equipment can be quickly and easily reconfigured by making changes to software. This allows modification and upgrading without the need to change any hardware. It also permits the rapid cloning of operational parameters such as frequency, channel spacing and selectivity.

Figures 3.35 and 3.36 respectively show the simplified arrangements of receivers and transmitters based on SDR technology. Note how

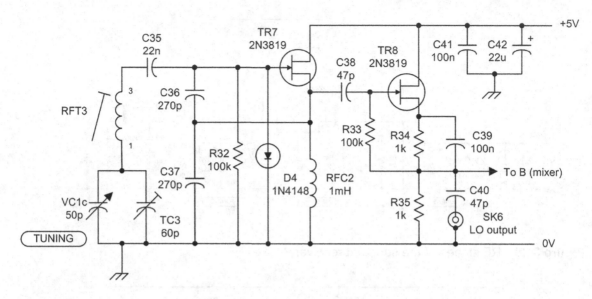

Figure 3.30   Local oscillator and buffer stages of the superhet receiver

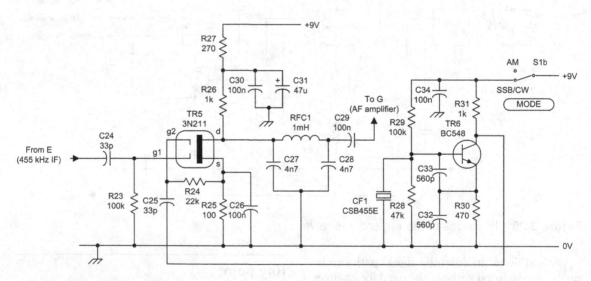

Figure 3.31   Product detector and BFO/CIO stages of the superhet receiver

analogue to digital conversion (ADC) is used in the receiver while digital to analogue conversion (DAC) is employed in the transmitter. Analogue circuitry is present in both the receiver and transmitter. The receiver uses analogue circuitry in the low-level RF amplifier and mixer stages while in the transmitter analogue technology is employed in the high-level driver and RF power amplifier stages.

In the SDR receiver arrangement shown in Figure 3.35 the RF sub-system (typically comprising band-pass filters and RF amplifiers) supplies an analogue signal to the splitter with identical in-phase signals applied to the two mixer stages. The local oscillator input to the two mixers is derived from a PLL VCO arrangement similar to those shown earlier in this chapter. The local oscillator input to one of the

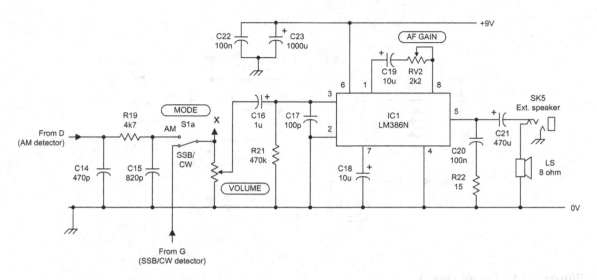

Figure 3.32 AF stages of the superhet receiver

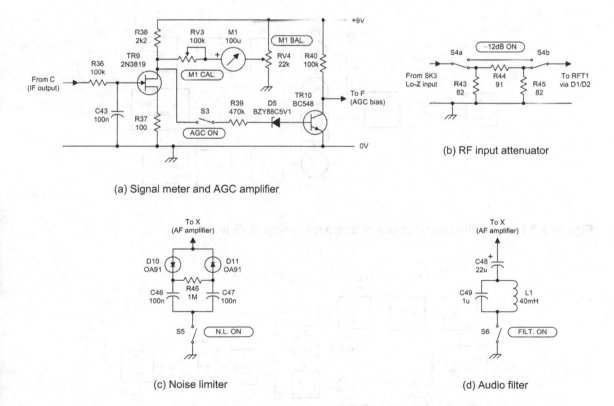

(a) Signal meter and AGC amplifier

(b) RF input attenuator

(c) Noise limiter

(d) Audio filter

Figure 3.33 Signal meter, AGC, RF input attenuator, noise limiter and audio filter stages

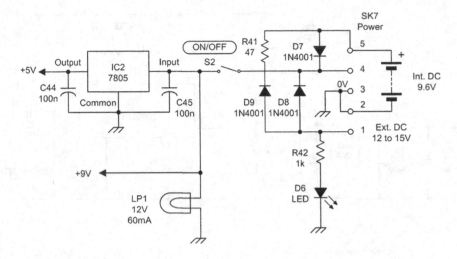

Figure 3.34   Power supply

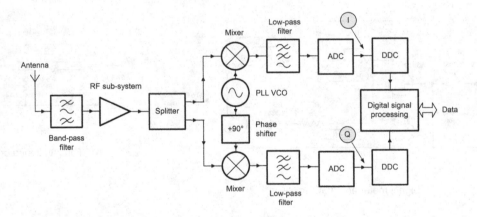

Figure 3.35   Simplified architecture of a receiver using SDR technology

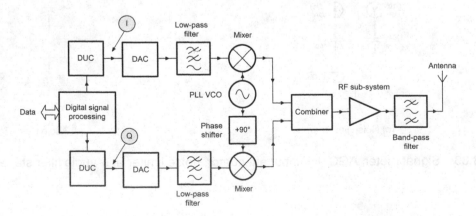

Figure 3.36   Simplified architecture of a transmitter using SDR technology

mixers is phase shifted by an angle of 90° so that the two local oscillators signals are in phase quadrature. The two mixer outputs are passed through low-pass filters before being applied to two analogue-to-digital converters (ADC) the outputs of which constitute in-phase, I, and quadrature, Q, components. The I and Q signals are then passed into a digital down converter (DDC) to reduces the sampling rate of the signal before it is passed to the digital signal processor (DSP) where the modulation (AM, FM or PSK data) is recovered from corresponding pairs of down-sampled quadrature data.

In the corresponding SDR transmitter arrangement shown in Figure 3.36 the digital data is processed before being applied to digital up-converters (DUC) from which the I and Q signals are derived. These are then fed to two DAC, each followed by low-pass filters. The mixing process (similar to that used in the SDR receiver shown in Figure 3.35) produces two quadrature signals that are combined before application to the RF sub-system typically comprising a pre-driver, driver and power amplifier. After band-pass filtering, the final output is applied to the antenna.

Key point

The digital back-end of an SDR contains an FPGA (or equivalent embedded processing) with on-board DSP functionality for modulation, demodulation, up-converting, down-converting, coding, decoding, and protocol handling. All of this is achieved in software rather than hardware.

Key point

Legacy radio platforms are non-digital, use analogue signal processing, and their functionality cannot be reconfigured by software.

Test your understanding 3.9

State two advantages of SDR technology compared with legacy (analogue) radio architecture.

Test your understanding 3.10

List four parameters that are usually configurable by software in an SDR.

Important note: Further details of the principles and operation of devices such as EPROM, FPGA, ADC and DAC, can be found in our companion volume, "Aircraft Digital Electronic and Computer Systems", 3rd edition published by Taylor Francis ISBN 9781032104805.

3.14 Multiple choice questions

1. A receiver in which selected signals of any frequency are converted to a single frequency is called a:
   (a) wideband TRF
   (b) multi-channel receiver
   (c) superhet receiver.

2. Delayed AGC:
   (a) maintains receiver sensitivity for very small signals
   (b) increases receiver sensitivity for very large signals
   (c) has no effect on receiver sensitivity.

3. A receiver with a high IF will successfully reject:
   (a) the image frequency
   (b) the adjacent frequency
   (c) the local oscillator frequency.

4. An IF amplifier consists of several stages. These are normally coupled using:
   (a) resistor/capacitor coupling
   (b) pure resistor coupling
   (c) transformer coupling.

5. SSB filters have a typical bandwidth of:
   (a) less than 300 Hz
   (b) 3 kHz to 6 kHz
   (c) more than 10 kHz.

6. The output signal of a diode detector comprises the modulated waveform, a small ripple and a DC component. The DC component is:

(a) independent of the carrier strength
(b) proportional to the carrier strength
(c) inversely proportional to the carrier strength.

7. What is the principal function of the RF stage in a superhet receiver?
   (a) To improve the sensitivity of the receiver
   (b) To reduce second channel interference
   (c) To reduce adjacent channel interference.

8. A receiver having an IF of 1.6 MHz is tuned to a frequency of 12.8 MHz. Which of the following signals could cause image channel interference?
   (a) 11.2 MHz
   (b) 14.5 MHz
   (c) 16.0 MHz.

9. In an FM transmitter, the modulating signal is applied to:
   (a) the final RF amplifier stage
   (b) the antenna coupling unit
   (c) the RF oscillator stage.

10. The response of two coupled tuned circuits appears to be 'double-humped'. This is a result of:
    (a) under-coupling
    (b) over-coupling
    (c) critical coupling.

11. A disadvantage of low-level amplitude modulation is the need for:
    (a) a high-power audio amplifier
    (b) a high-power RF amplifier
    (c) a linear RF power amplifier.

12. The function of an antenna coupling unit in a transmitter is:
    (a) to provide a good match between the RF power amplifier and the antenna
    (b) to increase the harmonic content of the radiated signal
    (c) to reduce the antenna SWR to zero.

13. In order to improve the stability of a local oscillator stage:
    (a) a separate buffer stage should be used
    (b) the output signal should be filtered
    (c) an IF filter should be used.

14. A dual conversion superhet receiver uses:
    (a) a low first IF and a high second IF
    (b) a high first IF and a low second IF
    (c) the same frequency for both first and second IF.

15. The majority of the gain in a superhet receiver is provided by:
    (a) the RF amplifier stage
    (b) the IF amplifier stage
    (c) the AF amplifier stage.

16. Image channel rejection in a superhet receiver is improved by:
    (a) using an IF filter
    (b) using a low IF
    (c) using a high IF.

17. In a receiver based on SDR technology the I and Q signals are derived from:
    (a) a frequency multiplier
    (b) a beat frequency oscillator
    (c) a quadrature sampling detector.

Chapter 4 — VHF communications

Very high frequency (VHF) radio has long been the primary means of communication between aircraft and the ground. The system operates in the frequency range extending from 118 MHz to 137 MHz and supports both voice and data communication (the latter becoming increasingly important). This chapter describes the equipment used and the different modes in which it operates.

VHF communication is used for various purposes including air traffic control (ATC), approach and departure information, transmission of meteorological information, ground handling of aircraft, company communications and also for the Aircraft Communications Addressing and Reporting System (ACARS).

4.1 VHF range and propagation

In the VHF range (30 MHz to 300 MHz) radio waves usually propagate as direct line-of-sight (LOS) waves (see Chapter 1). Sky wave propagation still occurs at the bottom end of the VHF range (up to about 50 MHz depending upon solar activity) but at the frequencies used for aircraft communication, reflection from the ionosphere is exceptionally rare.

Communication by strict line-of-sight paths, augmented on occasions by diffraction and reflection, imposes a limit on the working range that can be obtained. It should also be evident that the range will be dependent on the height of an aircraft above the ground; the greater this is the further the range will be.

The maximum LOS distance (see Figure 4.1) between an aircraft and a ground station, in nautical miles (nm), is given by the relationship:

$$d = 1.1\sqrt{h}$$

where h is the aircraft's altitude in feet above ground (assumed to be flat terrain).

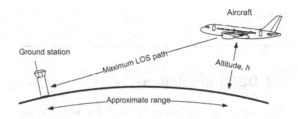

Figure 4.1 VHF line-of-sight range

Example 4.1

Determine the maximum line-of-sight distance when an aircraft is flying at a height of (a) 2500 feet, and (b) 25,000 feet.

In (a), h = 2500 hence:

$$d = 1.1\sqrt{2500} = 1.1 \times 50 = 55 \text{ nm}$$

In (b), h = 25,000 hence:

$$d = 1.1\sqrt{25,000} = 1.1 \times 158 = 174 \text{ nm}$$

The actual range obtained depends not only on the LOS distance but also on several other factors, including aircraft position, transmitter power and receiver sensitivity. However, the LOS distance usually provides a good approximation of the range that can be obtained between an aircraft and a ground station (see Table 4.1). The situation is slightly more complex when communication is from one aircraft to another; however, in such cases summing the two LOS distances will normally provide a guide as to the maximum range that can be expected.

Test your understanding 4.1

Determine the altitude of an aircraft that would provide a line-of-sight distance to a ground station located at a distance of 125 nm.

DOI: 10.1201/9781003411932-4

Table 4.1 Theoretical LOS range

Altitude (feet)	Approx. LOS range (nm)
100	10
1000	32
5000	70
10,000	100
20,000	141

4.2 DSB modulation

Amplitude modulation is used for voice com-munications as well as several of the **VHF data-link** (VDL) modes. The system uses **double sideband** (DSB) modulation and, because this has implications for the bandwidth of modulated signals, it is worth spending a little time explaining how this works before we look at how the available space is divided into channels.

Figure 4.2 shows the frequency spectrum of an RF carrier wave at 124.575 MHz amplitude modulated by a single pure sinusoidal tone with a frequency of 1 kHz. Note how the amplitude-modulated waveform comprises three separate components:

- an **RF carrier** at 124.575 MHz
- a **lower side frequency** (LSF) component at 124.574 MHz
- an **upper side frequency** (USF) component at 124.576 MHz.

Note how the LSF and USF are spaced away from the RF carrier by a frequency that is equal to that of the modulating signal (in this case 1 kHz). Note also from Figure 4.2 that the **bandwidth** (i.e. the range of frequencies occupied by the modulated signal) is *twice* the frequency of the modulating signal (i.e. 2 kHz).

Figure 4.3 shows an RF carrier modulated by a speech signal rather than a single sinusoidal tone. The **baseband** signal (i.e. the voice signal itself) typically occupies a frequency range extending from around 300 Hz to 3.4 kHz. Indeed, to improve intelligibility and reduce extraneous no ise, the frequency response of the microphone and speech amplifier is invariably designed to select this particular range of frequencies and reject any audio signals that lie

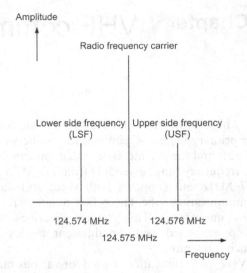

Figure 4.2 Frequency spectrum of an RF carrier using DSB modulation and a pure sinusoidal modulating signal

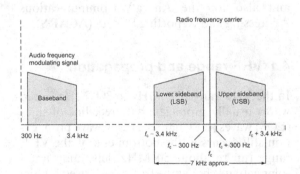

Figure 4.3 Frequency spectrum of a baseband voice signal (left) and the resulting DSB AM RF carrier (note that the bandwidth of the RF signal is approximately twice that of the highest modulating signal frequency)

outside it. From Figure 4.3 it should be noted that the bandwidth of the RF signal is approximately 7 kHz (i.e. twice that of the highest modulating signal).

Test your understanding 4.2

Determine the RF signal frequency components present in a DSB amplitude-modulated carrier wave at 118.975 MHz when the modulating signal comprises pure tones at 2 kHz and 5 kHz.

4.3 Channel spacing

VHF aircraft communications take place in a number of allocated channels. These channels were originally spaced at 200 kHz intervals throughout the VHF aircraft band. However, a relentless increase in air traffic coupled with the increasing use of avionic systems for datalink communication has placed increasing demands on the available frequency spectrum. In response to this demand, the spacing between adjacent channels in the band 118 MHz to 137 MHz has been successively reduced so as to increase the number of channels available for VHF communication (see Table 4.2).

Figure 4.4 shows the channel spacing for the earlier 25 kHz and current European 8.33 kHz VHF systems. Note how the 8.33 kHz system of channel spacing allows three DSB AM signals to occupy the space that was previously occupied by a single signal.

The disadvantage of narrow channel spacing is that the **guard band** of unused spectrum that previously existed with the 25 kHz system is completely absent and that receivers must be designed so that they have a very high degree of **adjacent channel rejection** (see page 66). Steps must also be taken to ensure that the bandwidth of the transmitted signal does not exceed the 7 kHz, or so, bandwidth required for effective voice communication. The penalty for not restricting the bandwidth is that signals from one channel may 'spill over' into the adjacent channels, causing interference and degrading communication (see Figure 4.6).

Table 4.2 Increase in the number of available VHF channels

Date	Frequency range	Channel spacing	Number of channels
1947	118 MHz to 132 MHz	200 kHz	70
1958	118 MHz to 132 MHz	100 kHz	140
1959	118 MHz to 136 MHz	100 kHz	180
1964	118 MHz to 136 MHz	50 kHz	360
1972	118 MHz to 136 MHz	25 kHz	720
1979	118 MHz to 137 MHz	25 kHz	760
1995	118 MHz to 137 MHz	8.33 kHz	2280

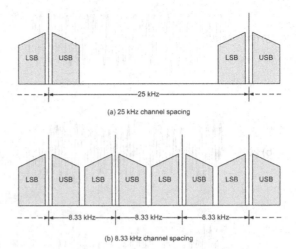

Figure 4.4 25 kHz and 8.33 kHz channel spacing

Test your understanding 4.3

How many channels at a spacing of 12.5 kHz can occupy the band extending from 118 MHz to 125 MHz?

Test your understanding 4.4

A total of 1520 data channels are to be accommodated in a band extending from 316 MHz to 335 MHz. What channel spacing must be used and what range of frequencies can the baseband signal have?

4.4 Depth of modulation

The depth of modulation of an RF carrier wave is usually expressed in terms of **percentage modulation**, as shown in Figure 4.5. Note that the level of modulation can vary between 0% (corresponding to a completely unmodulated carrier) and 100% (corresponding to a fully modulated carrier).

In practice, the intelligibility of a signal (i.e. the ability to recover information from a weak signal that may be adversely affected by noise and other disturbances) increases as the percentage modulation increases, and hence there is

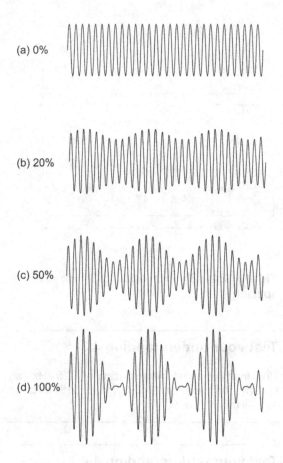

(a) 0%

(b) 20%

(c) 50%

(d) 100%

Figure 4.5   Different modulation depths

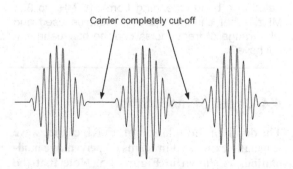

Carrier completely cut-off

Figure 4.6   Overmodulation

a need to ensure that a transmitted signal is fully modulated but without the attendant risk of overmodulation (see Figure 4.6). The result of overmodulation is excessive bandwidth, or 'splatter', causing adjacent channel interference, as shown in Figure 4.7.

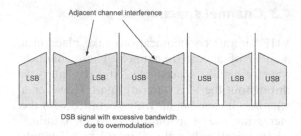

Adjacent channel interference

LSB   LSB   USB   USB   LSB   USB

DSB signal with excessive bandwidth
due to overmodulation

Figure 4.7   Adjacent channel interference cause by overmodulation

4.5 Compression

In order to improve the intelligibility of VHF voice communications, the speech amplifier stage of an aircraft VHF radio is invariably fitted with a **compressor** stage. This stage provides high gain for low amplitude signals and reduced gain for high amplitude signals. The result is an increase in the average modulation depth (see Figure 4.8).

Figure 4.9 shows typical speech amplifier characteristics with and without compression. Note that most aircraft VHF radio equipment provides adjustment both for the level of modulation and for the amount of compression that is applied (see Figure 4.10).

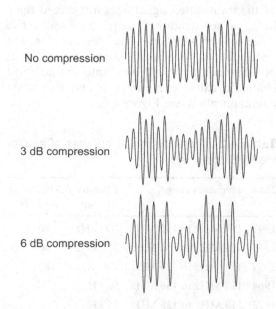

No compression

3 dB compression

6 dB compression

Figure 4.8   Modulated RF carrier showing different amounts of compression applied to the modulating signal

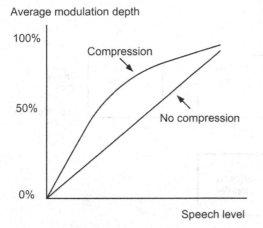

Figure 4.9 Effect of compression on average modulation depth

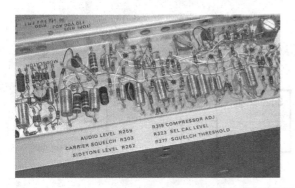

Figure 4.10 VHF radio adjustment points

4.6 Squelch

Aircraft VHF receivers invariably incorporate a system of muting the receiver audio stages in the absence of an incoming signal. This system is designed to eliminate the annoying and distracting background noise that is present when no signals are being received. Such systems are referred to as **squelch**, and the threshold at which this operates is adjusted (see Figure 4.10) so that the squelch 'opens' for a weak signal but 'closes' when no signal is present.

Two quite different squelch systems are used, but the most common (and easiest to implement) system responds to the amplitude of the received carrier and is known as **carrier operated squelch**. The voltage used to inhibit the receiver

audio can be derived from the receiver's AGC system and fed to the squelch gate (Figure 4.11).

The alternative (and somewhat superior) squelch system involves sensing the noise present at the output of the receiver's detector stage and using this to develop a control signal which is dependent on the signal-to-noise ratio of the received signal rather than its amplitude. This latter technique, which not only offers better sensitivity but is also less prone to triggering from general background noise and off-channel signals, is often found in FM receivers and is referred to as **noise operated squelch**.

4.7 Data modes

Modern aircraft VHF communications equipment supports both data communication as well as voice communication. The system used for the aircraft datalink is known as Aircraft Communications Addressing and Reporting System (ACARS). Currently, aircraft are equipped with three VHF radios, two of which are used for ATC voice communications and one of which is used for the ACARS datalink (also referred to as **airline operational control** communications).

A datalink terminal on board the aircraft (see Figure 4.12) generates **downlink** messages and processes **uplink** messages received via the VHF datalink. The downlink and uplink ACARS messages are encoded as plain ASCII text. In the Unites States, the ACARS ground stations are operated by ARINC, whilst in Europe, Asia and Latin America, the equivalent service is provided by SITA.

Initially each VHF ACARS provider was allocated a single VHF channel. However, as the use of VHF datalinks (VDL) has grown, the number of channels used in the vicinity of the busiest airports has increased to as many as four, and these are often operating at full capacity.

Unfortunately, due to the pressure for additional voice channels, it has not been possible to assign a number of additional VHF channels for ACARS datalink operation. As a result, several new data modes have recently been introduced that support higher data rates and make more efficient use of each 25 kHz channel currently assigned for datalink purposes.

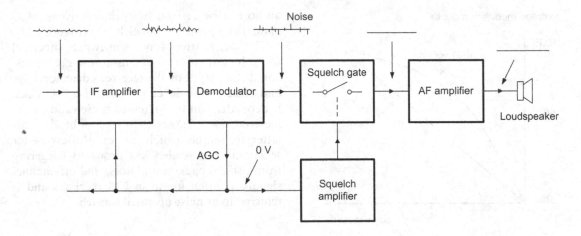

(a) No signal present (squelch gate open)

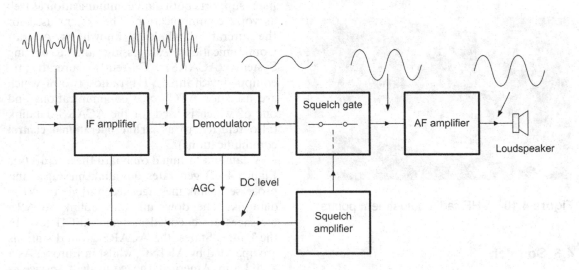

(b) Signal present (squelch gate closed)

Figure 4.11   Action of the squelch system

In addition, the FAA is developing a system that will permit the integration of ATC voice and data communications. This system uses digitally encoded audio rather than conventional analogue voice signals.

When operating in **VDL Mode 0**, the required datalink protocols are implemented in the ACARS **management unit** (see Figure 4.11). Data is transferred from the VHF radio to the management unit at a rate of 2400 bits per second (bps) by means of **frequency shift keying** (FSK). The FSK audio signal consists of two sinusoidal tones, one at 1.2 kHz and one at 2.4 kHz depending on whether the polarity of the information bit being transmitted is the same as that of the previous bit or is different. Note that the phase of the tones varies linearly and that there is no phase change on the transition between the two tones.

This type of modulation (in which the frequency spacing between the two audio tones is exactly half the data rate) is highly efficient in

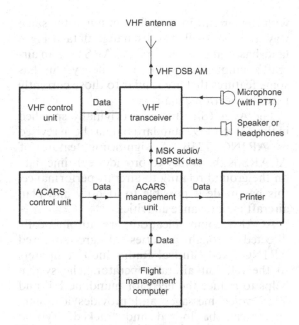

Figure 4.12 VHF radio data management

terms of bandwidth and is thus referred to as **minimum shift keying** (MSK). When data is transmitted, the MSK signal is used to modulate the amplitude of the VHF carrier (in much the same way as the voice signal). The resultant transmitted signal is then a double side-band (DSB) AM signal whose amplitude is modulated at 2400 bps. The RF carrier is then said to use DSB AM MSK modulation.

VHF carrier frequency selection and transmit/receive control are provided by the ACARS management unit working in conjunction with an ARINC 429 interface to the VHF radio (Figure 4.12). The channel access protocol employed is known as **carrier sense multiple**

access (CSMA). It consists of listening for activity on the channel (i.e. transmissions from other users) and transmitting only when the channel is free.

Operation in **VDL Mode A** is similar to Mode 0 except uplink and downlink ACARS data packets are transferred between the VHF radio and the ACARS management unit via a transmit/receive pair of 100 kbps ARINC 429 digital interfaces rather than the analogue audio interface used by Mode 0. The digital data is then used by the VHF radio to modulate the RF carrier at a rate of 2400 bps using the same DSB AM MSK modulation scheme used by VDL Mode 0.

Another difference between VDL Mode 0 and VDL Mode A is that, when using the latter, the VHF radio controls when to access the channel to transmit data using the same CSMA protocol employed by the management unit in VDL Mode 0. However, the selection of the frequency to be used is still controlled by the CMU or ATSU by means of commands issued via the same ARINC 429 interface used for data transfer. Note that, as far as the VHF datalink ground stations are concerned, there is no difference in the air/ground VDL Mode 0 or VDL Mode A transmissions.

Operation in **VDL Mode 2** is based on an improved set of data transfer protocols and, as a result, it provides a significant increase in data capacity. VDL Mode 2 has been designed to provide for the future migration of VDL to the aeronautical telecommunications network (ATN). This network will permit more efficient and seamless delivery of data messages and data files between aircraft and the ground computer systems used by airlines and air traffic control facilities.

Table 4.3 Summary of voice and data modes

Mode	Modulation	Channel spacing	Access method	Data rate	Type of traffic	Radio interface
Voice	DSB AM	25/8.33 kHz	PTT	Not applicable	Voice	Analogue
Data (Mode 0)	DSB AM MSK	25 kHz	CSMA	2400 bps	ACARS	Analogue
Data (Mode A)	DSB AM MSK	25 kHz	CSMA	2400 bps	ACARS	ARINC 429
Data (Mode 2)	D8PSK	25 kHz	CSMA	31,500 bps	ACARS and ATN	ARINC 429

ATN will be supported by a number of air/ground networks and ground/ground networks. The air/ground and ground/ground networks will be interconnected by means of ATN **routers** that implement the required protocols and will operate in much the same way as the Internet with which you are probably already familiar.

VDL Mode 2 employs a data rate of 31,500 bits per second over the air/ground link using a single 25 kHz channel. The increased utilisation of the 25 kHz channel is achieved by employing a system of modulation that is more efficient in terms of its use of bandwidth. This system is known as **differential eight phase shift keying** (D8PSK). In this system, an audio carrier signal is modulated by means of shift in phase that can take one of eight possible phases; 0°, 45°, 90°, 135°, 180°, 225°, 270° or 315°. These phase changes correspond to three bits of digital data as follows: 000, 001, 011, 010, 110, 111, 101 or 100. The D8PSK modulator uses the bits in the data message, in groups of three, to determine the carrier phase change at a rate of 10.5 kHz. Consequently, the bit rate will be three times this value, or 31.5 kbps. D8PSK modulation of the phase of the VHF carrier is accomplished using a **quadrature modulator**. Note that, in D8PSK modulation, groups of three bits are often referred to as **D8PSK symbols**.

VDL Mode 3 offers an alternative to the European solution of reducing the channel spacing to 8.33 kHz. VDL Mode 3 takes a 25 kHz frequency assignment and divides it into 120 ms frames with four 30 ms time slots (each of which constitutes a different channel). Thus Mode 3 employs **time division domain multiplexing** (TDM) rather than **frequency division domain multiplexing** (FDM) used in the European system. Note that VDL Mode 3 is the only planned VDL mode that is designed to support voice and data traffic *on the same frequency*.

4.8 ACARS

ACARS (Aircraft Communications Addressing and Reporting System) is a digital datalink system transmitted in the VHF range (118 MHz to 136 MHz). ACARS provides a means by which aircraft operators can exchange data with an aircraft without human intervention. This makes it possible for an airline to communicate with the aircraft in their fleet in much the same way as it is possible to exchange data using a land-based digital network. ACARS uses an aircraft's unique identifier, and the system has some features that are similar to those currently used for electronic mail.

The ACARS system was originally specified in the ARINC 597 standard but has been revised as ARINC 724B. A significant feature of ACARS is the ability to provide real-time data on the ground relating to aircraft performance; this has made it possible to identify and plan aircraft maintenance activities.

ACARS communications are automatically directed through a series of ground-based ARINC (Aeronautical Radio Inc.) computers to the relevant aircraft operator. The system helps to reduce the need for mundane HF and VHF voice messages and provides a system which can be logged and tracked. Typical ACARS messages are used to convey routine information such as:

- passenger loads
- departure reports
- arrival reports
- fuel data
- engine performance data.

This information can be requested by the company and retrieved from the aircraft at periodic intervals or on demand. Prior to ACARS this type of information would have been transferred via VHF voice.

ACARS uses a variety of hardware and software components including those that are installed on the ground and those that are present in the aircraft. The aircraft ACARS components include a **management unit** (see Figure 4.12) which deals with the reception and transmission of messages via the VHF radio transceiver and the **control unit** which provides the crew interface and consists of a display screen and printer. The ACARS **ground network** comprises the ARINC ACARS remote transmitting/receiving stations and a network of computers and switching systems. The ACARS **command, control and management subsystem** consists of the ground-based airline operations and associated functions including operations control, maintenance and crew scheduling.

There are two types of ACARS messages; **downlink** messages that originate from the aircraft and **uplink** messages that originate from ground stations (see Figures 4.13 to 4.17). Frequencies used for the transmission and reception of ACARS messages are in the band extending from 129 MHz to 137 MHz (VHF), as shown in Table 4.4. Note that different channels are used in different parts of the world. A typical ACARS message (see Figure 4.14) consists of:

* mode identifier (e.g. 2)
* aircraft identifier (e.g. G-DBCC)
* message label (e.g. 5U—a weather request)
* block identifier (e.g. 4)
* message number (e.g. M55A)
* flight number (e.g. BD01NZ)
* message content (see Figure 4.14).

```
ACARS mode: E Aircraft reg: N27015
Message label: H1 Block id: 3
Msg no: C36C
Flight id: CO0004
Message content:-
#CFBBY ATTITUDE INDICATOR
MSG 2820121 A 0051 06SEP06 CL H PL
DB FUEL QUANTITY PROCESSOR UNIT
MSG 3180141 A 0024 06SEP06 TA I 23
PL
DB DISPLAYS-2 IN LEFT AIMS
MSG 2394201 A 0005 06SEP06 ES H PL
MSG 2717018
```

Figure 4.13 Example of a downlink ACARS message sent from a Boeing 777 aircraft

```
ACARS mode: 2
Aircraft reg: G-DBCC
Message label: 5U
Block id: 4
Msg no: M55A
Flight id: BD01NZ
Message content:-
01 WXRQ 01NZ/05 EGLL/EBBR .G-DBCC
/TYP 4/STA EBBR/STA EBOS/STA EBCI
```

Figure 4.14 Example of an ACARS message (see text)

```
ACARS mode: 2 Aircraft reg: N788UA
Message label: RA Block id: L
Msg. no: QUHD
Flight id: QWDUA~
Message content:-
WEIGHT MANIFEST
UA930 SFOLHR
SFO
ZFW      383485
TOG      559485
MAC      40.1
TRIM     02.8
PSGRS    285
```

Figure 4.15 Example of aircraft transmitted data (in this case, a weight manifest)

```
ACARS mode: X Aircraft reg: N199XX
Message label: H1 Block id: 7
Msg no: F00M
Flight id: GS0000
Message content:-
#CFBER FAULT/WRG [SWPA2]
INTERFACE
TCAS FAIL ADVISORY
TERRAIN 1 FAIL ADVISORY
TERRAIN 1-2 FAIL ADVISORY
THROTTLE QUADRANT 1-2 FAIL ADVISORY
22-10 221009ATA1 OC=1
TQA FAULT [ATA1]
INTERFACE
22-10 221009ATA
```

Figure 4.16 Example of a failure advisory message transmitted from an aircraft

```
ACARS mode: R Aircraft reg: G-EUPR
Message label: 10 Block id: 8
Msg no: M06A
Flight id: BA018Z
Message content:-
FTX01.ABZKOBA
BA1304
WE NEED ENGINEERING TO DO PDC ON
NUMBER 2 IDG
CHEERS
ETL 0740 GMT
```

Figure 4.17 Example of a plain text message sent via ACARS

Table 4.4   ACARS channels

Frequency	ACARS service
129.125 MHz	USA and Canada (additional)
130.025 MHz	USA and Canada (secondary)
130.450 MHz	USA and Canada (additional)
131.125 MHz	USA (additional)
131.475 MHz	Japan (primary)
131.525 MHz	Europe (secondary)
131.550 MHz	USA, Canada, Australia (primary)
131.725 MHz	Europe (primary)
136.900 MHz	Europe (additional)

Test your understanding 4.5

Explain the need for (a) speech compression and (b) squelch in an aircraft VHF radio.

Test your understanding 4.6

Explain, with the aid of a block diagram, how data transfer is possible using an aircraft VHF radio.

Test your understanding 4.7

Explain the difference between MSK and D8PSK modulation. Why is the latter superior?

4.9 VHF radio equipment

The typical specification of a modern aircraft VHF data radio is shown in Table 4.5. This radio can be used with analogue voice as well as data in Modes 0, A and 2 (see page 87). Figures 4.18 to 4.20 show typical equipment and control locations in a passenger aircraft whilst

Table 4.5   Aircraft VHF radio specifications

Parameter	Specification
Frequency range	118.00 MHz to 136.99167 MHz
Channel spacing	8.33 kHz or 25 kHz
Operating modes	Analogue voice (ARINC 716); Analogue data 2400 bps AM MSK ACARS (external modem); ARINC 750 Mode A analogue data 2400 bps AM MSK ACARS; Mode 2 data 31.5 kbps D8PSK
Sensitivity	2 μV for 6 dB (S+N)/N
Selectivity (25 kHz channels)	6 dB max. attenuation at ±16 kHz 60 dB min. attenuation at ±34 kHz
Selectivity (8.33 kHz channels)	6 dB max. attenuation at ± 5.5 kHz 60 dB min. attenuation at ±14.7 kHz
Audio power output	Adjustable from less than 50 μW to 50 m into 600 Ω ± 20%
RF output power	25 W min. DSB AM operation 18 W min. D8PSK operation
Frequency stability	±0.005%
Modulation level	0.25 V RMS input at 1 kHz will modulate the transmitter at least 90%
Speech processing	Greater than 20 dB of compression
Mean time between failure	Greater than 40,000 hours

Figures 4.21 to 4.24 show internal and external views of a typical VHF radio. Finally, Figure 4.25 shows a typical VHF quarter-wave blade antenna fitted to an Airbus A380 aircraft.

Key point

Checks for satisfactory VHF radio communication should be carried out by carrying out systematic tests with a nearby ground station. Tests should only be made on approved frequencies.

Figure 4.18 Three VHF radios (on the extreme left) installed in the aircraft's avionic equipment bay

Figure 4.19 VHF communications frequency selection panel (immediately above the ILS panel)

Figure 4.20 ACARS control panel (immediately to the right of the VHF communications frequency selection panel)

Figure 4.21 Aircraft VHF radio removed from its rack mounting

Figure 4.22 Digital frequency synthesiser stages of the VHF radio. The quartz crystal controlled reference oscillator is at the bottom left corner and the frequency divider chain runs from left to right with the screened VCO at the top

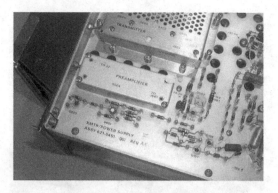

Figure 4.23 Screened receiver pre-amplifier and transmitter power amplifier stages (top)

Figure 4.24 RF power amplifier stages with the screening removed. There are three linear power stages and one driver (left)

Figure 4.25 Typical VHF antenna location on the Airbus A380 (see page 27 for the VHF antenna locations on a Boeing 757)

4.10 Multiple choice questions

1. The angle between successive phase changes of a D8PSK signal is:
   (a) 45°
   (b) 90°
   (c) 180°.

2. The method of modulation currently employed for aircraft VHF voice communication is:
   (a) MSK
   (b) D8PSK
   (c) DSB AM.

3. The channel spacing currently used in Europe for aircraft VHF voice communication is:
   (a) 8.33 kHz and 25 kHz
   (b) 12.5 kHz and 25 kHz
   (c) 25 kHz and 50 kHz.

4. Which one of the following gives the approximate LOS range for an aircraft at an altitude of 15,000 feet?
   (a) 74 nm
   (b) 96 nm
   (c) 135 nm.

5. The function of the compressor stage in an aircraft VHF radio is:
   (a) to reduce the average level of modulation
   (b) to increase the average level of modulation
   (c) to produce 100% modulation at all times.

6. The function of the squelch stage in an aircraft VHF radio is:
   (a) to eliminate noise when no signal is received
   (b) to increase the sensitivity of the receiver for weak signals
   (c) to remove unwanted adjacent channel interference.

7. Large passenger aircraft normally carry:
   (a) two VHF radios
   (b) three VHF radios
   (c) four VHF radios.

8. The typical bandwidth of a DSB AM voice
   signal is:
   (a) 3.4 kHz
   (b) 7 kHz
   (c) 25 kHz.

9. The disadvantage of narrow channel
   spacing is:
   (a) the need for increased receiver
       sensitivity
   (b) the possibility of adjacent channel
       interference
   (c) large amounts of wasted space between
       channels.

10. The standard for ACARS is defined in:
    (a) ARINC 429
    (b) ARINC 573
    (c) ARINC 724.

11. The frequency band currently used in
    Europe for aircraft VHF voice
    communication is:
    (a) 88 MHz to 108 MHz
    (b) 108 MHz to 134 MHz
    (c) 118 MHz to 137 MHz.

12. The typical output power of an aircraft
    VHF radio using voice mode is:
    (a) 25 W
    (b) 150 W
    (c) 300 W.

Chapter 5   HF communications

High frequency (HF) radio provides aircraft with an effective means of communication over long distance oceanic and trans-polar routes. In addition, global data communication has recently been made possible using strategically located HF datalink (HFDL) ground stations. These provide access to ARINC and SITA airline networks. HF communication is thus no longer restricted to voice and is undergoing a resurgence of interest due to the need to find a means of long distance data communication that will augment existing VHF and SATCOM datalinks.

An aircraft HF radio system operates on spot frequencies within the HF spectrum. Unlike aircraft VHF radio, the spectrum is not divided into a large number of contiguous channels, but aircraft allocations are interspersed with many other services, including short wave broadcasting, fixed point-to-point, marine and land-mobile, government and amateur services. This chapter describes the equipment used and the different modes in which it operates.

5.1 HF range and propagation

In the HF range (3 MHz to 30 MHz) radio waves propagate over long distances due to reflection from the ionised layers in the upper atmosphere. Due to variations in height and intensities of the ionised regions, different frequencies must be used at different times of day and night and for different paths. There is also some seasonal variation (particularly between winter and summer). Propagation may also be disturbed and enhanced during periods of intense solar activity. The upshot of this is that HF propagation has considerable vagaries and is far less predictable than propagation at VHF.

Frequencies chosen for a particular radio path are usually set roughly mid-way between the lowest usable frequency (LUF) and the maximum usable frequency (MUF). The daytime LUF is usually between 4 to 6 MHz during the day, falling rapidly after sunset to around 2 MHz. The MUF is dependent on the season and sunspot cycle but is often between 8 MHz and 20 MHz. Hence a typical daytime frequency for aircraft communication might be 8 MHz whilst this might be as low as 3 MHz during the night. Typical ranges are in the region of 500 km to 2500 km, and this effectively fills in the gap in VHF coverage (see Figure 5.1).

As an example of the need to change frequencies during a 24-hour period, Figure 5.2 shows how the service provided by the Santa Maria HF oceanic service makes use of different parts of the HF spectrum at different times of the day and night. Note the correlation between the

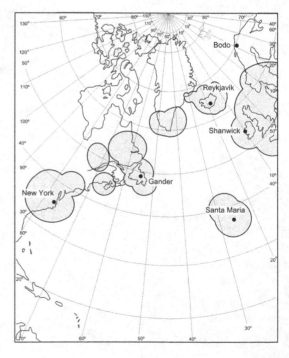

Figure 5.1 VHF aircraft coverage in the North Atlantic area

DOI: 10.1201/9781003411932-5

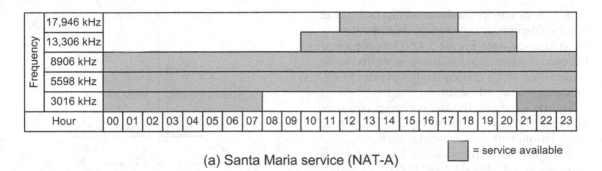

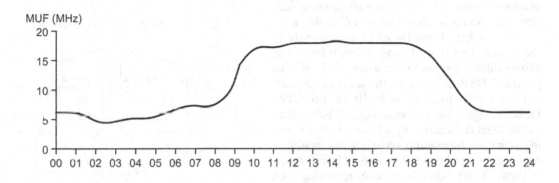

Figure 5.2 Santa Maria oceanic service (NAT-A) showing operational frequencies and times together with typical variation of MUF for a path from Madrid to New York

service availability chart shown in Figure 5.2(a) and the typical variation in maximum usable frequency (MUF) for the radio path between Madrid and New York.

The following HF bands are allocated to the aeronautical service:

- 2850 to 3155 kHz
- 3400 to 3500 kHz
- 4650 to 4750 kHz
- 5480 to 5730 kHz
- 6525 to 6765 kHz
- 8815 to 9040 kHz
- 10,005 to 10,100 kHz
- 11,175 to 11,400 kHz
- 13,200 to 13,360 kHz
- 15,010 to 15,100 kHz
- 17,900 to 18,030 kHz
- 21,870 to 22,000 kHz
- 23,200 to 23,350 kHz.

5.2 SSB modulation

Unfortunately, the spectrum available for aircraft communications at HF is extremely limited. As a result, steps are taken to restrict the bandwidth of transmitted signals, for both voice and data. **Double sideband** (DSB) amplitude modulation requires a bandwidth of at least 7 kHz, but this can be reduced by transmitting only one of the two sidebands. Note that either the **upper sideband** (USB) or the **lower sideband** (LSB) can be used because they both contain the same modulating signal information. In addition, it is possible to reduce (or 'suppress') the carrier as this, in itself, does not convey any information.

In order to demodulate a signal transmitted without a carrier it is necessary to reinsert the carrier at the receiving end (this is done in the demodulator stage where a beat frequency oscillator or **carrier insertion oscillator** replaces the missing carrier signal at the final intermediate frequency—see Figure 5.4). The absence of the carrier means that less power is wasted in the transmitter which consequently operates at significantly higher efficiency.

Figure 5.3 shows the frequency spectrum of an RF signal using different types of amplitude modulation, with and without a carrier.

In Figure 5.3(a) the mode of transmission is conventional **double sideband** (DSB) amplitude modulation with full-carrier. This form of modulation is used for VHF aircraft communications and was described earlier in Chapter 4.

Figure 5.3(b) shows the effect of suppressing the carrier. This type of modulation is known as **double sideband suppressed carrier** (DSB-SC). In practical DSB-SC systems the level of the carrier is typically reduced by 30 dB, or more. The DSB-SC signal has the same overall bandwidth as the DSB full-carrier signal, but the reduction in carrier results in improved efficiency as well as reduced susceptibility to heterodyne interference.

Figure 5.3(c) shows the effect of removing both the carrier and the upper sideband. The resulting signal is referred to as **single sideband** (SSB), in this case using only the **lower sideband** (LSB). Note how the overall bandwidth has been reduced to only around 3.5 kHz, i.e. half that of the comparable DSB AM signal shown in Figure 5.3(a).

Finally, Figure 5.3(d) shows the effect of removing the carrier and the lower sideband. Once again, the resulting signal is referred to as single sideband (SSB), but in this case we are using only the **upper sideband** (USB). Here again, the overall bandwidth has been reduced to around 3.5 kHz. Note that aircraft HF communication requires the use of the upper sideband (USB). DSB AM may also be available but is now very rarely used due to the superior performance offered by SSB.

Test your understanding 5.1

1. Explain why HF radio is used on transoceanic routes.
2. Explain why different frequencies are used for HF aircraft communications during the day and at night.
3. State two advantages of using SSB modulation for aircraft HF communications.

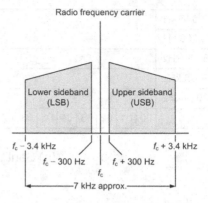

(a) Double sideband (DSB) full-carrier AM

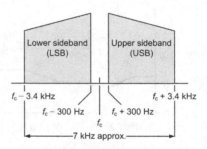

(b) Double sideband suppressed-carrier (DSB-SC)

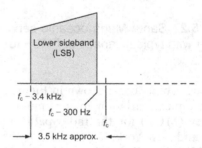

(c) Single sideband suppressed-carrier (SSB-SC)

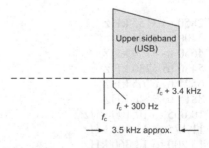

(d) Single sideband suppressed-carrier (SSB-SC)

Figure 5.3 Frequency spectrum of an RF carrier using DSB and SSB modulation

5.3 SELCAL

Selective calling (SELCAL) reduces the burden on the flight crew by alerting them to the need to respond to incoming messages. SELCAL is available at HF and VHF but the system is more used on HF. This is partly due to the intermittent nature of voice communications on long oceanic routes and partly due to the fact that squelch systems are more difficult to operate when using SSB because there is no transmitted carrier to indicate that a signal is present on the channel.

The aircraft SELCAL system is defined in Annex 10 to the Convention on International Civil Aviation Organization (ICAO), Volume 1, fourth edition of 1985 (amended 1987). The system involves the transmission of a short burst of audio tones.

Each transmitted code comprises two consecutive tone pulses, with each pulse containing two simultaneously transmitted tones. The pulses are of 1-second duration separated by an interval of about 0.2 seconds. To ensure proper operation of the SELCAL decoder, the frequency of the transmitted tones must be held to an accuracy of better than 0.15%.

SELCAL codes are uniquely allocated to particular aircraft by air traffic control (ATC). As an example, a typical transmitted SELCAL code might consist of a one-second burst of 312.6 Hz and 977.2 Hz followed by a pause of about 0.2 seconds and a further one-second burst of tone comprising 346.7 Hz and 977.2 Hz. Table 5.1 indicates that the corresponding transmitted SELCAL code is 'AM-BM', and only the aircraft with this code would then be alerted to the need to respond to an incoming message.

The RF signal transmitted by the ground radio station should contain (within 3 dB) equal amounts of the two modulating tones, and the combination of tones should result in a modulation envelope having a nominal modulation percentage as high as possible (and in no case less than 60%).

The transmitted tones are made up from combinations of the tones listed in Table 5.1. Note that the tones have been chosen so that they are not harmonically related (thus avoiding possible confusion within the SELCAL decoder when harmonics of the original tone frequencies might be present in the demodulated waveform).

Table 5.1   SELCAL tone frequencies

Character	Frequency
A	312.6 Hz
B	346.7 Hz
C	384.6 Hz
D	426.6 Hz
E	473.2 Hz
F	524.8 Hz
G	582.1 Hz
H	645.7 Hz
J	716.1 Hz
K	794.3 Hz
L	881.0 Hz
M	977.2 Hz
P	1083.9 Hz
Q	1202.3 Hz
R	1333.5 Hz
S	1479.1 Hz

5.4 HF datalink

ARINC's global high frequency datalink (HFDL) coverage provides a highly cost-effective datalink capability for carriers on remote oceanic routes, as well as the polar routes at high latitudes where SATCOM coverage is unavailable. HFDL is lower in cost than SATCOM, and many carriers are using HFDL instead of satellite services, or as a backup system. HFDL is still the only datalink technology that works over the North Pole, providing continuous, uninterrupted datalink coverage on the popular polar routes between North America and eastern Europe and Asia.

The demand for HFDL has grown steadily since ARINC launched the service in 1998, and today HFDL avionics are offered as original equipment by all the major airframe manufacturers. HFDL offers a cost-effective solution for global datalink service. The demand for HFDL service is currently growing by more than several hundred aircraft per year.

Advantages of HFDL can be summarised as:

- wide coverage due to the extremely long range of HF signals

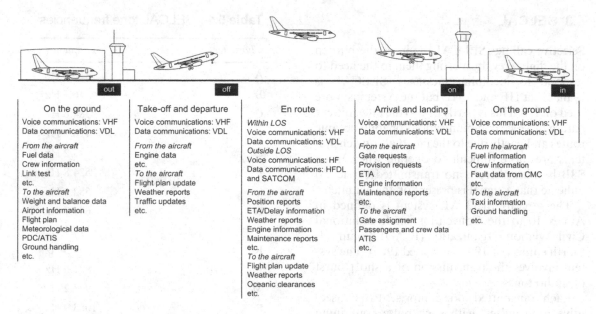

Figure 5.4 Aircraft operational control at various 'out-off-on-in' (OOOI) stages

* simultaneous coverage on several bands and frequencies (currently 60)
* multiple ground stations (currently 14) at strategic locations around the globe
* relatively simple avionics using well-tried technology
* rapid network acquisition
* exceptional network availability.

Disadvantages of HFDL are:

* very low data rates (making the system unsuitable for high-speed wideband communications).

As a result of the above, the vast majority of HFDL messages are related to **airline operational control** (AOC) (see Figure 5.5), but HFDL is also expected to play an important part in **future air navigation systems** (FANS; see Chapter 21) where it will provide a further means of data linking with an aircraft, supplementing VDL, GPS and SATCOM systems. Note that SATCOM can support much faster data rates, but it can also be susceptible to interruptions and may not be available at high latitudes.

HFDL uses **phase shift keying** (PSK) at data rates of 300, 600, 1200 and 1800 bps. The rate

used is dependent on the prevailing propagation conditions. HFDL is based on **frequency division (domain) multiplexing** (FDM) for access to ground station frequencies and **time division (domain) multiplexing** (TDM) within individual communication channels. Figure 5.6 shows how the frequency spectrum of a typical HFDL signal at 300 bps compares with an HF voice signal.

Figure 5.7 shows typical HFDL messages sent from the four aircraft shown in Figure 5.8 to the Shannon HFDL ground station using the same communications channel. The radio path from one of the aircraft (LH8409) is illustrated in Figure 5.9. The first two of the messages shown in Figure 5.7 are **log-on requests**, and the maximum bit rate is specified in the header. In each log-on request, the aircraft is identified by its unique 24-bit **ICAO address**. Once logged on, the aircraft is allocated an 8-bit **address code** (AF hex in the case of the third message and AD hex in the case of the fourth message). Each aircraft also transmits its current location data (longitude and latitude).

The system used for HFDL data exchange is specified in ARINC 635. Each ground station transmits a frame called a 'squitter' every 32 seconds. The **squitter frame** informs aircraft of the system status, provides a timing reference and

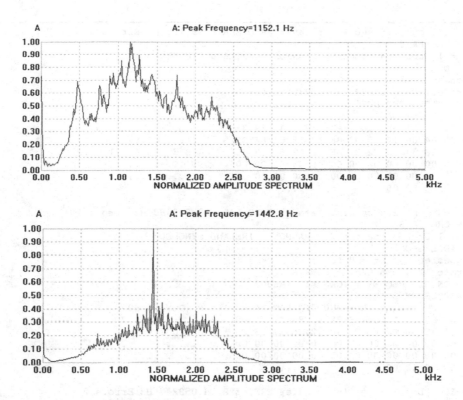

Figure 5.5 Frequency spectra of voice (upper trace) and HFDL signals (lower-trace)

provides protocol control. Each ground station has a time offset for its squitters. This allows aircraft to jump between ground stations finding the best one before logging on. When passing traffic, dedicated TDM time slots are used. This prevents two aircraft transmitting at the same time causing **data collisions**.

5.5 HF radio equipment

The block schematic of a simple HF transmitter/receiver is shown in Figure 5.4. Note that, whilst this equipment uses a single intermediate frequency (IF), in practice most modern aircraft HF radios are much more complex and use two or three intermediate frequencies.

On transmit mode, the DSB suppressed carrier (Figure 5.3(b)) is produced by means of a **balanced modulator** stage. The balanced modulator rejects the carrier and its output just comprises the upper and lower sidebands. The DSB signal is then passed through a multiple-stage crystal or mechanical filter. This filter has a very narrow pass-band (typically 3.4 kHz) at the

intermediate frequency (IF) and this rejects the unwanted sideband. The resulting SSB signal is then mixed with a signal from the digital frequency synthesiser to produce a signal on the wanted channel. The output from the mixer is then further amplified before being passed to the output stage. Note that, to avoid distortion, all of the stages must operate in linear mode.

When used on receive mode, the incoming signal frequency is mixed with the output from the digital frequency synthesiser in order to produce the intermediate frequency signal. Unwanted adjacent channel signals are removed by means of another multiple-stage crystal or mechanical filter which has a pass-band similar to that used in the transmitter. The IF signal is then amplified before being passed to the demodulator.

The (missing) carrier is reinserted in the demodulator stage. The carrier signal is derived from an accurate crystal controlled carrier oscillator which operates at the IF frequency. The recovered audio signal from the demodulator is then passed to the audio amplifier where it is

```
Preamble 300  bps 1.8 sec Interleaver FREQ ERR 5.398116 Hz Errors 0
[MPDU AIR CRC PASS]
Nr LPDUs = 1 Ground station ID SHANNON - IRELAND SYNCHED
Aircraft ID LOG-ON
Slots Requested medium = 0 Low = 0
Max Bit rate 1800 bps U(R) = 0 UR(R)vect = 0
[LPDU LOG ON DLS REQUEST] ICAO AID 0A123C
[HFNPDU FREQUENCY DATA]
14:45:24  UTC  Flight ID = AB3784  LAT 39 37 10  N  LON 0 21 20  W
07 87 FF 00 04 00 14 85 92 BF 3C 12 0A FF D5      ...............
41 42 33 37 38 34 C8 C2 31 BF FF C2 67 88 8C      A B 3 7 8 4 ..1 ...g ..
00 00 00 00 00 00 00 00 00 00 00 00 00 00 00      ...............
00 00 00 00 00 00 00 00 00 00 00 00 00 00 00      ...............
00 00 00 00 00 00 00                              .......

Preamble 300  bps 1.8 sec Interleaver FREQ ERR -18.868483 Hz Errors 19
[MPDU AIR CRC PASS]
Nr LPDUs = 1 Ground station ID SHANNON - IRELAND SYNCHED
Aircraft ID LOG-ON
Slots Requested medium = 0 Low = 0
Max Bit rate 1200 bps U(R) = 0 UR(R)vect = 0
[LPDU LOG ON DLS REQUEST] ICAO AID 4A8002
[HFNPDU FREQUENCY DATA]
14:45:30  UTC  Flight ID = SU0106  LAT 54 42 16  N  LON 25 50 42  E
07 87 FF 00 03 00 14 80 1E BF 02 80 4A FF D5      ............J ..
53 55 30 31 30 36 6A 6E F2 60 12 C5 67 33 FB      S U 0 1 0 6 j n ....g 3 .
00 00 00 00 00 00 00 00 00 00 00 00 00 00 00      ...............
00 00 00 00 00 00 00 00 00 00 00 00 00 00 00      ...............
00 00 00 00 00 00 00                              .......

Preamble 300  bps 1.8 sec Interleaver FREQ ERR 15.059247 Hz Errors 2
[MPDU AIR CRC PASS]
Nr LPDUs = 1 Ground station ID SHANNON - IRELAND SYNCHED
Aircraft ID AF
Slots Requested medium = 0 Low = 0
Max Bit rate 1200 bps U(R) = 0 UR(R)vect = 0
[LPDU UNNUMBERED DATA]
[HFNPDU PERFORMANCE]
14:45:30  UTC  Flight ID = LH8409  LAT 46 42 34  N  LON 21 22 55  E
07 87 AF 00 03 00 31 4D 1D 0D FF D1 4C 48 38      ......1 M ....L H 8
34 30 39 73 13 82 34 0F C5 67 01 36 03 02 02      4 0 9 s ..4 ..g .6 ...
00 B6 00 00 00 00 00 00 00 00 03 00 00 00 00      ...............
02 00 00 00 00 00 01 00 00 00 01 01 D3 EA 00      ...............
00 00 00 00 00 00 00                              .......

Preamble 300  bps 1.8 sec Interleaver FREQ ERR 8.355845 Hz Errors 0
[MPDU AIR CRC PASS]
Nr LPDUs = 1 Ground station ID SHANNON - IRELAND SYNCHED
Aircraft ID AD
Slots Requested medium = 0 Low = 0
Max Bit rate 1200 bps U(R) = 0 UR(R)vect = 0
[LPDU UNNUMBERED DATA]
[HFNPDU PERFORMANCE]
14:43:30  UTC  Flight ID = LH8393  LAT 52 37 27  N  LON 16 46 41  E
07 87 AD 00 03 00 31 C5 0B 0D FF D1 4C 48 38      ......1 .....L H 8
33 39 33 BF 56 62 EE 0B 89 67 01 8A 07 01 B8      3 9 3 .V b ...g .....
00 7E 00 00 00 00 00 00 00 00 06 0F 00 00 00 00   ...............
2E 00 00 00 00 00 05 00 00 00 05 07 08 27 00      ...............
00 00 00 00 00 00 00                              .......
```

Figure 5.6 Examples of aircraft communication using HFDL

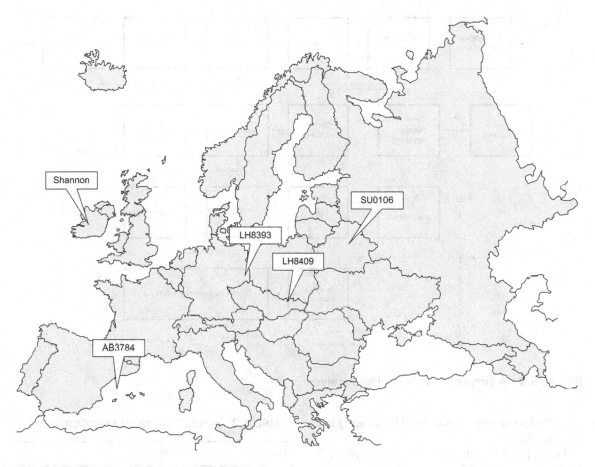

Figure 5.7 Ground station and aircraft locations for the HFDL communications in Figure 5.6

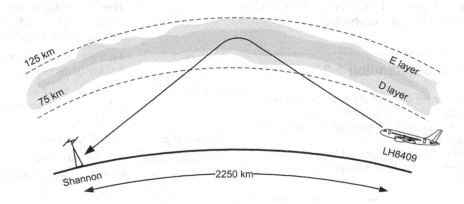

Figure 5.8 Radio path for LH8409

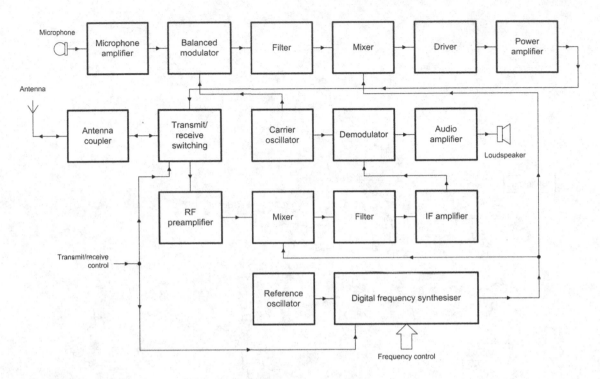

Figure 5.9   A simple SSB transmitter/receiver

amplified to an appropriate level for passing to a loudspeaker.

The typical specification for an aircraft HF radio is shown in Table 5.2. One or two radios of this type are usually fitted to a large commercial aircraft (note that at least one HF radio is a requirement for *any* aircraft following a transoceanic route). Figure 5.10 shows the flight deck location of the **HF radio controller**.

Test your understanding 5.2

1. Explain how HF datalink (HFDL) differs from VH F d ata l i nk (V DL). Un de r wh a t circumstances is HFDL used and what advantages does it offer?
2. Explain briefly how an aircraft logs on to the HFDL system. How are data collisions avoided?

Table 5.2   Aircraft HF radio specifications

Parameter	Specification
Frequency range	2.0000 MHz to 29.9999 MHz
Tuning steps	100 Hz
Operating modes	SSB SC analogue voice (ARINC 719) and analogue data (ARINC 753 and ARINC 635) at up to 1800 bps;DSB AM (full carrier)
Sensitivity	1 μ V for 10 dB (S+N)/N SSB;4 μ V for 10 dB (S+N)/N AM
Selectivity	6 dB max. attenuation at +2.5 kHz 60 dB min. attenuation at +3.4 kHz
Audio output	50 mW into 600 Ω
SELCAL output	50 mW into 600 Ω
RF output power	200 W pep min. SSB;50 W min. DSB AM
Frequency stability	±20 Hz
Audio response	350 Hz to 2500 Hz at −6 dB
Mean time between failure	Greater than 50,000 hours

Figure 5.10 HF radio control unit

5.6 HF antennas and coupling units

External wire antennas were frequently used on early aircraft. Such antennas would usually run from the fuselage to the top of the vertical stabiliser, and they were sufficiently long to permit resonant operation on one or more of the aeronautical HF bands. Unfortunately this type of antenna is unreliable and generally unsuitable for use with a modern high-speed passenger aircraft. The use of a large probe antenna is unattractive due to its susceptibility to static discharge and lightning strike. Hence an alternative solution in which the HF antenna is protected within the airframe is highly desirable. Early experiments (see Figure 5.13) showed that the vertical stabiliser (tail fin) would be a suitable location and is now invariably used to house the HF antenna and its associated coupling unit on most large transport aircraft—see Figures 5.11 and 5.12.

Due to the restriction in available space (which mitigates against the use of a resonant antenna such as a quarter-wave Marconi antenna—see page 36) the HF antenna is based on a notch which uses part of the airframe in order to radiate effectively. The notch itself has a very high-Q factor and its resistance and reactance varies very widely over the operating frequency range (i.e. 3 MHz to 24 MHz). The typical variation of **standing wave ratio** (SWR—see page 45) against frequency for an HF notch antenna is shown in Figure 5.14. For comparison, the variation of SWR with frequency for a typical quarter-wave VHF blade antenna is shown in Figure 5.15.

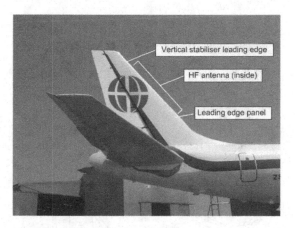

Figure 5.11 HF antenna location

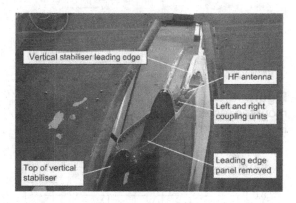

Figure 5.12 View from the top of the vertical stabiliser (leading edge panel removed)

From Figures 5.14 and 5.15 it should be obvious that the HF antenna, whilst well matched at 21 MHz, would be severely mismatched to a conventional 50 Ω feeder/transmitter at most other HF frequencies. Because of this, and because the notch antenna is usually voltage fed, it is necessary to use an HF coupling/tuning unit between the HF radio feeder and the notch antenna. This unit is mounted in close proximity to the antenna, usually close to the top of the vertical stabiliser (see Figure 5.12). Figure 5.16 shows the effect of using a coupling/tuning unit on the SWR-frequency characteristic of the same notch antenna that was used in Figure 5.14. Note how the SWR has been reduced to less than 2:1 for most (if not all) of the HF range.

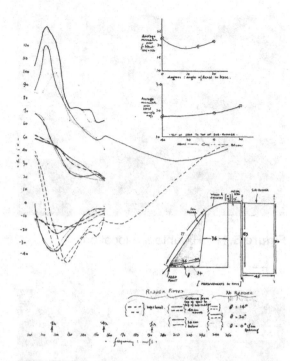

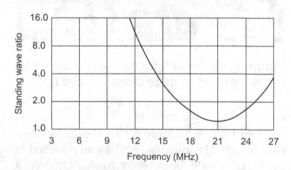

Figure 5.13 Original sketches for a tail-mounted antenna from work carried out by E. H. Tooley in 1944

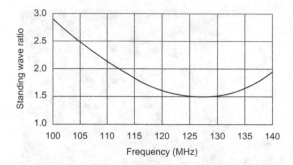

Figure 5.15 Variation of SWR with frequency for a VHF quarter-wave blade antenna (note the linear scale used for SWR)

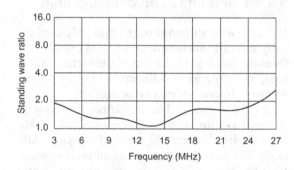

Figure 5.16 Variation of SWR with frequency for an HF notch antenna fitted with an antenna coupling/tuning unit

Figure 5.14 Variation of SWR with frequency for an HF notch antenna (note the logarithmic scale used for SWR)

The tuning adjustment of an HF antenna coupler is entirely automatic and only requires a brief signal from the transmitter to retune to a new HF frequency. The HF antenna coupler unit incorporates an SWR bridge (see page 47) and a feedback control system (see Figure 5.17) to adjust a roller coaster inductor (L1) and high-voltage vacuum variable capacitor (C1) together

with a number of switched high-voltage capacitors (C1 to C4). The internal arrangement of a typical HF antenna coupler is shown in Figures 5.18 and 5.19. The connections required between the HF antenna coupler, HF radio and control unit are shown in Figure 5.20.

Voltages present in the vicinity of the HF antenna (as well as the field radiated by it) can be extremely dangerous. It is therefore **essential** to avoid contact with the antenna and to maintain a safe working distance from it (at least 5 meters) whenever the HF radio system is 'live'.

Test your understanding 5.3

Explain the function of an HF antenna coupler. What safety precautions need to be observed when accessing this unit?

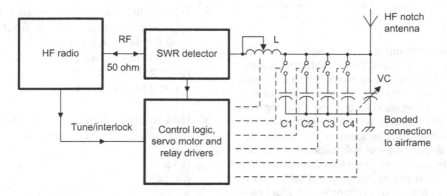

Figure 5.17   Typical feedback control system used in an HF antenna coupler

Figure 5.18   Interior view of an HF antenna coupler showing the roller coaster inductor (top) and vacuum variable capacitor (bottom). The high-voltage antenna connector is shown in the extreme right

Figure 5.19   SWR bridge circuit incorporated in the HF antenna coupler. The output from the SWR bridge provides the error signal input to the automatic feedback control system

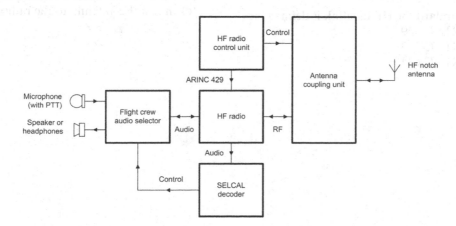

Figure 5.20   Connections to the HF radio, control unit and antenna coupling unit

Key point

Safety precautions must be observed since very high voltages are present on the antenna system with the resulting danger of electric shock or arcing. No personnel should be in the vicinity of the antenna when transmitting, nor should fuelling operations be in progress. Note that with many HF systems a change of frequency will invariably result in transmission to allow automatic antenna tuning. Checks for satisfactory HF radio communication should be carried out by carrying out tests with a remote ground station on an approved frequency.

5.7 Multiple choice questions

1. The typical bandwidth of an aircraft HF SSB signal is:
   (a) 3.4 kHz
   (b) 7 kHz
   (c) 25 kHz.

2. The principal advantage of SSB over DSB AM is:
   (a) reduced bandwidth
   (b) improved frequency response
   (c) faster data rates can be supported.

3. HF datalink uses typical data rates of:
   (a) 300 bps and 600 bps
   (b) 2400 bps and 4800 bps
   (c) 2400 bps and 31,500 bps.

4. The standard for HF datalink is defined in:
   (a) ARINC 429
   (b) ARINC 573
   (c) ARINC 635.

5. Which one of the following gives the approximate range of audio frequencies used for SELCAL tones?
   (a) 256 Hz to 2048 Hz
   (b) 312 Hz to 1479 Hz
   (c) 300 Hz to 3400 Hz.

6. How many alphanumeric characters are transmitted in a SELCAL code?
   (a) 4
   (b) 8
   (c) 16.

7. How many bits are used in an ICAO aircraft address?
   (a) 16
   (b) 24
   (c) 32.

8. The typical RF output power from an aircraft HF transmitter is:
   (a) 25 W pep
   (b) 50 W pep
   (c) 400 W pep.

9. An HF radio is required for use on oceanic routes because:
   (a) VHF coverage is inadequate
   (b) higher power levels can be produced
   (c) HF radio is more reliable.

10. The function of an HF antenna coupler is to:
    (a) reduce static noise and interference
    (b) increase the transmitter output power
    (c) match the antenna to the radio.

Chapter 6 · Flight-deck audio systems

As well as systems for communication with ground stations, modern passenger aircraft require a number of facilities for local communication within the aircraft. In addition, there is a need for communications with those who work on the aircraft when it is being serviced on the ground.

Systems used for local communications need to consist of nothing more than audio signals, suitably amplified, switched and routed, and incorporating a means of alerting appropriate members of the crew and other personnel.

These flight-deck audio systems include:

* passenger address (PA) system
* service interphone system
* cabin interphone system
* ground crew call system
* flight interphone system.

The **passenger address system** provides the flight crew and cabin crew with a means of making announcements and distributing music to passengers through cabin speakers. Circuits in the system send chime signals to the cabin speakers.

The **service interphone system** provides the crew and ground staff with interior and exterior communication capability. Circuits in the system connect service interphone jacks to the flight compartment.

The **cabin interphone system** provides facilities for communication among cabin attendants and between the flight compartment crew members and attendants. The system can be switched to the input of the passenger address system for PA announcements.

The **ground crew call system** provides a signalling capability (through the ground crew call horn) between the flight compartment and nose landing gear area.

The **flight interphone system** provides facilities for interphone communication among flight compartment crew members and provides the means for them to receive, key and transmit using the various aircraft radio systems. The flight interphone system also extends communication to ground personnel at the nose gear interphone station and allows flight compartment crew members to communicate and to make passenger address announcements. The flight interphone system also incorporates amplifiers and mixing circuits in the audio accessory unit, audio selector panels, cockpit speakers, microphone/headphone jacks and press-to-talk (PTT) switches.

In addition to the audio systems used for normal operation of the aircraft, large commercial aircraft are also required to carry a **cockpit voice recorder** (CVR). This device captures and stores information derived from a number of the aircraft's audio channels. Such information may later become invaluable in the event of a crash or malfunction.

6.1 Flight interphone system

The flight interphone system provides the essential connecting link between the aircraft's communication systems, navigation receivers and flight-deck crew members. The flight interphone system also extends communication to ground personnel at external stations (e.g. the nose gear interphone station). It also provides the means by which members of the flight crew can communicate with the cabin crew and also make passenger address announcements. The flight interphone system comprises a number of subsystems including amplifiers and mixing circuits in the audio accessory unit, audio selector panels, cockpit speakers, microphone/headphone jacks and PTT switches.

The flight interphone components provided for the captain and first officer usually comprise the following components:

* audio selector panel
* headset, headphone and hand microphone jack connectors

DOI: 10.1201/9781003411932-6

- audio selector panel and control wheel PTT
- switches
- cockpit speakers.

Note that, where a third (or fourth) seat is provided on the flight deck, a third (or fourth) set of flight interphone components will usually be available for the observer(s) to use. In common with other communication systems fitted to the aircraft, the flight interphone system normally derives its power from the aircraft's 28 V DC battery bus through circuit breakers on the overhead panel.

The simplified block schematic diagram of a typical flight interphone system is shown in Figure 6.1. Key subsystem components are the captain's and first officer's audio selector panels

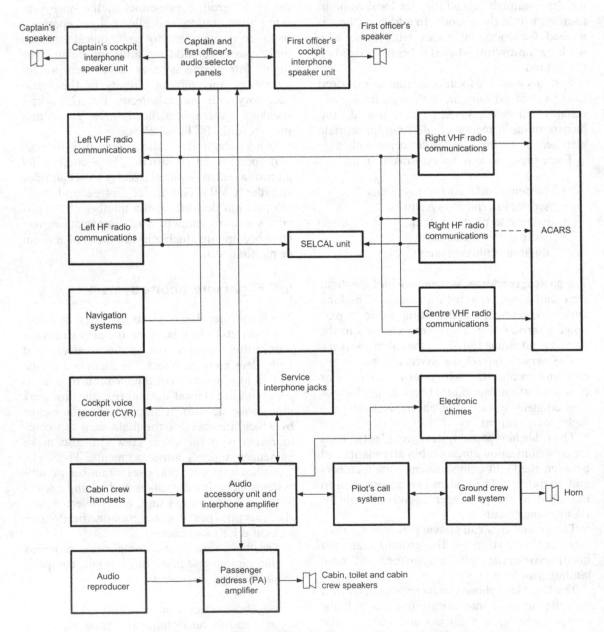

Figure 6.1 Simplified block schematic diagram of a typical flight interphone system

and the **audio accessory unit** that provides a link from the flight deck audio system to the passenger address, cabin and service interphones and ground crew call systems. It is also worth noting from Figure 6.1 that the audio signals (inputs and outputs) from the HF and VHF radio communications equipment as well as the audio signals derived from the navigation receivers (outputs only) are also routed via the audio selector panels. This arrangement provides a high degree of configuration flexibility together with a degree of redundancy sufficient to cope with failure of individual subsystem components. Finally, it should be emphasised that the arrangement depicted in Figure 6.1 is typical and that minor variations can and do exist. For example, most modern aircraft incorporate SATCOM facilities (not shown in Figure 6.1).

The **flight interphone amplifier** is usually located in the audio accessory unit in the main avionic equipment rack. The amplifier receives low-level microphone inputs and provides audio to all flight interphone stations. The amplifier has preset internal adjustments for compression, squelch and volume.

Audio selector panels are located in the flight compartment within easy of reach of the crew members. Audio selector panels are provided for the captain and first officer as well as any observers that may be present on the flight deck. Depending on aircraft type and flight deck configuration, audio selector panels may be fitted in the central pedestal console or in the overhead panels. Typical examples of cockpit audio selector panel layouts are shown in Figures 6.2 and 6.3. Each audio selector panel contains microphone selector switches which connect microphone circuits to the interphone systems, to the radio communication systems or to the passenger address system. The PTT switch on the audio selector panels can be used to key the flight compartment microphones. Volume control is provided by switches on each audio selector panel.

Two **cockpit speaker units** are usually fitted in the flight compartment. These are usually located in the sidewall panels adjacent to the captain's and first officer's stations. Each cockpit speaker unit contains a loudspeaker, amplifier, muting circuits and a **volume control**. The speakers receive all audio signals provided to

Figure 6.2 First officer's audio selector panel (top) and radio panel (bottom) fitted in the overhead panel of an A320 aircraft

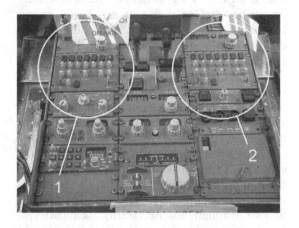

Figure 6.3 Captain's audio selector (1) and first officer's audio selector (2) fitted in the central console of a Boeing 757 aircraft

the audio selector panels. The speakers are muted whenever a PTT switch is pushed at the captain's or first officer's station.

Several **jack panels** are provided for a headset with integral boom microphone for the captain, first officer and observer. Hand microphones may also be used. PTT switches are located at all flight interphone stations. The hand

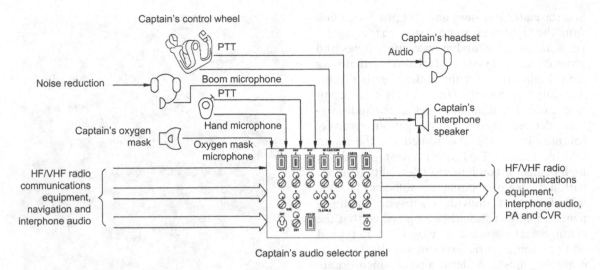

Captain's audio selector panel

Figure 6.4 Typical arrangement for the captain's audio selector. A similar arrangement is used for the first officer's audio selector as well as any supernumerary crew members that may be present on the flight deck

microphone, control wheel and audio selector panels all have PTT switches. The switch must be pushed before messages are begun or no transmission can take place. Audio and control circuits to the audio selector panel are completed when the PTT switch is operated.

The flight interphone system provides common microphone circuits for the communications systems and common headphone and speaker circuits for the communications and navigation systems.

Figure 6.4 shows a typical arrangement for the captain's audio selector panel (note that the flight interphone components and operation are identical for both the captain and first officer). Similar (though not necessarily identical) systems are available for use by the observer and any other supernumerary crew members (one obvious difference is the absence of a control wheel push-to-talk switch and cockpit speaker). Switches are provided to select boom microphone and hand microphone (where available) as well as microphones located in the oxygen masks (for emergency use). Outputs can be selected for use with the headset or cockpit loudspeakers.

Amplifiers, summing networks and filters in the audio selector panel provide audio signals from the interphone and radio communication

systems to the headphones and speakers. Audio signals from the navigation receivers are also monitored through the headphones and speakers. Reception of all audio signals is controlled by the volume switches. The captain's INT microphone switch is illuminated when active. Note that this switch is interlocked with the other microphone switches so that only one at a time can be pushed.

The navigation system's (ADF, VOR, ILS, etc.) audio is also controlled by switches on the audio selector panel. The left, centre or right (L, C, R) switches control selection and volume of the desired receiver. The VOICE-BOTH-RANGE switch acts as a filter that separates voice signals and range signals. The filter switch can also combine both voice and range signals. All radio communication, interphone and navigation outputs are received and recorded by the **cockpit voice recorder** (CVR).

A typical procedure for checking that the microphone audio is routed to the radio communication, interphone or passenger address system is as follows:

1. Push the microphone select switch on the audio selector panel to select the required communication system.

2. If a handheld microphone is used, push the PTT switch on the microphone and talk.
3. If a boom microphone or oxygen mask microphone is used, select MASK or BOOM with the toggle switch on the audio selector panel and push the audio selector panel or control wheel PTT switch and talk.

The following procedure is used to listen to navigation and communication systems audio:

1. For communications systems, adjust the volume control switch on the audio selector panel and listen to the headset.
2. For navigation systems audio, select desired left-centre-right (L-C-R) and filter (VOICE-BOTH-RANGE) positions on the audio selector panel, adjust volume control switch and listen to headset.
3. The captain's and first officer's cockpit speakers (see Figure 6.5) can be used to listen to navigation as well as communication system audio. A control in the centre of the cockpit speaker (Boeing aircraft) or on an adjacent panel (Airbus) adjusts the speaker volume to the desired level.
4. External interphone panels (as appropriate to the aircraft type—see Figures 6.9 and 6.10) should be similarly tested by connecting a headset or handset (as appropriate) to each interphone jack.

Figures 6.5 to 6.10 show examples of some typical flight deck audio communications equipment used on modern passenger aircraft.

Figure 6.6 Captain's headset and boom microphone in a Boeing 757 aircraft. The press-to-talk (PTT) switch can be seen on the left-hand section of the control wheel

Figure 6.5 First officer's loudspeaker (centre) in a Boeing 757 aircraft (the volume control is mounted in the centre of the loudspeaker panel)

Figure 6.7 First officer's headset and boom microphone in an A320 aircraft

Figure 6.8 Headsets and boom micro-phones in a four-seat rotorcraft

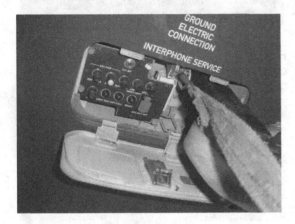

Figure 6.9 Ground staff interphone jack connector

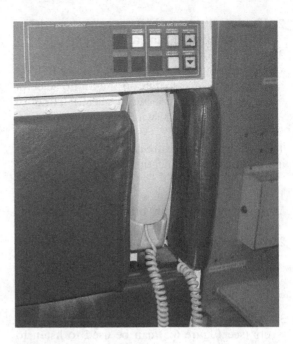

Figure 6.10 Cabin interphone/passenger address handset

Key point

Appropriate audio levels should be checked follow-ing established maintenance procedures. In most cases the master volume control is first set to an arbitrary level before selecting and tuning local VHF communication ground stations, navigation aids, etc. Levels should then be adjusted for flight deck speaker(s) and headsets. The pilot's micro-phone level is set by operating the press-to-talk switch and listening to the VHF side-tone from the flight deck speaker(s) and headsets.

Test your understanding 6.1

Explain the differences between (a) the flight inter-phone and (b) the cabin interphone systems.

Test your understanding 6.2

1. Explain the function of the audio selector panels used by members of the flight crew.
2. List three different examples of inputs to an audio selector panel and three different examples of outputs from an audio selector panel.

6.2 Cockpit voice recorder

The cockpit voice recorder (CVR) can provide valuable information that can later be analysed in the event of an accident or serious malfunc-tion of the aircraft or any of its systems. The voice recorder preserves a continuing record of typically between 30 and 120 minutes of the most recent flight crew communications and conversations.

The storage medium used with the CVR fitted to modern aircraft is usually based on one or more solid-state memory devices whereas on older aircraft the CVR is usually based on a continuous loop of magnetic tape.

The CVR storage unit must be recoverable in the event of an accident. This means that the entire recorder unit including storage media must be mounted in an enclosure that can withstand severe mechanical and thermal shock as well as the high pressure that exists when a body is immersed at depth in water.

The CVR is usually fitted with a test switch, headphone jack, status light (green) and an externally mounted **underwater locator beacon** (ULB) to facilitate undersea recovery. The ULB is a self-contained device (invariably attached to the front panel of the CVR) that emits an ultrasonic vibration (typically at 37.5 kHz) when the water-activated switch is activated as a result of immersion in either sea water or fresh water. A label on the ULB indicates the date by which the internal battery should be replaced. A typical specification for a ULB is shown in Table 6.1. An external view of a CVR showing its externally mounted ULB is shown in Figure 6.11.

The audio input to the CVR is derived from the captain, first officer, observer (where present) and also from an **area microphone** in the

Figure 6.11 FDR/CVR fitted with an underwater locator beacon (ULB)

flight compartment which is usually mounted in the overhead panel and thus collects audio input from the entire flight-deck area.

In order to improve visibility and aid recovery, the external housing of the CVR is painted bright orange. The unit is thermally insulated and hermetically sealed to prevent the ingress of water. Because of the crucial nature of the data preserved by the flight, the unit should only be opened by authorised personnel following initial recovery from the aircraft.

Magnetic CVRs use a multi-track tape transport mechanism. This normally comprises a tape drive, four recording heads, a single (full-width) erase head, a monitor head and a bulk erase coil. The bias generator usually operates at around 65 kHz and an internal signal (at around 600 Hz) is often provided for test purposes. Bulk erase can be performed by means of an erase switch (which is interlocked so that bulk tape erasure can only be performed when the aircraft is on the ground and the parking brake is set). The erase current source is usually derived directly from the aircraft's 115 V AC 400 Hz supply. The magnetic tape (a continuous loop) is usually 308 ft in length and ¼-in wide.

More modern solid-state recording media uses no moving parts (there is no need for a drive mechanism) and is therefore much more reliable. Erasure can be performed electronically and there is no need for a separate erase coil and AC supply. Finally, it is important to note that the CVR is

Table 6.1 Typical ULB specification

Parameter	Specification
Operating frequency	37.5 kHz (1 kHz)
Acoustic output	160 dB relative to 1 μPa at 1 m
Pulse repetition rate	0.9 pulses per sec
Pulse duration	10 ms
Activation	Immersion in either salt water or fresh water
Power source	Internal lithium battery
Battery life	6 years standby (shelf-life)
Beacon operating life	30 days
Operating depth	20,000 ft (6,096 m)
Housing material	Aluminium
Length	3.92 in (9.95 cm)
Diameter	1.3 in (3.3 cm)
Weight	6.7 oz (190 g)

usually mounted in the aft passenger cabin ceiling. This location offers the greatest amount of protection for the unit in the event of a crash.

Test your understanding 6.3

1. Explain the function and principle of operation of the underwater locator beacon (ULB) fitted to a cockpit voice recorder (CVR).
2. Explain why the CVR is located in the ceiling of the aft passenger cabin.

6.3 Multiple choice questions

1. Audio selector panels are located:
   (a) in the main avionic equipment bay
   (b) close to the pilot and first officer stations
   (c) in the passenger cabin for use by cabin crew members.

2. When are the flight-deck speaker units muted?
   (a) when a PTT switch is operated
   (b) when a headset is connected
   (c) when a navigation signal is received.

3. Input to the captain's interphone speaker unit is derived from:
   (a) the audio selector panel
   (b) the passenger address system
   (c) the audio accessory unit and interphone amplifier.

4. The microphone PTT system is interlocked in order to prevent:
   (a) unwanted acoustic feedback
   (b) more than one switch being operated at any time
   (c) loss of signal due to parallel connection of microphones.

5. Bulk erasure of the magnetic tape media used in a CVR is usually carried out:
   (a) immediately after take-off
   (b) as soon as the aircraft has touched down
   (c) on the ground with the parking brake set.

6. The typical bias frequency used in a magnetic CVR is:
   (a) 3.4 kHz
   (b) 20 kHz
   (c) 65 kHz.

7. The typical frequency emitted by a ULB is:
   (a) 600 Hz
   (b) 3.4 kHz
   (c) 37.5 kHz.

8. Which one of the following is a suitable audio tone frequency for testing a CVR?
   (a) 60 Hz
   (b) 600 Hz
   (c) 6 kHz.

9. A ULB is activated:
   (a) automatically when immersed in water
   (b) manually when initiated by a crew member
   (c) when the unit is subjected to a high impact mechanical shock.

10. The CVR flight deck area microphone is usually mounted:
    (a) on the overhead panel
    (b) on the left-side flight deck floor
    (c) immediately behind the jump seat.

11. The typical pulse rate for a ULB is:
    (a) 0.9 pulses per sec
    (b) 10 pulses per sec
    (c) 60 pulses per sec.

12. The CVR is usually located:
    (a) on the flight deck
    (b) in the avionic equipment bay
    (c) in the ceiling of the aft passenger cabin.

13. What colour is used for the external housing of a CVR?
    (a) red
    (b) green
    (c) orange.

14. A ULB usually comprises:
    (a) a separate externally fitted canister
    (b) an internally fitted printed circuit module
    (c) an external module that derives its power from the CVR.

15. A ULB will operate:
    (a) only in salt water
    (b) only in fresh water
    (c) in either salt water or fresh water.

16. The typical shelf-life of the battery fitted to
    a ULB is:
    (a) six months
    (b) 18 months
    (c) six years

Chapter 7   Emergency locator transmitters

The detection and location of an aircraft crash is vitally important to the search and rescue (SAR) teams and to potential survivors. Studies show that whilst the initial survivors of an aircraft crash have less than a 10% chance of survival if rescue is delayed beyond two days, the survival rate is increased to over 60% if the rescue can be accomplished within eight hours. For this reason, emergency locator transmitters (ELT) are required for most general aviation aircraft. ELT are designed to emit signals on the VHF and UHF bands, thereby helping search crews locate aircraft and facilitating the timely rescue of survivors. This chapter provides a general introduction to the types and operating principles of ELT fitted to modern passenger aircraft.

7.1 Types of ELT

Several different types of ELT are in current use. These include the older (and simpler) units that produce a modulated RF carrier on one or both of the two spot VHF frequencies used for distress beacons (121.5 MHz and its second harmonic 243.0 MHz). Note that the former frequency is specified for civil aviation use whilst the latter is sometimes referred to as the military aviation distress frequency. Simultaneous transmission on the two frequencies (121.5 MHz and 243.0 MHz) is easily possible and only requires a frequency doubler and dual-band output stage.

Simple VHF ELT devices generate an RF carrier that is modulated by a distinctive sirenlike sound. This sweeps downwards at a repetition rate of typically between 2 and 4 Hz. This signal can be readily detected by Sarsat and Cospas satellites (see p. 111), or by any aircraft monitoring 121.5 MHz or 243.0 MHz.

More modern ELT operate on a spot UHF frequency (460.025 MHz). These devices are much more sophisticated and also operate at a significantly higher power (5 W instead of the 150 mW commonly used at VHF). Unlike the simple amplitude modulation used with their VHF counterparts, 460 MHz ELT transmit digitally encoded data which incorporates a code that is unique to the aircraft that carries them.

Provided they have been properly maintained, most ELT are capable of continuous operation for up to 50 hours. It is important to note that ELT performance (and, in particular, the operational range and period for which the signal is maintained) may become seriously impaired when the batteries are out of date. For this reason, routine maintenance checks are essential and any ELT which contains outdated batteries should be considered unserviceable.

The different types of ELT are summarised in Table 7.1. These are distinguished by application and by the means of activation. Modern passenger aircraft may carry several different types of ELT. Figure 7.1 shows a typical example of the Type-W (water-activated) survival ELT carried on a modern transport aircraft.

Most ELT in general aviation aircraft are of the automatic type. Fixed automatic units contain a crash activation sensor, or G-switch, which is designed to detect the deceleration characteristics of a crash and automatically activate the transmitter.

With current Sarsat and Cospas satellites now in orbit, ELT signals will usually be detected within 90 minutes, and the appropriate search and rescue (SAR) agencies alerted. Military aircrew monitor 121.5 MHz or 243.0 MHz and they will also notify ATS or SAR agencies of any ELT transmissions they hear.

It is worth noting that the detection ranges for Type-W and Type-S ELT can be improved if the ELT is placed upright, with the antenna vertical, on the highest nearby point with any accessible metal surface acting as a ground plane. Doubling the height will increase the range by about 40%.

DOI: 10.1201/9781003411932-7

Table 7.1 Types of ELT

Type	Class	Description
A or AD	Automatic ejectable or automatic deployable	This type of ELT automatically ejects from the aircraft and is set in operation by inertia sensors when the aircraft is subjected to a crash deceleration force acting through the aircraft's flight axis. This type is expensive and is seldom used in general aviation.
F or AF	Fixed (non-ejectable) or automatic fixed	This type of ELT is fixed to the aircraft and is automatically set in operation by an inertia switch when the aircraft is subjected to crash deceleration forces acting in the aircraft's flight axis. The transmitter can be manually activated or deactivated and in some cases may be remotely controlled from the cockpit. Provision may also be made for recharging the ELT's batteries from the aircraft's electrical supply. Most general aviation aircraft use this ELT type, which must have the function switch placed to the ARM position for the unit to function automatically in a crash (see Figure 7.5).
AP	Automatic portable	This type of ELT is similar to Type-F or AF except that the antenna is integral to the unit for portable operation.
P	Personnel activated	This type of ELT has no fixed mounting and does not transmit automatically. Instead, a switch must be manually operated in order to activate or deactivate the ELT's transmitter.
W or S	Water activated or Survival	This type of ELT transmits automatically when immersed in water (see Figure 7.1). It is waterproof, floats and operates on the surface of the water. It has no fixed mounting and should be tethered to survivors or life rafts by means of the supplied cord.

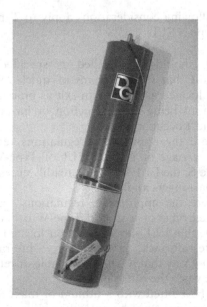

Figure 7.1 Type-W ELT with attachment cord secured by water-soluble tape (the antenna has been removed).

7.2 Maintenance and testing of ELT

ELT should be regularly inspected in accordance with the manufacturer's recommendations. The ELT should be checked to ensure that it is secure, free of external corrosion and that antenna connections are secure. It is also important to ensure that the ELT batteries have not reached their expiry date (refer to external label) and that only approved battery types are fitted.

Air testing normally involves first listening on the beacon's output frequency (e.g. 121.5 MHz), checking first that the ELT is not transmitting before activating the unit and then checking the radiated signal. Simple air tests between an aircraft and a ground station (or between two aircraft) can sometimes be sufficient to ensure that an ELT is functional; however, it is important to follow manufacturer's instructions when testing an ELT. Two-station air testing (in conjunction with a nearby ground station) is usually preferred because, due to the proximity of the transmitting and receiving antennae, a test carried out with the aircraft's own VHF receiver may not reveal a fault condition in which the ELT's RF output has become reduced.

To avoid unnecessary SAR missions, all accidental ELT activations should be reported to the appropriate authorities (e.g. the nearest rescue coordination centre), giving the location of the transmitter and the time and duration of the accidental transmission. Promptly notifying the appropriate authorities of an accidental ELT transmission can be instrumental in preventing the launch of a search aircraft. Any testing of an ELT must be conducted only during the first five minutes of any UTC hour and restricted in duration to not more than five seconds.

7.3 ELT mounting requirements

In order to safeguard the equipment and to ensure that it is available for operation should the need arise, various considerations should be observed when placing and mounting an ELT and its associated antenna system in an aircraft. The following requirements apply to Type-F, AF and AP ELT installations in fixed-wing aircraft and rotorcraft:

1. When installed in a fixed-wing aircraft, ELT should be mounted with its sensitive axis pointing in the direction of flight
2. When installed in a rotorcraft an ELT should be mounted with its sensitive axis pointing approximately 45° downward from the normal forward direction of flight
3. ELT should be installed to withstand ultimate inertia forces of 10 g upward, 22.5 g downward, 45 g forward and 7.5 g sideways
4. The location chosen for the ELT should be sufficiently free from vibration to prevent involuntary activation of the transmitter
5. ELT should be located and mounted so as to minimise the probability of damage to the transmitter and antenna by fire or crushing as a result of a crash impact
6. ELT should be accessible for manual activation and deactivation. If it is equipped with an antenna for portable operation, the ELT should be easily detachable from inside the aircraft

7. The external surface of the aircraft should be marked to indicate the location of the ELT
8. Where an ELT has provision for remote operation it is important to ensure that appropriate notices are displayed.

The antenna used by a fixed type of ELT should conform to the following:

1. ELT should not use the antenna of another avionics system
2. ELT antenna should be mounted as far away as possible from other very high frequency (VHF) antennas
3. The distance between the transmitter and antenna should be in accordance with the ELT manufacturer's installation instructions or other approved data
4. The position of the antenna should be such as to ensure essentially omnidirectional radiation characteristics when the aircraft is in its normal ground or water attitude
5. The antenna should be mounted as far aft as possible
6. ELT antenna should not foul or make contact with any other antennas in flight.

The following considerations apply to Type-W and Type-S ELTs:

1. ELT should be installed as specified for Type-F but with a means of quick release and located as near to an exit as practicable without being an obstruction or hazard to aircraft occupants
2. Where the appropriate regulations require the carriage of a single ELT of Type-W or Type-S, the ELT should be readily accessible to passengers and crew
3. Where the appropriate regulations require the carriage of a second Type-W or Type-S ELT, that ELT should be either located near a life raft pack, or attached to a life raft in such a way that it will be available or retrievable when the raft is inflated
4. An ELT fitted with a lithium or magnesium battery must not be packed inside a life raft in an aircraft.

7.4 Typical ELT

Figures 7.1 to 7.4 show the external and internal construction of a basic Type-W ELT. The unit is hermetically sealed at each end in order to prevent the ingress of water. The procedure for disassembling the ELT usually involves withdrawing the unit from one end of the cylindrical enclosure. When reassembling an ELT, care must be taken to reinstate the hermetic seals at each end of the enclosure.

The specification for a modern Type-AF ELT is shown in Table 7.2. This unit provides outputs on all three ELT beacon frequencies; 121.5 MHz,

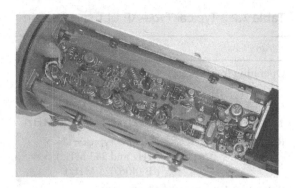

Figure 7.3 ELT transmitter and modulator printed circuit board (the crystal oscillator is located on the right with the dual-frequency output stages on the left).

Figure 7.4 ELT test switch and test light (the antenna base connector is in the centre of the unit).

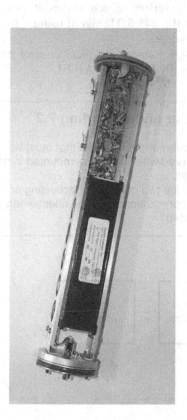

Figure 7.2 Interior view of the ELT shown in Figure 7.1. Note how the battery occupies approximately 50% of the internal volume.

Test your understanding 7.1

Distinguish between the following types of ELT: (a) Type-F, (b) Type-AF and (c) Type-W.

Figure 7.5 Type-AF ELT control panel (note the three switch positions marked ON, ARMED and TEST/RESET).

Table 7.2 Typical Type-AF ELT specification

Parameter	Specification
Operating frequencies	121.5 MHz, 243 MHz and 406.025 MHz
Frequency tolerance	0.005% (121.5 MHz and 243 MHz); kHz (406.025 MHz)
RF output power	250 mW typical (121.5 MHz and 243 MHz); 5 W dB (406.025 MHz)
Pulse duration	10 ms
Activation	G-switch
Power source	Internal lithium battery
Battery life	5 years (including effects of monthly operational checks)
Beacon operating life	50 hours
Digital messagerepetition period (406.025 MHz only)	Every 50 s
Modulation	AM (121.5 MHz and 243 MHz); phase modulation (406.025 MHz)
Housing material	Aluminium alloy

243 MHz and 406.025 MHz. The ELT uses amplitude modulation (AM) on the two VHF frequencies (121.5 MHz and 243 MHz) and phase modulation (PM) on the UHF frequency (406.025 MHz). The AM modulating signal consists of an audio tone that sweeps downwards from 1.5 kHz to 500 Hz with three sweeps every second. The modulation depth is greater than 85%.

The block schematic diagram for a simple Type-W ELT is shown in Figure 7.6. The supply is connected by means of a water switch (not shown in Figure 7.6). The unit shown in Figure 7.6 only provides outputs at VHF (121.5 MHz and 243 MHz). These two frequencies are harmonically related which makes it possible to generate the 243 MHz signal using a frequency doubler stage.

Figure 7.5 shows a typical control panel for an internally fixed Type-AF ELT.

Test your understanding 7.2

1. State three requirements that must be observed when an ELT is mounted in an aircraft.
2. Describe two methods of activating an ELT.
3. What precautions must be taken when an ELT is tested?

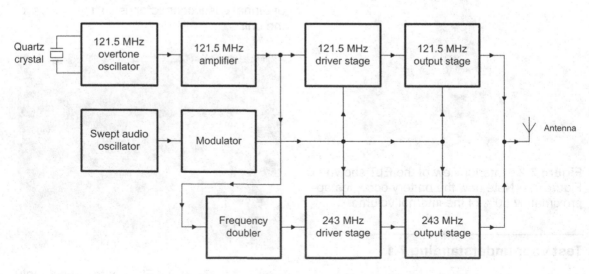

Figure 7.6 Block schematic diagram for a Type-W ELT.

7.5 Cospas–Sarsat satellites

Cospas–Sarsat is a satellite system designed to supply alert and location information to assist search and rescue operations. The Russian Cospas stands for 'space system for the search of vessels in distress' whilst Sarsat stands for 'search and rescue satellite-aided tracking'.

The system uses satellites and ground stations to detect and locate signals from ELT operating at frequencies of 121.5 MHz, 243 MHz and/or 406 MHz. The system provides worldwide support to organizations responsible for air, sea or ground SAR operations.

The basic configuration of the Cospas–Sarsat system features:

- ELT that transmit VHF and/or UHF signals in case of emergency
- Instruments on board geostationary and low-orbiting satellites detecting signals transmitted by the ELT
- **Local user terminals** (LUT) which receive and process signals transmitted via the satellite downlink to generate distress alerts
- **Mission control centres** (MCC) which receive alerts from LUTs and send them to a **rescue coordination centre** (RCC)
- **Search and rescue** (SAR) units.

There are two Cospas–Sarsat systems. One operates at 121.5 MHz (VHF) whilst the other operates at 406 MHz (UHF). The Cospas–Sarsat 121.5 MHz system uses low earth orbit (LEO) polar-orbiting satellites together with associated ground receiving stations. The basic system is shown in Figure 7.7.

The signals produced by ELT beacons are received and relayed by Cospas–Sarsat LEO-SAR satellites to Cospas–Sarsat LUTs that process the signals to determine the location of the ELT. The computed position of the ELT transmitter is relayed via an MCC to the appropriate RCC or **search and rescue point of contact (SPOC)**.

The Cospas–Sarsat system uses Doppler location techniques (using the relative motion between the satellite and the distress beacon) to accurately locate the ELT. The carrier frequency transmitted by the ELT is reasonably stable during the period of mutual beacon-satellite visibility. Doppler performance is enhanced due to the low-altitude near-polar orbit used by the Cospas–Sarsat satellites. However, despite this it is important to note that the location accuracy of the 121.5 MHz system is not as good as the accuracy that can be achieved with the 406 MHz system. The low-altitude orbit also makes it possible for the system to operate with very low uplink power levels.

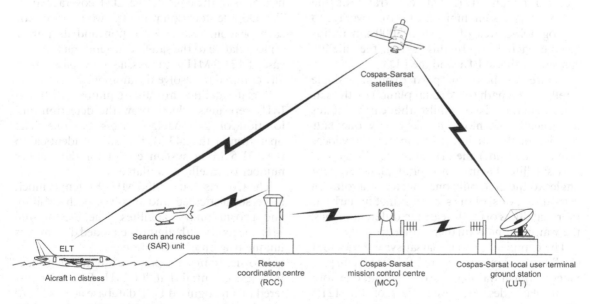

Figure 7.7 The Cospas–Sarsat system in operation.

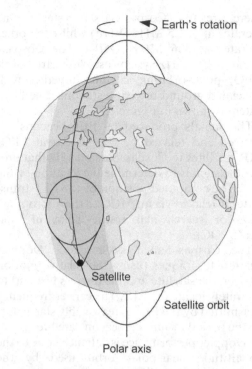

Figure 7.8 Polar orbit for a low earth orbit (LEO) search and rescue (SAR) satellite.

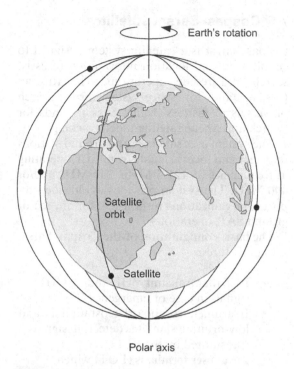

Figure 7.9 The constellation of four LEO–SAR satellites.

A near polar orbit could provide full global coverage but 121.5 MHz can only be produced if the uplink signals from the ELT are actually received by an LUT. This constraint of the 121.5. MHz system limits the useful coverage to a geographic area of about a 3000 km radius around each LUT. In this region, the satellite can see' both the ELT and the LUT.

Figure 7.8 shows the polar orbit of a single satellite. The path (or 'orbital plane') of the satellite remains fixed, whilst the earth rotates underneath. At most, it takes only one half rotation of the earth (i.e. 12 hours) for any location to pass under the orbital plane. With a second satellite, having an orbital plane at right angles to the first, only one quarter of a rotation is required, or six hours maximum. Similarly, as more satellites orbit the earth in different planes, the waiting time is further reduced.

The complete Cospas–Sarsat system uses four satellites, as shown in Figure 7.9. The system provides a typical waiting time of less than one hour at mid-latitudes. However, users of the 121.5 MHz system have to wait for a satellite pass which provides for a minimum of four minutes, simultaneous visibility of an ELT and an LUT. This additional constraint may increase the waiting time to several hours if the transmitting beacon is at the edge of the LUT coverage area. The Doppler location provides two positions for each beacon: the true position and its mirror image relative to the satellite ground track. In the case of 121.5 MHz beacons, a second pass is usually required to resolve the ambiguity.

Sarsat satellites are also equipped with 243 MHz repeaters which allow the detection and location of 243 MHz distress beacons. The operation of the 243 MHz system is identical to the 121.5 MHz system except for the smaller number of satellites available.

The Cospas–Sarsat 406 MHz System is much more sophisticated and involves both orbiting and geostationary satellites. The use of 406 MHz beacons with digitally encoded data allows unique beacon identification.

In order to provide positive aircraft identification, it is essential that 406 MHz ELT are registered in a recognised ELT database accessible to search and rescue authorities. The information held in the database includes data on the ELT, its owner and the aircraft on which the ELT is mounted. This information can be invaluable in a SAR operation.

The unique coding of a UHF ELT is imbedded in the final stage of manufacture using aircraft data supplied by the owner or operator. The ELT data is then registered with the relevant national authorities. Once this has been done, the data is entered into a database available for interrogation by SAR agencies worldwide.

Key point

It is essential to observe published safety precautions when testing an ELT. This includes:

- self-test
- operational test
- transmission test.

All checks must follow the approved test procedures.

7.6 Multiple choice questions

1. ELT transmissions use:
   (a) Morse code and high-power RF at HF
   (b) pulses of acoustic waves at 37.5 kHz
   (c) low-power RF at VHF or UHF.

2. A Type-P ELT derives its power from:
   (a) aircraft batteries
   (b) internal batteries
   (c) a small hand-operated generator.

3. Transmission from an ELT is usually initially detected by:
   (a) low-flying aircraft
   (b) one or more ground stations
   (c) a satellite.

4. The operational state of an ELT is tested using:
   (a) a test switch and indicator lamp
   (b) immersion in a water tank for a short period
   (c) checking battery voltage and charging current.

5. A Type-W ELT needs checking. What is the first stage in the procedure?
   (a) Inspect and perform a load test on the battery
   (b) Open the outer case and inspect the hermetic seal
   (c) Read the label on the ELT to determine the unit's expiry date.

6. If bubbles appear when an ELT is immersed in a tank of water, which one of the following statements is correct?
   (a) This is normal and can be ignored
   (b) This condition indicates that the internal battery is overheating and producing gas
   (c) The unit should be returned to the manufacturer.

7. The air testing of an ELT can be carried out:
   (a) at any place or time
   (b) only after notifying the relevant authorities
   (c) only at set times using recommended procedures.

8. On which frequencies do ELTs operate?
   (a) 125 MHz and 250 MHz
   (b) 122.5 MHz and 406.5 MHz
   (c) 121.5 MHz and 406.025 MHz.

9. A Type-W ELT is activated by:
   (a) a member of the crew
   (b) immersion in water
   (c) a high G-force caused by deceleration.

10. The location accuracy of a satellite-based beacon locator system is:
    (a) better on 121.5 MHz than on 406 MHz
    (b) better on 406 MHz than on 121.5 MHz
    (c) the same on 121.5 MHz as on 406 MHz.

11. An ELT fitted with a lithium battery is:
    (a) safe for packing in a life raft
    (b) unsafe for packing in a life raft
    (c) not suitable for use with a Type-F ELT.

12. A Type-W or Type-S ELT will work better
    when the antenna is:
    (a) held upright
    (b) slanted downwards slightly
    (c) carefully aligned with the horizontal.

13. The satellites used by the Cospas–Sarsat
    121.5 MHz system are:
    (a) in high earth orbit
    (b) in low earth orbit
    (c) geostationary.

Chapter 8

Aircraft navigation

Navigation is the science of conducting journeys over land and/or sea. Whether the journey is to be made across deserts or oceans, we need to know the ultimate destination and how the journey's progress will be checked along the way. Finding a position on the earth's surface and deciding on the direction of travel can be simply made by observations or by mathematical calculations. Aircraft navigation is no different, except that the speed of travel is much faster! Navigation systems for aircraft have evolved with the nature and role of the aircraft itself. Starting with visual references and the basic compass, leading on to radio ground aids and self-contained systems, many techniques and methods are employed.

Although the basic requirement of a navigation system is to guide the crew from point A to point B, increased traffic density and airline economics means that more than one aircraft is planning a specific route. Flight planning takes into account such things as favourable winds, popular destinations and schedules. Aircraft navigation is therefore also concerned with the management of traffic and safe separation of aircraft. This chapter reviews some basic features of the earth's geometry as it relates to navigation and introduces some basic aircraft navigation terminology. The chapter concludes by reviewing a range of navigation systems used on modern transport and military aircraft (a full description of these systems follows in subsequent chapters).

8.1 The earth and navigation

Before looking at the technical aspects of navigation systems, we need to review some basic features of the earth and examine how these features are employed for aircraft navigation purposes. Although we might consider the earth to be a perfect sphere, this is not the case. There is a flattening at both the poles such that the earth

is shaped more like an orange. For short distances, this is not significant; however, for long-range (i.e. global) navigation we need to know some accurate facts about the earth. The mathematical definition of a sphere is where the distance (radius) from the centre to the surface is equidistant. This is not the case for the earth, where the actual shape is referred to as an oblate spheroid.

8.1.1 Position

To define a unique two-dimensional position on the earth's surface, a coordinate system using imaginary lines of **latitude** and **longitude** is drawn over the globe, see Figure 8.1. Lines of longitude join the poles in **great circles** or **meridians**. A great circle is defined as the intersection of a sphere by a plane passing through the centre of the sphere; this has a radius measured from the centre to the surface of the earth. These north–south lines are spaced around the globe and measured in angular distance from the **zero (or prime) meridian**, located in Greenwich, London. Longitude referenced to the prime-meridian extends east or west up to 180 **degrees**. Note that the distance between lines of longitude converges at the poles. Latitude is the angular distance north or south of the equator; the poles are at latitude 90 degrees.

For accurate navigation, the degree (symbol ° after the value, e.g. 90° north) is divided by 60 giving the unit of **minutes** (using the symbol ' after numbers), e.g. one half of a degree will be 30'. This can be further refined into smaller units by dividing again by 60 giving the unit of **seconds** (using the symbol " after numbers), e.g. one half of a minute will be 30". A second of latitude (or longitude at the equator) is approximately 31 meters, just over 100 feet. Defining a unique position on the earth's surface, e.g. Land's End in Cornwall, UK, using latitude and longitude, is written as:

DOI: 10.1201/9781003411932-8

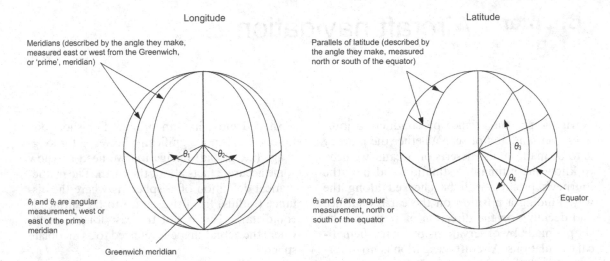

Longitude

Meridians (described by the angle they make,
measured east or west from the Greenwich,
or 'prime', meridian)

θ_1 and θ_2 are angular
measurement, west or
east of the prime
meridian

Greenwich meridian

Latitude

Parallels of latitude (described by
the angle they make, measured
north or south of the equator)

θ_3 and θ_4 are angular
measurement, north or
south of the equator

Equator

Figure 8.1　Longitude and latitude

Latitude N 50° 04′ 13″ Longitude W 5° 42′ 42″

Key point

Horizontal magnetic lines of flux are only of use to
a compass between latitudes of 70 degrees. Between latitudes of 50 degrees, the strength of horizontal field lines have decreased by up to 50%.

8.1.2 Direction

Direction to an observed point (**bearing**) can be
referenced to a known point on the earth's surface, e.g. **magnetic north**. Bearing is defined as
the angle between the vertical plane of the reference point through to the vertical plane of the
observed point. Basic navigational information
is expressed in terms of **compass points** from
zero referenced to north through 360° in a
clockwise direction, see Figure 8.2. For practical navigation purposes, north has been taken
from the natural feature of the earth's magnetic
field; however; magnetic north is not at 90° latitude; the latter defines the position of **true
north**. The location of magnetic north is in the
Canadian Arctic, approximately 83° latitude
and 115° longitude west of the prime meridian,
see Figure 8.3.

Magnetic north is a natural feature of the
earth's geology; it is slowly drifting across the

Canadian Arctic at approximately 40 km north-
west per year. Over a long period of time, magnetic north describes an elliptical path. The
Geological Survey of Canada keeps track of
this motion by periodically carrying out magnetic surveys to redetermine the pole's location.
In addition to this long-term change, the earth's
magnetic field is also affected on a random basis
by the weather, i.e. electrical storms.

Navigation charts based on magnetic north
have to be periodically updated to consider this
gradual drift. Compass-based systems are referenced to magnetic north; since this is not at 90°
latitude, there is an angular difference between
magnetic and true north. This difference will be
zero if the aircraft's position happens to be on
the same longitude as magnetic north, and maximum at longitudes 90° either side of this longitude. The angular difference between magnetic
north and true north is called **magnetic variation**. It is vital that when bearings or headings
are used, we are clear on what these are referenced to.

The imaginary lines of latitude and longitude
described above are curved when superimposed
on the earth's surface; they also appear as
straight lines when viewed from above. The
shortest distance between points A and B on a
given route is a straight line. When this route is
examined, the projection of the path (the **track**)
flown by the aircraft over the earth's surface is
described by a great circle.

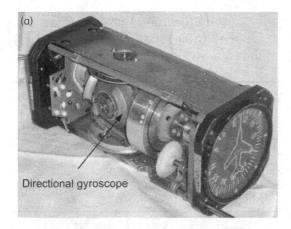

Directional gyroscope

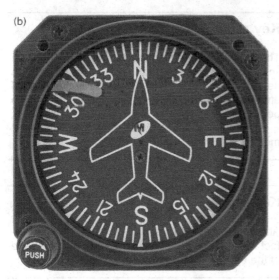

Figure 8.2 Directional gyro (DG) instrument

Figure 8.3 Approximate location of magnetic north

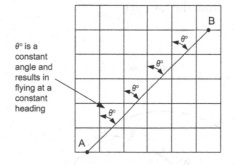

(a) Local meridians and the rhumb line

$\theta°$ is a constant angle and results in flying at a constant heading

The shortest distance between A and B is defined by a great circle

Rhumb line intersects each meridian at the same angle

(b) Great circle and the rhumb line

Figure 8.4 Flying a constant heading

Flying in a straight line implies that we are maintaining a constant heading, but this is not the case. Since the lines of longitude converge, travelling at a constant angle at each meridian yields a track that actually curves as illustrated in Figure 8.4. A track that intersects the lines of longitude at a constant angle is referred to as a **rhumb line**. Flying a rhumb line is readily achieved by reference to a fixed point, e.g. magnetic north. The great circle route; however, requires that the direction flown (with respect to the meridians) changes at any given time, a role more suited to a navigation computer.

8.1.3 Distance and speed

The standard unit of measurement for distance used by most countries around the world (the exceptions being the UK and USA) is the kilometre (km). This quantity is linked directly to the earth's geometry; the distance between the poles and equator is 10,000 km. The equatorial radius of the earth is 6378 km; the polar radius is 6359 km.

For aircraft navigation purposes, the quantity of distance used is the **nautical mile (nm)**. This quantity is defined by distance represented by one minute of arc of a great circle (assuming the earth to be a perfect sphere). The nautical mile (unlike the statute mile) is therefore directly linked to the geometry of the earth. Aircraft speed, i.e. the rate of change of distance with respect to time, is given by the quantity '**knots**'; nautical miles per hour.

Calculating the great circle distance between two positions defined by an angle is illustrated in Figure 8.5. The distance between two positions defined by their respective latitudes and longitudes, (*lat1*, *lon1*) and (*lat2*, *lon2*), can be calculated from the formula:

$$d = \cos^{-1}\left(\begin{array}{l}\sin(lat_1)\sin(lat_2)\\ +\cos(lat_1)\times\cos(lat_2)\times\cos(lon_1-lon_2)\end{array}\right)$$

Test your understanding 8.1

Explain each of the following terms:

1. Latitude
2. Longitude
3. Great circle
4. Rhumb line.

Key point

The nautical mile (unlike the statute mile) is directly linked to the geometry of the earth. This quantity is defined by distance represented by one degree of arc of a great circle (assuming the earth to be a perfect sphere).

One nautical mile = 1.15 statute miles; 1852 meters; 6076 feet

---

Key point

Although we might consider the earth to be a perfect sphere, this is not the case. The actual shape of the earth is referred to as an oblate spheroid.

Key point

Longitude referenced to the prime-meridian extends east or west up to 180 degrees. Latitude is the angular distance north or south of the equator; the poles are at a latitude of 90 degrees.

Key point

Both latitude and longitude are angular quantities measured in degrees. For accurate navigation, degrees can be divided by 60, giving the unit of 'minutes'; this can be further divided by 60, giving the unit of 'seconds'.

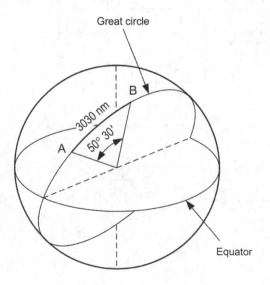

50° 30' = (50 × 60) + 30 = 3030' = 3030 nm

Figure 8.5 Calculation of great circle distances

8.2 Dead reckoning

Estimating a position by extrapolating from a known position and then keeping note of the direction, speed and elapsed time is known as **dead reckoning**. An aircraft passing over a given point on a heading of 90° at a speed of 300 knots will be five miles due east of the given point after one minute. If the aircraft is flying in zero wind conditions, this simple calculation holds true. In realistic terms, the aircraft will almost certainly be exposed to wind at some point during the flight and this will affect the navigation calculation. With our aircraft flying on a heading of 90° at a speed of 300 knots, let's assume that the wind is blowing from the south at 10 knots, see Figure 8.6. In a one-hour time period, the air that the aircraft is flying in will have moved north by ten nautical miles. This means that the aircraft's path (referred to as its **track**) over the earth's surface is not due east. In other words, the aircraft track is not the same as the direction in which the aircraft is heading. This leads to a horizontal displacement (**drift**) of the aircraft from the track it would have followed in zero wind conditions.

The angular difference between the heading and track is referred to as the **drift angle** (quoted as being to port/left or starboard/right of the heading). If the wind direction were in the same direction as the aircraft heading, i.e. a tail wind, the aircraft speed of 300 knots through the air would equate to a ground speed of 310 knots. Likewise, if the wind were from the east (a headwind) the ground speed would be 290 knots.

Knowledge of the wind direction and speed allows the crew to steer the aircraft into the

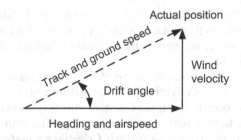

Figure 8.7 Resolving actual position

wind such that the wind actually moves the aircraft onto the desired track. For dead reckoning purposes, we can resolve these figures in mathematical terms and determine a position by triangulation as illustrated in Figure 8.7. Although the calculation is straightforward, the accuracy of navigation by dead reckoning will depend on up to date knowledge of wind speed and direction. Furthermore, we need accurate measurements of speed and direction. Depending on the accuracy of measuring these parameters, positional error will build up over time when navigating by dead reckoning. We therefore need a means of checking our calculated position on a periodic basis; this is the process of **position fixing**.

Key point

Dead reckoning is used to estimate a position by extrapolating from a known position and then keeping note of the direction and distance travelled.

Key point

The angular difference between the heading and track is referred to as the drift angle.

8.3 Position fixing

When travelling short distances over land, natural terrestrial features such as rivers, valleys, hills, etc., can be used as direct observation to keep a check on (**pinpointing**) the journey's progress. If the journey is by sea, then we can use the coastline and specific features such as

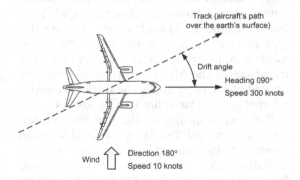

Figure 8.6 Effect of crosswind

lighthouses to confirm our position. If the journey is now made at night or out of sight of the coast, we need other means of fixing our position.

The early navigators used the sun, stars and planets very effectively for navigation purposes; if the position of these celestial objects is known, then the navigator can confirm a position anywhere on the earth's surface. **Celestial navigation** (or **astronavigation**) was used very effectively in the early days of long distance aircraft navigation. Indeed, it has a number of distinct advantages when used by the military: the aircraft does not radiate any signals; navigation is independent of ground equipment; the references cannot be jammed; and navigation references are available over the entire globe.

The disadvantage of celestial navigation for aircraft is that the skies are not always clear and it requires a great deal of skill and knowledge to be able to fix a position whilst travelling at high speed. Although automated celestial navigation systems were developed for use by the military, they are expensive; modern avionic equipment has now phased out the use of celestial navigation for commercial aircraft.

The earliest ground-based references (**navigation aids**) developed for aircraft navigation are based on radio beacons. These beacons can provide angular and/or distance information; when using this information to calculate a position fix, the terms are referred to mathematically as theta (θ) and rho (ρ). By utilising the directional properties of radio waves, the intersection of signals from two or more navigation aids can be used to fix a position (**theta–theta**), see Figure 8.8. Alternatively, if we know the distance and direction (bearing) to a navigation aid, the aircraft position can be confirmed (**rho–theta**). Finally, we can establish our position if we know the aircraft's distance (**rho–rho**) from any two navigation aids, i.e. without knowledge of the bearing.

8.4 Maps and charts

Maps provide the navigator with a representative diagram of an area that contains a variety of physical features, e.g. cities, roads and topographical information. Charts contain lines of latitude and longitude together with essential data such as the location of navigation aids. Creating charts and maps requires that we transfer distances and geographic features from the earth's spherical surface onto a flat piece of paper. This is not possible without some kind of compromise in geographical shape, surface area, distance or direction. Many methods of producing charts have been developed over the centuries; the choice of projection depends on the intended purpose.

In the sixteenth century Gerhardus Mercator, the Flemish mathematician, geographer and cartographer, developed what was to become the standard chart format for nautical navigation: the **Mercator projection**. This is a cylindrical map projection where the lines of latitude and longitude are projected from the earth's centre, see Figure 8.9. Imagine a cylinder of paper wrapped around the globe and a light inside the globe; this projects the lines of latitude and longitude onto the paper. When the cylinder is unwrapped, the lines of latitude appear incorrectly as having equal length. Directions and the shape of geographic features remain true; however, distances and sizes become distorted. The advantage of using this type of chart is that the navigator sets a constant heading to reach the destination. The meridians of the Mercator projected chart are crossed at the same angle; the track followed is referred to as a **rhumb line** (see Figure 8.4).

For aircraft navigation the Mercator projection might be satisfactory; however, if we want to navigate by great circle routes then we need true directions. An alternative projection used for aircraft navigation, and most popular maps and charts, is the **Lambert** azimuthal equal-area

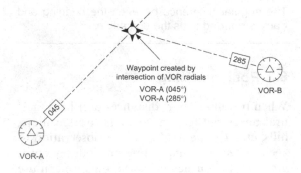

Figure 8.8   Position fixing

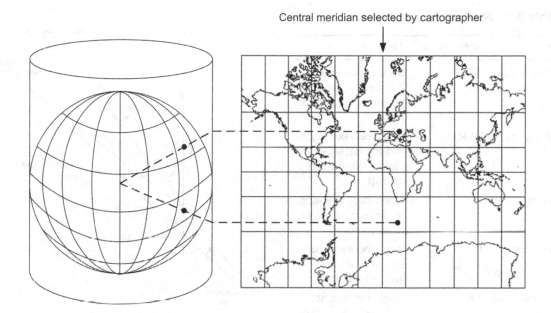

Figure 8.9 Mercator projection

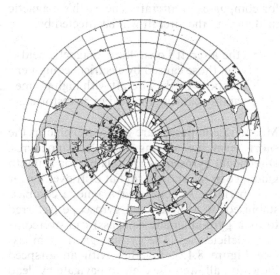

Figure 8.10 Lambert projection (viewed from true north)

projection. This projection was developed by Johann Heinrich Lambert (1728–77) and is particularly useful in high latitudes. The projection is developed from the centre point of the geographic feature to be surveyed and charted. Using true north as an example, Figure 8.10 illustrates the Lambert projection.

8.5 Navigation terminology

The terms shown in Table 8.1 are used with numerous navigaton systems including INS and RNAV; computed values are displayed on a control display unit (CDU) and/or primary flight instruments. These terms are illustrated in Figure 8.11, all terms are referenced to true north.

8.6 Navigation systems evolution

This section provides a brief overview of the evolution of increasingly sophisticated navigation systems used on aircraft.

8.6.1 Gyro-magnetic compass

The early aviators used visual aids to guide them along their route; these visual aids would have included rivers, roads, rail tracks, coastlines, etc. This type of navigation is not possible at high altitudes or in low visibility and so the earth's magnetic field was used as a reference, leading to the use of simple **magnetic compasses** in aircraft. We have seen that magnetic variation has to be taken into account for navigation; there are additional considerations to be addressed

Table 8.1 Navigation terminology

Term	Abbreviation	Description
Cross track distance	XTK	Shortest distance between the present position and desired track
Desired track angle	DSRTK	Angle between north and the intended flight path of the aircraft
Distance	DIS	Great circle distance to the next waypoint or destination
Drift angle	DA	Angle between the aircraft's heading and ground track
Ground track angle	TK	Angle between north and the flight path of the aircraft
Heading	HDG	Horizontal angle measured clockwise between the aircraft's centreline (longitudinal axis) and a specified reference
Present position	POS	Latitude and longitude of the aircraft's position
Track angle error	TKE	Angle between the actual track and desired track (equates to the desired track angle minus the ground track angle)
Wind direction	WD	Angle between north and the wind vector
True airspeed	TAS	True airspeed measured in knots
Wind speed	WS	Measured in knots
Ground speed	GS	Measured in knots

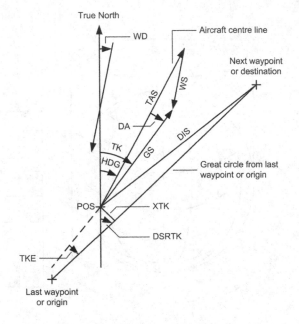

Figure 8.11 Navigation terminology

for compasses in aircraft. The earth's magnetic field around the aircraft will be affected by:

- the aircraft's own 'local' magnetic fields, e.g. those caused by electrical equipment
- sections of the aircraft with high permeability causing the field to be distorted.

Magnetic compasses are also unreliable in the short term, i.e. during turning manoeuvres. **Directional gyroscopes** are reliable for azimuth guidance in the short term but drift over longer time periods. A combined magnetic compass stabilised by a directional gyroscope (referred to as a **gyro-magnetic compass**) can overcome these deficiencies. The gyro-magnetic compass (see Figure 8.12), together with an airspeed indicator, allowed the crew to navigate by dead reckoning, i.e. estimating their position by extrapolating from a known position and then keeping note of the direction and distance travelled.

In addition to directional references, aircraft also need an attitude reference for navigation, typically from a vertical gyroscope. Advances in sensor technology and digital electronics have led to combined attitude and heading reference systems (AHRS) based on laser gyros

Figure 8.12 Directional gyroscope (DG) instrument

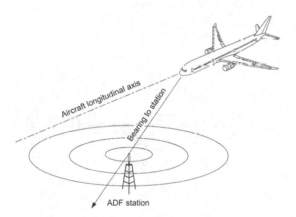

Figure 8.13 ADF radio navigation

and microelectromechanical sensors (see Chapter 17).

Instrumentation errors inevitably lead to deviations between the aircraft's actual and calculated positions; these deviations accumulate over time. Crews therefore need to be able to confirm and update their position by means of a fixed ground-based reference, e.g. a radio navigation aid.

8.6.2 Radio navigation

Early airborne navigation systems using ground-based navigation aids consisted of a loop antenna in the aircraft tuned to amplitude modulation (AM) commercial radio broadcast stations transmitting in the low/medium frequency (LF/MF) bands. Referring to Figure 8.13, pilots

would know the location of the radio station (indeed, it would invariably have been located close to or even in the town/city that the crew wanted to fly to), and this provided a means of fixing a position. Although technology has moved on, these **automatic direction finder** (ADF) systems are still in use today.

Operational problems are encountered using low frequency (LF) and medium frequency (MF) transmissions. During the mid to late 1940s, it was evident to the aviation world that an accurate and reliable short-range navigation system was needed. Since radio communication systems based on very high frequency (VHF) were being successfully deployed, a decision was made to develop a radio navigation system based on VHF. This system became the **VHF omnidirectional range** (VOR) system, see Figure 8.14; a system that is in widespread use throughout the world today. VOR is the basis of the current network of 'airways' that are used in navigation charts.

The advent of radar in the 1940s led to the development of a number of navigation aids including **distance measuring equipment** (DME). This is a short/medium-range navigation system, often used in conjunction with the VOR system to provide accurate navigation fixes. The system is based on secondary radar principles, see Figure 8.15.

Navigation aids such as automatic direction finder (ADF), VOR and DME are used to define airways for en route navigation, see Figure 8.16. They are also installed at airfields to assist with approaches to those airfields. These navigation aids cannot, however, be used for precision approaches and landings. The standard approach and landing system installed at airfields around the world is the **instrument landing system** (ILS), see Figure 8.17. The ILS uses a combination of VHF and UHF radio waves and has been in operation since 1946. There are a number of shortcomings with ILS; in 1978 the **microwave landing system** (MLS) was adopted as the long-term replacement. The system is based on the principle of time referenced scanning beams and provides precision navigation guidance for approach and landing. MLS provides threedimensional approach guidance, i.e. azimuth, elevation and range. The system provides multiple approach angles for both

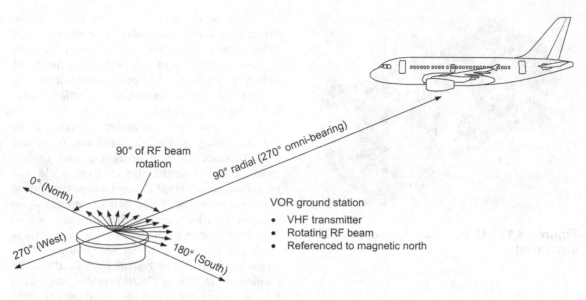

Figure 8.14   VOR radio navigation

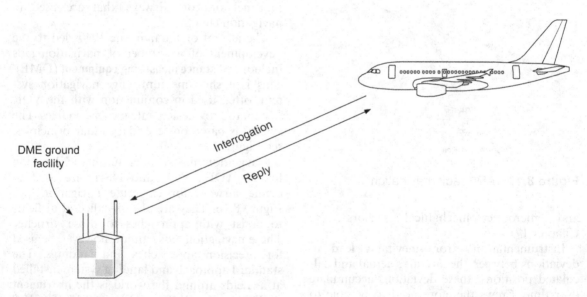

Figure 8.15   Distance measuring equipment (DME)

azimuth and elevation guidance. Despite the advantages of MLS, it has not yet been introduced on a worldwide basis for commercial aircraft. MLS is now superseded by the introduction of satellite-based augmentation systems (SBAS). Military operators of MLS often use mobile equipment that can be deployed within hours. The chapter for MLS has been deleted from this book and archived in the author's website (refer to Preface).

The aforementioned radio navigation aids have one disadvantage in that they are land based and only extend out beyond coastal regions. Long-range radio navigation systems based on **hyperbolic navigation** were introduced in the 1940s to provide for en route operations over oceans and unpopulated areas. Several hyperbolic systems have been developed since, including Decca, Omega and Loran. The operational use of Omega and Decca navigation

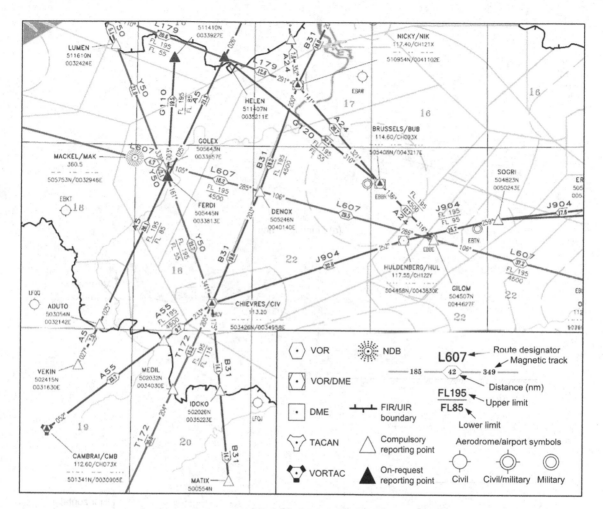

Figure 8.16 Airways defined by navigation aids

systems ceased in 1997 and 2000 respectively. Loran systems are still very much available today as stand-alone systems; they are also being proposed as a complementary navigation aid for global navigation satellite systems. The **Loran-C** system is based on a master station and a number of secondary stations; the use of VLF radio provides an increased area of coverage, see Figure 8.18. Hyperbolic navigation is now considered redundant for commercial aircraft; the chapter for this subject has been deleted from this book and archived in the author's website (refer to Preface).

The advent of computers, in particular the increasing capabilities of integrated circuits using digital techniques, has led to a number of advances in aircraft navigation. One example of

this is the **area navigation system** (RNAV); this is a means of combining, or filtering, inputs from one or more navigation sensors and defining positions that are not necessarily colocated with ground-based navigation aids. Typical navigation sensor inputs to an RNAV system can be from external ground-based navigation aids such as VHF omni directional range (VOR) and distance measuring equipment (DME), see Figure 8.19.

8.6.3 Dead reckoning systems

Dead reckoning systems require no external inputs or references from ground stations. Doppler navigation systems were developed in the mid-1940s and introduced in the mid-1950s

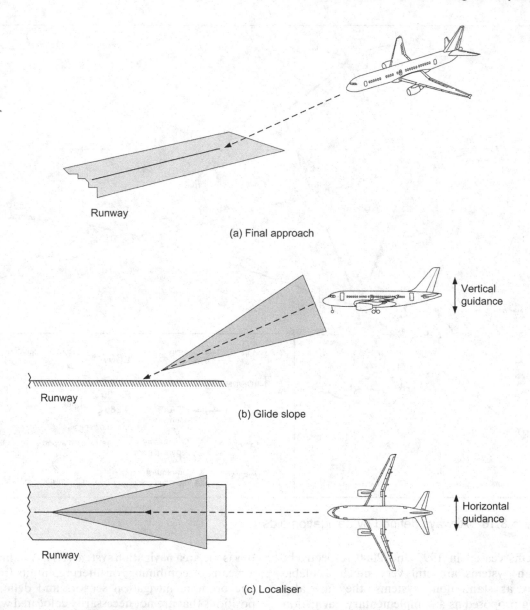

Runway

(a) Final approach

Runway

(b) Glide slope

Runway

(c) Localiser

Figure 8.17   Instrument landing system

as a primary navigation system. Ground speed and drift can be determined using a fundamental scientific principle called **Doppler shift**. Being self-contained, the system can be used for long-distance navigation over oceans and undeveloped areas of the globe.

A major advance in aircraft navigation came with the introduction of the **inertial navigation system** (INS). This is an autonomous dead reckoning system, i.e. it requires no external inputs or references from ground stations. The system was developed in the 1950s for use by the US military and subsequently the space programmes. INS were introduced into commercial aircraft service during the early 1970s. The system is able to compute navigation data such as present position, distance to waypoint, heading, ground speed, wind speed, wind direction, etc. It does not need radio navigation inputs and it does not transmit radio frequencies. Being

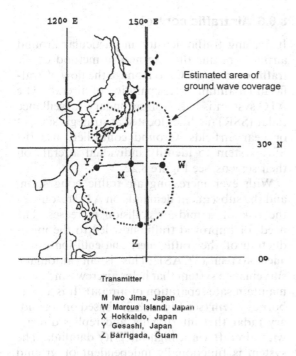

Figure 8.18 Loran-C oceanic coverage using VLF transmissions

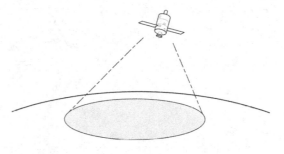

(a) Single satellite describes a circle on the earth's surface

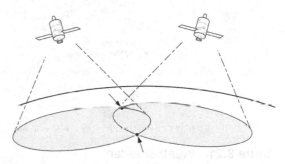

(b) Two satellites define two unique positions. A third satellite defines a unique position

Figure 8.20   Satellite navigation

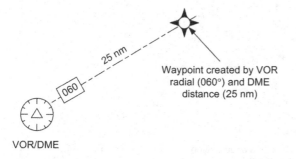

Figure 8.19   Area navigation

self-contained, the system can be used for long distance navigation over oceans and undeveloped areas of the globe.

8.6.4 Satellite navigation

Navigation by reference to the stars and planets has been employed since ancient times; commercial aircraft used to have periscopes to take celestial fixes for long distance navigation. An artificial constellation of navigation aids was initiated in 1973 and referred to as **Navstar** (navigation system with timing and ranging). The global positioning system (GPS) was developed for use by the US military; the first satellite was launched in 1978 and the full constellation was in place and operating by 1994. GPS is now widely available for use by many applications including aircraft navigation; the system calculates the aircraft position by triangulating the distances from a number of satellites, see Figure 8.20.

8.6.5 Radar navigation

The planned journey from A to B could be affected by adverse weather conditions. Radar was introduced onto passenger aircraft during the 1950s to allow pilots to identify weather conditions (see Figure 8.21) and subsequently reroute around these conditions for the safety and comfort of passengers. A secondary use of weather radar is a terrain-mapping mode that allows the pilot to identify features on the ground, e.g. rivers, coastlines and mountains.

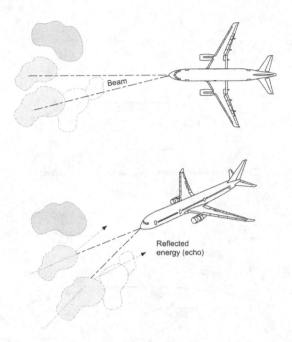

Figure 8.21   Weather radar

8.6.6 Air traffic control

Increasing traffic density, in particular around airports, means that we need a method of **air traffic control** (ATC) to manage the flow of traffic and maintain safe separation of aircraft. The ATC system is based on secondary surveillance radar (SSR) facilities located at strategic sites, at or near airfields. Ground controllers use the SSR system to identify individual aircraft on their screens, see Figure 8.22.

With ever increasing air traffic congestion, and the subsequent demands on ATC resources, the risk of a mid-air collision increases. The need for improved traffic flow led to the introduction of the **traffic alert and collision avoidance system** (TCAS). This is an automatic surveillance system that helps aircrews and ATC maintain safe separation of aircraft. It is an airborne system (see Figure 8.23) based on secondary radar that interrogates and replies directly with aircraft via a high-integrity datalink. The system is functionally independent of ground

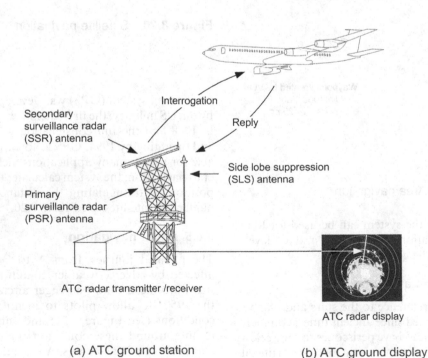

(a) ATC ground station   (b) ATC ground display

Figure 8.22   Secondary surveillance radar

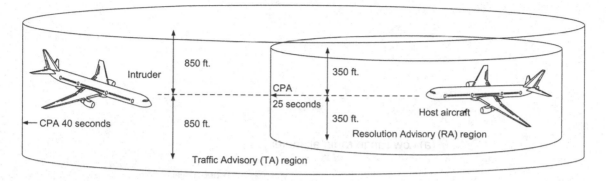

Figure 8.23   Traffic alert and collision avoidance system

stations and alerts the crew if another aircraft comes within a predetermined time to a potential collision. It is important to note that TCAS is a backup system, i.e. it provides warnings when other navigation systems (including ATC) have failed to maintain safe separation of aircraft that could lead to a collision.

8.6.7 Instrument approaches

An instrument approach comprises a series of predetermined manoeuvres by reference to the on-board navigation systems and flight instruments. Approaches are documented in hard or electronic format, informing the pilot with specific information, e.g. the transition from en route navigation to the landing, and (if the landing cannot be completed) a missed approach procedure. There are three classes of instrument approach, as defined by ICAO:

* non-precision approach (NPA)
* approach procedures vertical guidance (APV)
* precision approach (PA).

NPA provides lateral guidance only, i.e. no vertical information, using VOR, ADF and DME or GPS. APV provides both lateral and vertical guidance information, using a localizer directional aid (LDA) with glide path. LDA is an ILS type installation that does not meet precision approach standards, e.g. because the localiser is offset. PA provides lateral and vertical guidance information, using an instrument landing system (ILS), or GNSS landing system (GLS).

8.6.8 Low range radio altimeter

The low range radio altimeter (LRRA) is a self-contained, vertically directed primary radar system operating in the 4.2 to 4.4 GHz band. Airborne equipment comprises a transmitting/receiving antenna, transmitter/receiver and a flight deck indicator/display. Radar energy is directed via a transmitting antenna to the ground; some of this energy is reflected back from the ground and is collected in the receiving antenna, see Figure 8.24(a). A radio altimeter is more accurate and responsive than an air pressure (barometric) altimeter at low altitudes, typically 2500 feet. Radio altitude is either incorporated into an electronic display, as in Figure 8.24(b), or displayed on a dedicated instrument, as in Figure 8.24(c).

Two types of LRRA methods are used to determine the aircraft's radio altitude. The pulse modulation method measures the elapsed time taken for the signal to be transmitted and received; this time delay is directly proportional to altitude. The frequency-modulated, continuous wave (FM/CW) method uses a changeable FM signal where the rate of change is fixed. A proportion of the transmitted signal is mixed with the received signal; the resulting beat signal frequency is proportional to altitude.

Radio altimeter systems perform a critical function in an aircraft's operation, providing a direct measurement of the aircraft's clearance over the terrain. This function is used as a sensor in many aircraft systems, including the instrument landing system (ILS), terrain awareness and collision avoidance (TCAS) and

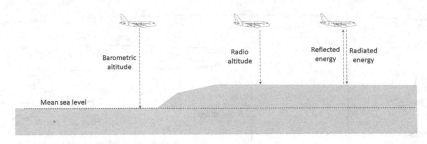

(a) Low range radio altimeter

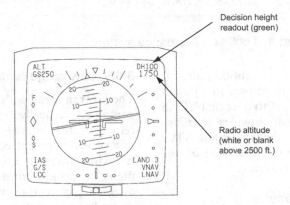

(b) LRRA integrated display

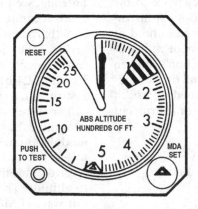

(c) LRRA dedicated display

Figure 8.24   Low range radio altimeter (LRRA)

terrain awareness system (TAWS). The ILS and TCAS are described in more detail in this book; TAWS is described in another book title in the series, Aircraft electrical and electronic systems (AEES). LRRA can also be a self-contained system for low-level rotorcraft operations. A secondary safety system derived from the LRRA is the automatic voice alert device (AVAD). This is a voice alerting system that stores pre-recorded human speech in digital form, e.g. 'Low height' or 'One hundred feet'.

*TAWS is covered in a separate title in this series, Aircraft Electrical and Electronic Systems (AEES).

8.6.9 Reduced vertical separation minimums

The majority of flights over the North Atlantic are by aircraft operating between flight levels FL290 – FL410 (29,000 – 41,000 feet). To ensure adequate airspace capacity and provide for safe vertical separations, reduced vertical separation minimums (RVSM) is applied throughout the region. RVSM allows aircraft to fly with a vertical separation of 1000 feet between FL290 and FL410 inclusive. Entry/exit and movement along these tracks is controlled by Oceanic Control Centres air traffic controllers to maintain aircraft separation.

To fly in this airspace, an aircraft must be equipped with at least two fully serviceable long range navigation systems (LRNS); these will be either an inertial navigation system (INS) or global navigation satellite system (GNSS).

8.6.10 Minimum navigation performance specifications

Minimum navigation performance specifications (MNPS) are required for remote areas, e.g. in the North Atlantic. The requirements are specified in ICAO document North Atlantic Operations and Airspace Manual (NAT Doc 007).

The North Atlantic airspace linking Europe and North America is the busiest oceanic airspace in the world; in 2012 approximately 460,000 flights crossed the North Atlantic. For the majority of this airspace, direct controller pilot communications (DCPC) and air traffic system (ATS) surveillance are unavailable. Aircraft separation therefore requires the highest standards of horizontal and vertical navigation performance and accuracy together with operating discipline.

8.6.11 Electronic flight bag

An electronic flight bag (EFB) is an electronic display system, replacing information and data traditionally based on paper documents and manuals, e.g. navigation charts. The EFB may also support other functions that have no paper equivalent, e.g. data communication systems. A basic EFB can perform flight planning calculations and display a variety of navigational charts, operations manuals, aircraft checklists, etc. Advanced EFBs are fully certified and integrated with aircraft systems. EFBs are classified in one of three ways; Class 1, 2 or 3.

The Class 1 EFB is typically a standard commercial off-the-shelf (COTS) item such as a laptop or handheld electronic device. These devices are deemed as portable or loose equipment and are typically stowed during critical phases of flight. Class 2 EFB can either be COTS or designed for the purpose. They are typically mounted in the aircraft with the display being viewable to the pilot during all phases of flight. Under certain conditions, they can be certified to interface with aircraft systems. They do not share any display or other input/output device (e.g. keyboard, pointing device) with certified aircraft systems. Class 2 EFBs do not require any installation/removal tools. Class 3 EFBs are neither Class 1 nor 2; they are installed items of equipment with certified operating system software. A Class 3 EFB is part of the certified aircraft configuration including installed applications, database, resources, etc.

A typical Class 1/2 electronic flight bag (EFB) is shown in Figure 8.25, on the left of the glareshield. These are typically high-definition sunlight-readable display screens, featuring detailed electronic charts and weather data. EFBs reduce the use of paper charts in the cockpit. They can display victor airways, jet routes, minimum en route altitudes and leg distance, standard terminal arrival routes (STARs), approach charts and airport diagrams replicating traditional en route charts.

The latest versions of avionic systems provide remote access to aircraft systems via an electronic flight bag (EFB) and cloud connectivity. This allows a variety of tasks to be completed remotely, e.g., remote loading and retrieval of flight plans.

8.7 Navigation systems summary

Navigation systems for aircraft have evolved with the nature and role of the aircraft itself. These individual systems are described in detail in the following chapters. Each system has been developed to meet specific requirements within the available technology and cost boundaries.

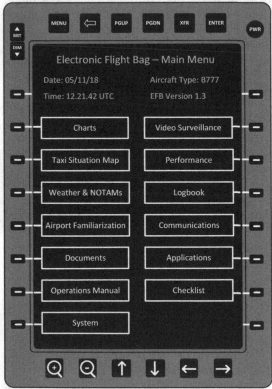

Figure 8.25   Electronic flight bag (EFB)

Whatever the requirement, all navigation systems are concerned with several key factors:

- **Accuracy**: conformance between calculated and actual position of the aircraft
- **Integrity**: ability of a system to provide timely warnings of system degradation
- **Availability**: ability of a system to provide the required function and performance

- **Continuity**: probability that the system will be available to the user
- **Coverage**: geographic area where each of the above are satisfied at the same time.

Test your understanding 8.2

The nautical mile is directly linked to the geometry of the earth; how is a nautical mile defined?

Test your understanding 8.3

Explain the difference between dead reckoning and position fixing.

Test your understanding 8.4

For a given airspeed, explain how tailwinds and headwinds affect groundspeed.

Test your understanding 8.5

Explain the following terms: accuracy, integrity, availability.

Test your understanding 8.6

Describe three ways that bearings and ranges can be used for position fixing.

Test your understanding 8.7

1. Explain the difference between Mercator and Lambert projections.
2. Where on the earth's surface is the difference between a rhumb line and great circle route the greatest?

8.8 Multiple choice questions

1. Longitude referenced to the prime
   meridian extends:
   (a) north or south up to 180°
   (b) east or west up to 180°
   (c) east or west up to 90°

2. Latitude is the angular distance:
   (a) north or south of the equator
   (b) east or west of the prime meridian
   (c) north or south of the prime meridian.

3. The distance between lines of longitude
   converges at the:
   (a) poles
   (b) equator
   (c) great circle.

4. Lines of latitude are always:
   (a) converging
   (b) parallel
   (c) the same length.

5. Degrees of latitude can be divided by 60,
   giving the unit of:
   (a) longitude
   (b) minutes
   (c) seconds.

6. The location of magnetic north is
   approximately:
   (a) 80° latitude and 110° longitude, east of
   the prime meridian
   (b) 80° longitude and 110° latitude, west of
   the prime meridian
   (c) 80° latitude and 110° longitude, west of
   the prime meridian.

7. One minute of arc of a great circle
   defines a:
   (a) nautical mile
   (b) kilometre
   (c) knot.

8. The angular difference between magnetic
   north and true north is called the:
   (a) magnetic variation
   (b) great circle
   (c) prime meridian.

9. Mercator projections produce parallel
   lines of:
   (a) the earth's magnetic field
   (b) longitude
   (c) great circle routes.

10. With respect to the polar radius, the
    equatorial radius of the earth is:
    (a) equal
    (b) larger
    (c) smaller.

11. Dead reckoning is the process of:
    (a) fixing the aircraft's position
    (b) correcting the aircraft's position
    (c) estimating the aircraft's position.

12. The angle between the aircraft's heading
    and ground track is known as the:
    (a) drift angle
    (b) cross track distance
    (c) wind vector.

13. Magnetic compasses are unreliable in the:
    (a) long-term, flying a constant heading
    (b) short-term, during turning
    manoeuvres.
    (c) equatorial regions.

14. The angle between north and the flight
    path of the aircraft is the:
    (a) ground track angle
    (b) drift angle
    (c) heading.

15. When turning into a 25-knot headwind at
    constant indicated airspeed, the ground
    speed will:
    (a) increase by 25 knots
    (b) remain the same
    (c) decrease by 25 knots.

Chapter 9 — Automatic direction finder

Radio waves have directional characteristics as we have seen from earlier chapters. This is the basis of the automatic direction finder (ADF); one of earliest forms of radio navigation that is still in use today. ADF is a short/medium-range (200 nm) navigation system providing directional information; it operates within the frequency range 190–1750 kHz, i.e. low and medium frequency bands. The term 'automatic' refers to the introduction of electromechanical equipment to replace manually operated devices. In this chapter we will look at the historical background to radio navigation, review typical ADF hardware and conclude with some practical aspects associated with the operational use of ADF.

9.1 Introducing ADF

The early aviators used visual aids to guide them along their route; these visual aids would have included rivers, roads, rail tracks, coastlines, etc. This type of navigation is not possible in low visibility and so magnetic compasses were introduced. Magnetic compasses are somewhat unreliable in the short term, i.e. during turning manoeuvres. Directional gyroscopes are reliable in the short term, but drift over longer time periods. A combined magnetic compass stabilised by a directional gyroscope (referred to as a **gyro-magnetic compass**) can overcome these deficiencies. The gyro-magnetic compass, together with an airspeed indicator, allowed the crew to navigate by dead reckoning, i.e. estimating their position by extrapolating from a known position and then keeping note of the direction and distance travelled. Instrumentation errors inevitably lead to deviations between the aircraft's actual and calculated positions; these deviations accumulate over time. Crews therefore need to be able to confirm and update their position by means of a fixed ground-based reference.

The early airborne navigation systems using ground-based navigation aids consisted of a fixed-loop antenna in the aircraft tuned to an amplitude modulation (AM) commercial radio broadcast station. Pilots would know the location of the radio station (indeed, it would invariably have been located close to or even in the town/city that the crew wanted to fly to). The fixed-**loop antenna** was aligned with the longitudinal axis of the aircraft, with the pilot turning the aircraft until he received the minimum signal strength (null reading). By maintaining a null reading, the pilot could be sure that he was flying towards the station. This constant turning was inefficient in terms of fuel consumption and caused inherent navigation problems in keeping note of the aircraft's position during these manoeuvres! The effects of crosswind complicated this process since the aircraft's heading is not aligned with its track.

9.2 ADF principles

The introduction of an 'automatic' direction finder (ADF) system addresses this problem. A loop antenna that the pilot could rotate by hand solves some of these problems; however, this still requires close attention from the crew. Later developments of the equipment used an electrical motor to rotate the loop antenna. The received signal strength is a function of the angular position of the loop with respect to the aircraft heading and bearing to the station, see Figures 9.1(a) and (b). If a plot is made of loop angle and signal strength, the result is a sine wave, as shown in Figure 9.1(c). The **null point** is easier to determine than the maximum signal strength since the rate of change is highest. Rotating the antenna (rather than turning the aircraft) to determine the null reading from the radio station was a major advantage of the

DOI: 10.1201/9781003411932-9

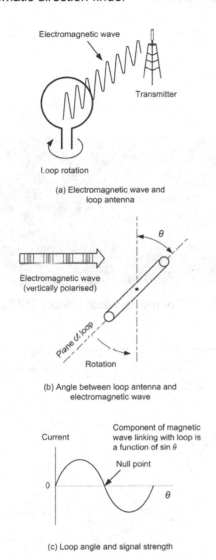

(a) Electromagnetic wave and loop antenna

(b) Angle between loop antenna and electromagnetic wave

(c) Loop angle and signal strength

Figure 9.1 Loop antenna output

system. The pilot read the angular difference between the aircraft's heading and the direction of the radio station, see Figure 9.2(a), from a graduated scale, and a bearing to the station could then be determined. The industry drive towards solid-state components, i.e. with no moving parts, has led to the equipment described in Section 9.3.

Navigation based on ADF (using AM commercial radio stations broadcasting in the frequency range 540–1620 kHz) became an established method of travelling across country. With the growth of air travel, dedicated radio navigation aids were installed along popular air transport routes. These radio stations, known as **non-directional beacons** (NDB), gradually supplemented the commercial radio stations and a network of NDBs sprang up in the nations developing their aeronautical infrastructure. These NDBs broadcast in the low frequency (LF) range 190–415 kHz and medium frequency (MF) range 510–535 kHz. As the quantity of NDBs increased, air navigation charts were produced and the NDBs were identified by a two- or three-letter alpha code linked to the location and frequency. In Figure 9.2(c), the NDB located at Mackel in Belgium transmits on 360.5 kHz and is identified as MAK; note the Morse code, latitude and longitude details on the chart. Beacons are deployed with varying power outputs classified as high (2 kW), medium (50W to 2 kW) and low (less than 50 W).

Table 9.1 provides a list of typical NDBs associated with airports and cities in a typical European country (note that these are provided for illustration purposes only). Beacons marked with an asterisk in this table are referred to as locator beacons; they are part of the final approach procedures for an airfield (see Chapter 12).

9.3 ADF equipment

9.3.1 Antenna

The rotating loop antenna was eventually replaced with a fixed antenna consisting of two loops combined into a single item; one aligned with the centreline of the fuselage, the other at right angles, as shown in Figure 9.3. This orthogonal antenna is still referred to as the 'loop' antenna. Measuring the signal strength from each of the loops, and deriving an angular position in a dedicated ADF receiver, determines the direction to the selected beacon (or commercial radio station). The loop antenna resolves the directional signal; however, this can have two possible solutions 180 degrees apart. A second **'sense' antenna** is therefore required to detect non-directional radio waves from the beacon; this signal is combined with the(a) Using an ADF system for navigation directional signals from the loop antenna to produce a single directional solution. The polar diagram for a

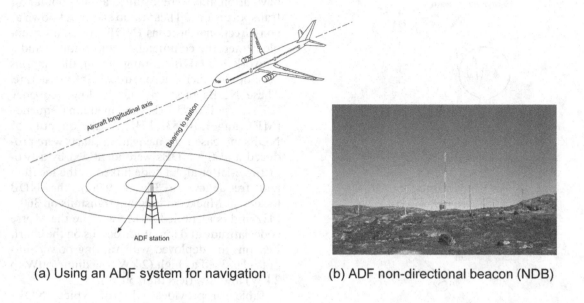

(a) Using an ADF system for navigation   (b) ADF non-directional beacon (NDB)

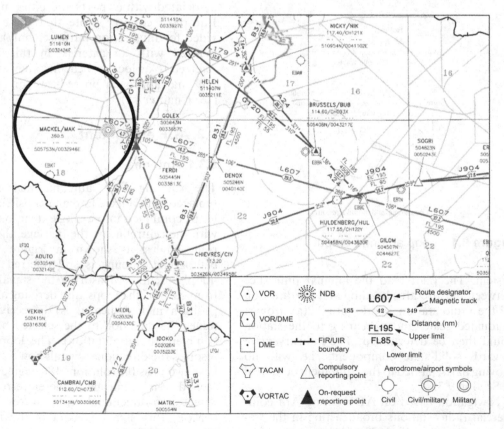

(c) MACKEL NDB shown on a navigation chart

Figure 9.2 Navigation by non-directional beacons NDB

Table 9.1 Examples of NDB codes and frequencies

Name	Identification code	Frequency (kHz)
Eelde	SO	330.00
Eindhoven	EHN	397.00
Eindhoven	PH	316.00
Gull	GUL	383.50
Maastricht	NW	373.00
Maastricht *	ZL	339.00
Rotterdam	ROT	350.50
Rotterdam *	PS	369.00
Rotterdam *	RR	404.50
Schiphol	CH	388.50
Schiphol *	NV	332.00
Stad	STD	386.00
Stadskanaal	STK	315.00
Thorn	THN	434.00
Twenthe	TWN	335.50

* Locator beacons

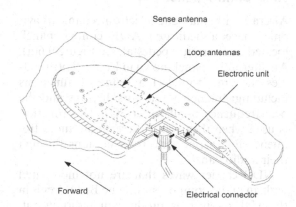

Figure 9.3 ADF antenna

loop and sense antenna is shown in Figure 9.4; when the two patterns are combined, it forms a **cardioid**. Most commercial transport aircraft are fitted with two independent ADF systems, typically identified as left and right systems; the antenna locations for a typical transport aircraft are shown in Figure 9.5.

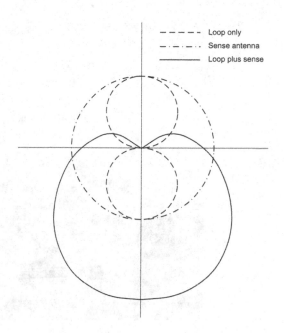

Figure 9.4 Polar diagram for the ADF loop and sense antennas

Key point

When the ADF loop is at right angles to the transmitter, Figure 9.1, the RF wave is detected simultaneously by both sides of the loop antenna, resulting in zero current, i.e. a null point. When the loop is at an angle to the beacon, the RF wave is detected by one side of the loop first; the other side of the loop will be out of phase, thereby generating a current. The null signal point is therefore used to determine the direction of the beacon's RF wave. The loop and sense components are combined algebraically, with the magnitude and polarity of the sense antenna arranged to be identical to the loop antenna. The resulting polar diagram, Figure 9.4, is a cardioid, i.e. it only has one null point.

9.3.2 Receiver

ADF receivers are located in the avionic equipment bay. The signal received at the antenna is coupled to the receiver in three ways:

- The sense signal
- A loop signal proportional in amplitude to the cosine of the relative angle of the aircraft centreline and received signal

Figure 9.5   Location of left and right ADF antennas on a typical transport aircraft

- A loop signal proportional in amplitude to the sine of the relative angle of the aircraft centreline and received signal.

The sense antenna signal is processed in the receiver via a superhet receiver (see page 63) which allows weak signals to be identified together with discrimination of adjacent frequencies. The output of the superhet receiver is then integrated into the aircraft's audio system. Loop antenna signals are summed with the sense antenna signal; this forms a phase-modulated (PM) carrier signal. The superhet intermediate frequency (IF) is coupled with the PM signal into a coherent demodulator stage that senses the presence of a sense antenna signal from the IF stage. The PM component of the signal is recovered from the voltage-controlled oscillator (VCO) phase lock circuit (see page 70). This recovered signal contains the bearing information received by the antenna and is compared to a reference modulation control signal.

Receivers based on analogue technology send bearing data to the flight deck displays using synchro systems. Digital receivers transmit bearing data to the displays using a data bus system, typically ARINC 429. The ADF receiver is often incorporated into a multi-mode receiver along with other radio navigation systems.

9.3.3 Control panel

Aircraft with analogue (electromechanical) avionics have a dedicated ADF control panel, located on the centre pedestal, see Figure 9.6(a). An alternative panel shown in Figure 9.6(b) enables the crew to select a range of functions including: frequency selectors/displays and the beat frequency oscillator (BFO). This function is used when they want to create an audio frequency for carrier wave transmissions through their audio panel.

NDB carrier waves that are not modulated with an audio component use the BFO circuit in the ADF receiver. To produce an audio output, the receiver heterodynes (beats) the carrier wave signal with a separate signal derived from an oscillator in the receiver.

Some ADF panels have an ADF/ANT switch where 'ADF' selects normal operation, i.e. combined sense and loop antennas; and 'ANT' selects the sense antenna by itself so that the crew can confirm that a station is broadcasting, i.e. without seeking a null. General aviation products combine the control panel and receiver

(a) Location of ADF control panel

(b) Typical ADF control panel

(c) ADF panel/receiver for general aviation

Figure 9.6   ADF control panels

into a single item, see Figure 9.6(c). A change-over switch is used to select the active and standby frequencies.

9.3.4 ADF bearing display

The output from the ADF receiver is transmitted to a display that provides the pilot with both magnetic heading and direction to the tuned NDB; this can either be a dedicated ADF instrument, as shown in Figure 9.7(a), or a **radio magnetic indicator** (RMI), see Figure 9.7(b).

In the RMI, two bearing pointers (coloured red and green) are associated with the two ADF systems and allow the crew to tune into two different NDBs at the same time. RMIs can have a dual purpose; pilots use a switch on the RMI to select either ADF and/or VHF omnidirectional range (VOR) bearings (see Chapter 10 for the latter). Referring to Figure 9.7(c), some aircraft have a bearing source indicator (located adjacent to the RMI) that confirms ADF or VOR selection.

The evolution of digital electronics, together with integration of other systems, has led to the introduction of the flight management system (FMS: see Chapter 19) control display unit (CDU) which is used to manage the ADF system. Aircraft fitted with electronic flight instrument systems (EFIS) have green NDB icons displayed on the electronic horizontal situation indicator (EHSI), as shown in Figure 9.7(e).

Key point

ADF is a short/medium-range (200 nm) navigation system operating within the frequency range 190 to 1750 kHz, i.e. low and medium frequency bands. The ADF system uses an orthogonal antenna consisting of two loops; one aligned with the centreline of the fuselage, the other at right angles.

Test your understanding 9.1

Why does the ADF system seek a null rather than the maximum signal strength from a transmitting station?

Test your understanding 9.2

Explain the function of the ADF/ANT switch that is present on some ADF panels.

(a) ADF bearing indicator

(b) RMI with two bearing indications

(c) Location of RMI and source indicator

(d) RMI and source indicator

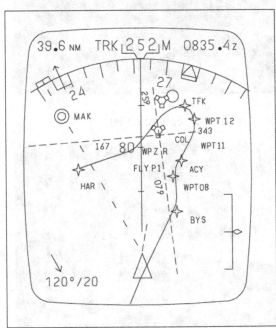

(e) EHSI with an NDB icon (shown as MAK on the upper left of the display)

Figure 9.7   ADF displays

Test your understanding 9.3

Explain the purpose of a beat frequency oscillator (BFO) and why it is needed in an ADF receiver.

9.4 Operational aspects of ADF

ADF radio waves are propagated as ground waves and/or sky waves. Problems associated with ADF are inherent in the frequency range that the system uses. ADF transmissions are susceptible to errors from:

- **Atmospheric conditions:** the height and depth of the ionosphere will vary depending on solar activity. The sky waves (see Figure 9.8) will be affected accordingly since their associated skip distances will vary due to refraction in the ionosphere. This is particularly noticeable at sunrise and sunset.
- **Physical aspects of terrain:** mountains and valleys will reflect the radio waves, causing multi-path reception.
- **Coastal refraction:** low frequency waves that are propagated across the surface of the earth as ground waves will exhibit different characteristics when travelling over land versus water. This is due to the attenuation of the ground wave being different over land and water. The direction of a radio wave across land will change (see Figure 9.9) when it reaches the coast and then travels over water. This effect depends on the angle between the radio wave and the coast.
- **Quadrantal error (QE):** many parts of the aircraft structure, e.g. the fuselage and wings, are closely matched in physical size to the wavelength of the ADF radio transmissions. Radiated energy is absorbed in the airframe and re-radiated, causing interference; this depends on the relative angle between the direction of travel, the physical aspects of the aircraft and the location of the ADF transmitter. Corrections can be made for QE in the receiver.

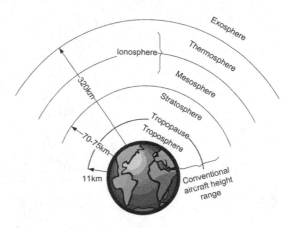

(a) Atmospheric layers

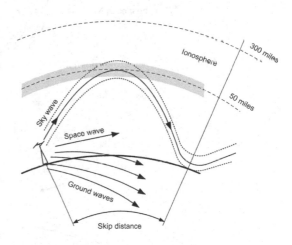

(b) Ionosphere and skip distance

Figure 9.8   Sky waves and the ionosphere

- **Interference:** this can arise from electrical storms, other radio transmissions, static build-up/discharges and other electrical equipment on the aircraft.

The accuracy of an ADF navigation system is in the order of ±5 degrees for locator beacons and ±10 degrees for en route beacons. Any of the above conditions will lead to errors in the bearing information displayed on the RMI. If these conditions occur in combination then the navigation errors will be significant. Pilots cannot

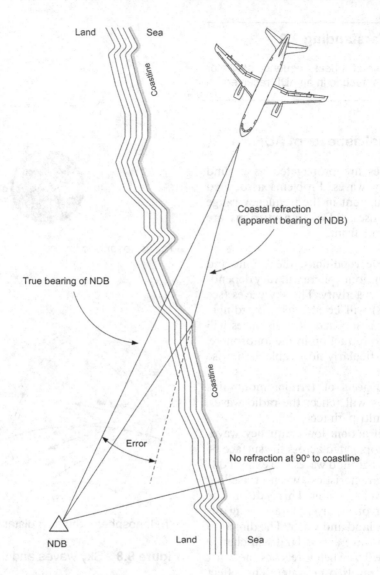

Figure 9.9   Effect of coastal refraction

use ADF for precision navigation due to these limitations.

The increased need for more accuracy and reliability of navigation systems led to a new generation of en route radio navigation aids; this is covered in the next chapter. In the meantime, ADF transmitters remain installed throughout the world and the system is used as a secondary radio navigation aid. The equipment remains installed on modern aircraft, albeit integrated with other radio navigation systems.

Key point

ADF radio waves are propagated as ground waves and/or sky waves. Problems associated with ADF are inherent in the frequency range that the system uses.

Key point

ADF cannot be used for precision navigation due to inherent performance limitations; it remains as a backup to other navigation systems.

Test your understanding 9.4

Why do ADF antennas need a sense loop?

Test your understanding 9.5

How are NDBs identified on navigation charts?

Test your understanding 9.6

Where would locator beacons be found?

Test your understanding 9.7

Why are there two pointers on the RMI?

Test your understanding 9.8

Describe how ground and sky waves are affected by:

(a) local terrain
(b) the ionosphere
(c) attenuation over land and water
(d) electrical storms.

Test your understanding 9.9

Explain, in relation to an ADF system, what is meant by quadrantal error. What steps can be taken to reduce this error?

9.5 ADF homing

Homing to an ADF radio station (non-directional beacon—NDB) simply means flying the aircraft in the direction of the station. This is achieved by steering the aircraft directly towards the beacon (Figure 9.10(a)) with the ADF needle pointing to the top of the indicator,

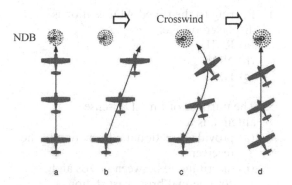

Figure 9.10 ADF homing

i.e. with a null reading. In the event of a crosswind, Figure 9.10(b), the aircraft is driven off-course, away from the intended flight path. To correct for this, the pilot has to constantly change the aircraft heading to maintain a direct course, resulting in a curved path to the NDB (Figure 9.10(c)). The crosswind requires a constant change of heading to maintain the ADF null reading; however, this has displaced the aircraft from the intended track. This is unacceptable for IFR navigation because the aircraft deviates too far from the intended course, potentially leading to close proximity with other aircraft, terrain and/or obstacles. Homing can be avoided by establishing a wind-correction angle that compensates for the drift caused by crosswind, Figure 9.10(d).

9.6 Multiple choice questions

1. ADF antennas are used to determine what aspect of the transmitted signal?
   (a) wavelength
   (b) null signal strength
   (c) maximum signal strength.

2. The ADF antennas include:
   (a) one sense loop and two directional loops
   (b) two sense loops and two directional loops
   (c) two sense loops and one directional loop.

3. ADF operates in the following frequency range:
   (a) MF to VHF
   (b) LF to MF
   (c) VLF.

4. Bearing to the tuned ADF station is
   displayed on the:
   (a) RMI
   (b) NDB
   (c) HSI.

5. The purpose of an ADF sense
   antenna is to:
   (a) provide directional information to the
       receiver
   (b) discriminate between NDBs and
       commercial broadcast stations
   (c) combine with the loop antenna to
       determine a station bearing.

6. The RMI has two pointers coloured red
   and green; these are used to indicate:
   (a) two separately tuned ADF stations
   (b) AM broadcast stations (red) and
       NDBs (green)
   (c) heading (red) and ADF bearing
       (green).

7. The bearing source indicator adjacent to
   the RMI confirms:
   (a) ADF or VOR selection
   (b) the NDB frequency
   (c) the NDB bearing.

8. NDBs on navigation charts can be
   identified by:
   (a) five letter codes
   (b) two/three letter codes
   (c) triangles.

9. Morse code is used to confirm the NDB's:
   (a) frequency
   (b) name
   (c) bearing.

10. During sunrise and sunset, ADF
    transmissions are affected by:
    (a) coastal refraction
    (b) static build-up in the airframe
    (c) variations in the ionosphere.

11. NDBs associated with the final approach
    to an airfield are called:
    (a) locator beacons
    (b) reporting points
    (c) en-route navigation aids.

12. Quadrantal error (QE) is associated
    with the:
    (a) ionosphere
    (b) physical aspects of terrain
    (c) physical aspects of the aircraft
        structure

13. ADF ground waves are affected by:
    (a) the ionosphere
    (b) coastal refraction and terrain
    (c) solar activity.

14. ADF sky waves are affected by:
    (a) the ionosphere
    (b) coastal refraction
    (c) the local terrain.

15. A BFO can be used to establish:
    (a) the non-directional output of an NDB
    (b) which loop antenna is receiving a null
    (c) an audio tone for an NDB.

16. Referring to Figure 9.11, which icon is for
    the NDB?
    (a) HAR
    (b) MAK
    (c) WPT08.

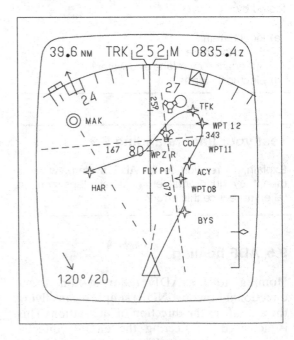

Figure 9.11   See Question 16

Chapter 10 VHF omnidirectional range

The early use of aircraft radio navigation was based on low/medium frequency (LF/MF) transmissions. During the mid to late 1940s, it was evident that increased accuracy and reliability were needed. Since radio communication systems based on very high frequency (VHF) were being successfully deployed, a decision was made to develop a radio navigation system based on VHF. This system became the VHF omnidirectional range (VOR) system; a system that is in widespread use throughout the world today. VOR navigation was the basis for the network of navigation 'airways'

10.1 VOR principles

10.1.1 Overview

VOR is a short/medium-range navigation system operating in the 108–117.95 MHz range of frequencies. This means that the radio waves are now propagated as space waves. The problems that were encountered with ground and sky waves in the LF and MF ranges are no longer present with a VHF system. VOR navigation aids are identified by unique three-letter codes (derived from their name, e.g. London VOR is called LON, Dover VOR is called DVR, etc.). The code is modulated onto the carrier wave as a 1020 Hz tone that the crew can listen to as a Morse code signal. Some VOR navigation aids have an automatic voice identification announcement that provides the name of the station; this alternates with the Morse code signal. The location of the VOR navigation aids (specified by latitude and longitude), together with their carrier wave frequencies, is provided on navigation charts as with ADF.

VOR operates in the same frequency range as the instrument landing system (ILS), described in Chapter 12. Although the two systems are completely independent and work on totally different principles, they often share the same receiver. The two systems are differentiated by their frequency allocations within this range. ILS frequencies are allocated to the odd tenths of each 0.5 MHz increment, e.g. 109.10 MHz, 109.15 MHz, 109.30 MHz, etc. VOR frequencies are allocated to even tenths of each 0.5 MHz increment, e.g. 109.20 MHz, 109.40 MHz, 109.60 MHz, etc. Table 10.1 provides an illustration of how these frequencies are allocated within the 109 MHz range.

10.1.2 VOR features

In addition to the inherently improved system performance and navigation reliability, VOR has another feature that makes it extremely useful for air navigation. The VOR system has the ability to transmit specific bearing information, referred to as a 'radial', see Figure 10.1(a).

Table 10.1 Allocation of ILS and VOR frequencies

ILS frequency (MHz)	VOR frequency (MHz)
	109.00
109.10	
109.15	
	109.20
109.30	
109.35	
	109.40
109.50	
109.55	
	109.60
109.70	
109.75	
	109.80
109.90	
109.95	

DOI: 10.1201/9781003411932-10

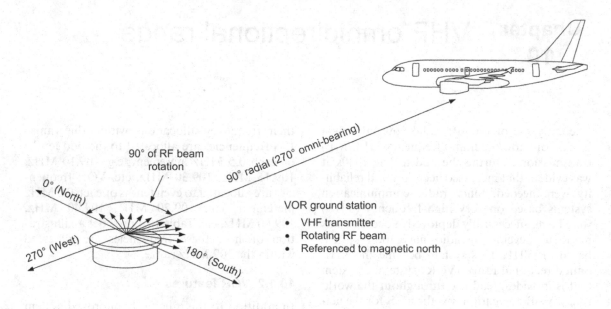

(a) VHF omni directional range (VOR) overview

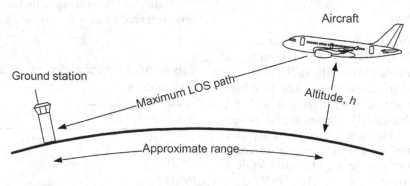

(b) VHF omni directional range—line of sight

Figure 10.1   VOR overview

The pilot can select any radial from a given VOR navigation aid and fly to or from that aid. Since the system is 'line of sight', i.e. receiving signals as space waves, the altitude of the aircraft will have a direct relationship with the range within which the system can be used, see Figure 10.1(b).

Using VHF navigation aids imposes a limit on the theoretical working range that can be obtained. The maximum theoretical line of sight (LOS) distance between an aircraft and the ground station is given by the relationship:

$$d = 1.1\sqrt{h}$$

where d is the distance in nautical miles, and h is the altitude in feet above ground level (assumed to be flat terrain). The theoretical LOS range for altitudes up to 20,000 feet is given in Table 10.2.

At higher altitudes, it is possible to receive VOR signals at greater distances but with

Table 10.2 Theoretical LOS range

Altitude (feet)	Range (nm)
100	10
1,000	32
5,000	70
10,000	100
20,000	141

Table 10.3 Navigation aid classifications

Classification	Altitude (feet)	Range (nm)
Terminal	1,000–12,000	25
Low altitude	1,000–18,000	40
High altitude	18,000–45,000	130

reduced signal integrity. Although the actual range also depends on transmitter power and receiver sensitivity, the above relationship provides a good approximation. In practice, navigation aids have a designated standard service volume (SSV); this defines the reception limits within an altitude envelope, as shown in Table 10.3.

Key point

VOR radials are referenced to magnetic north; they are the basis of airways for en route navigation.

Key point

VOR transmissions are 'line of sight'; therefore, range increases with increased altitude.

10.1.3 Conventional VOR (CVOR)

Conventional VOR (CVOR) stations radiate two signals: omnidirectional and directional. The omnidirectional (or reference) signal is the carrier wave frequency of the station, with AM and FM modulations to produce a sub-carrier

wave. The directional signal is radiated as a cardioid pattern, electronically rotating at 30 revolutions per second.

The directional signal is arranged to be in phase with the reference signal when the aircraft is due north (magnetic) of the VOR station. As the cardioid pattern rotates around the station, the two signals become out of phase on a progressive basis, see Figure 10.2. The bearing between any given angle and magnetic north is determined by the receiver as the phase angle difference between the reference and directional signals. This difference in phase angle is resolved in the aircraft receiver and displayed to the crew as a radial from the VOR station, see Figure 10.3.

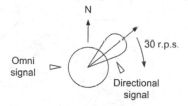

(a) Conventional VOR reference

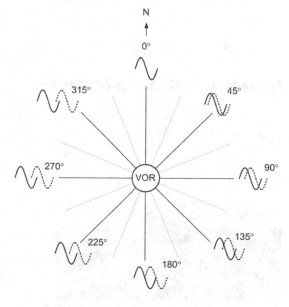

(b) Variable signal phase relationship

Figure 10.2 Conventional VOR (CVOR)

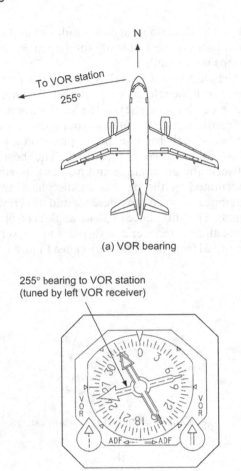

(a) VOR bearing

255° bearing to VOR station
(tuned by left VOR receiver)

(b) RMI

Figure 10.3 VOR bearings and displays

Locations of conventional VOR (CVOR) ground stations have to be carefully planned to take into account local terrain and obstacles. Mountains and trees can cause multi-path reflections, resulting in distortion (known as (siting errors) of the radiated signal. These errors can be overcome with an enhanced second-generation system known as Doppler VOR (DVOR).

10.1.4 Doppler VOR (DVOR)

Doppler is usually associated with self-contained navigation systems, and this subject is described in a separate chapter. The Doppler effect is also applied to the second-generation version of VOR ground transmitters. The Doppler effect can be summarised here as: '…the frequency of a wave apparently changes as its source moves closer to, or farther away from an observer'.

The DVOR ground station has an omnidirectional transmitter in the centre, amplitude modulated at 30 Hz; this is the reference phase. The directional signal is derived from a 44-foot diameter circular array comprising up to 52 individual antennas, see Figures 10.4(a) and 10.4(b). Each antenna transmits in turn to simulate a rotating antenna.

Consider two aircraft using the DVOR station as illustrated in Figure 10.4(c). The effect of the rotating 9960 Hz signal is to produce a Doppler shift; aircraft A will detect a decreased frequency, aircraft B will detect an increased frequency. Doppler shift creates a frequency

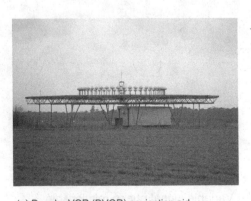

(a) Doppler VOR (DVOR) navigation aid

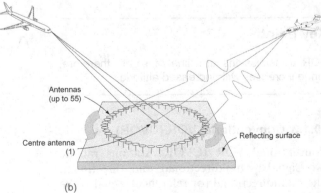

(b)

Figure 10.4 Doppler VOR (DVOR)   (Continued)

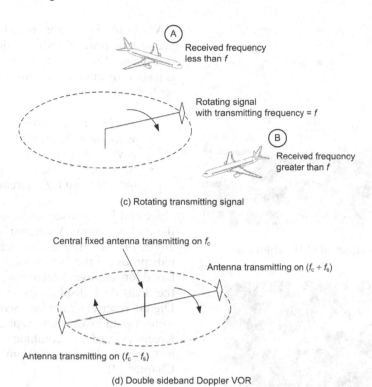

(c) Rotating transmitting signal

(d) Double sideband Doppler VOR

Figure 10.4 (Continued)

modulated (FM) signal in the aircraft receiver over the range 9960 Hz 480 Hz varying at 30 Hz in a sine wave. Note that the perceived frequency will be 9960 Hz when the aircraft are in the positions shown in Figure 10.4(c). The phase difference measured in the airborne equipment depends on the bearing of the aircraft relative to the station. Since the FM variable signal is less prone to interference, DVOR performance is superior to CVOR.

DVOR actually uses two rotating patterns, as shown in Figure 10.4(d). These patterns (diagonally opposite each other) rotate at 30 revolutions per second; one pattern is 9960 Hz above the reference, the other is 9960 Hz below the reference frequency. The diameter of the array, together with the speed of pattern rotation, creates a Doppler shift of 480 Hz (at VOR frequencies).

10.2 Airborne VOR equipment

Modern transport aircraft have two VOR systems, often designated left and right; note that airborne equipment is the same for conventional

and Doppler VOR. Radio frequency (RF) signals from the antenna are processed in the receiver as determined by frequency and course selections from the control panel; outputs are sent to various displays.

10.2.1 Antennas

The VOR antenna is a horizontally polarised, omnidirectional half-wave dipole, i.e. a single conductor with a physical length equal to half the wavelength of the VOR signals being received. Two such antennas are formed into a single package and usually located in the aircraft fin, as indicated in Figure 10.5. They are packaged within a composite fairing for aerodynamic streamline purposes. The antennas are connected to the receivers via coaxial cables.

10.2.2 Receivers

VOR receivers are often combined with other radio navigation functions, e.g. the instrument landing system; receivers are located in the avionic equipment bay, see Figure 10.6.

Figure 10.5   Location of VOR antennas

VOR receivers are based on the super-heterodyne principle with tuning from the control panel. The received radio frequency signal is passed through an amplitude modulation filter to separate out the:

- 30 Hz tone from the rotating pattern
- voice identification (if provided from the navigation aid)
- Morse code tone; 9960 Hz signal FM by 480 Hz at 30 Hz reference tone.

Voice and Morse code tones are integrated with the audio system. A comparison of the phase angles of the variable and reference 30 Hz signals produces the VOR radial signal. Receivers based on analogue technology interface with the flight deck displays using synchro systems. Digital receivers interface with other systems using a data bus system, typically ARINC 429. Receivers usually combine both VOR and instrument landing system functions (see Chapter 12).

10.2.3 Control panels

Control panels identified as 'VHF NAV' can be located on the glare shield (as shown in Figure 10.7) or centre pedestal. This panel is used to select the desired course and VOR frequencies. General aviation products have a combined VHF navigation and communications panel— see Figure 10.6(b)—this can be integrated with the GPS navigation panel.

(a) VOR receiver (remotely located in the aircraft's avionic equipment bay)

(b) Integrated GA navigation/ communication control panel

Figure 10.6   VOR receivers

Figure 10.7   Typical VOR control panel

10.2.4 VOR displays

The bearing to a VOR navigation aid (an output from the receiver) can be displayed on the radio magnetic indicator (RMI); this is often shared with the ADF system, as discussed in the previous chapter. The RMI (Figure 10.8(a)) provides the pilot with both magnetic heading and direction to the tuned VOR navigation aid. The two bearing pointers (coloured red and green) are associated with the two VOR systems and allow the crew to tune into two different VOR navigation aids at the same time. On some instruments, a switch on the RMI is used to select either ADF or VOR bearing information, see Figure 10.8(b). RMIs therefore have a dual purpose; pilots use a switch on the RMI to select either ADF and/or VOR bearings (see Chapter 9 for ADF). Some aircraft have a bearing source indicator adjacent to the RMI to confirm ADF or VOR selection, see Figure 10.8(c).

In order to fly along an airway, first it has to be intercepted. This is achieved by flying towards the desired radial on a specified heading; the method of displaying the selected VOR radial depends on the type of avionic fit. Electromechanical instruments include the omni-bearing selector (OBS), course deviation indicator (CDI) and horizontal situation indicator (HSI), see Figure 10.9. These instruments vary slightly in their design and layout; however, they have some common features.

A selector is used to manually rotate the course card. This card is calibrated from 0 to 360° and indicates the VOR bearing selected to fly TO or FROM (indicated by a flag). The deviation pointer moves to the left or right to guide

(a) RMI with two bearing pointers

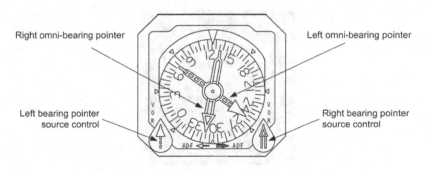

(b) RMI source control (VOR/ADF)

Figure 10.8 VOR displays and indicators (*Continued*)

(c) RMI and bearing source annunciator

Figure 10.8   (Continued)

(a) Omni-bearing selector (OBS)

(b) Course deviation indicator (CDI)

(c) Horizontal situation indicator (HSI)

Figure 10.9   Omni-bearing selector and course deviation indicator

the pilot in the required direction to maintain the selected course. Each dot on the scale represents a 2° deviation from the selected course. A red OFF flag indicates when the:

- VOR navigation aid is beyond reception range
- pilot has not selected a VOR frequency
- VOR system is turned off, or is inoperative.

Note that these indicators can also be used with the ILS navigation system; refer to Chapter 12 for a detailed description of this system.

For aircraft with electronic flight instruments, the desired radial is displayed on the electronic horizontal situation indicator (EHSI). This EHSI display can either be selected to show a conventional compass card (Figure 10.10(a)) or expanded display (Figure 10.10(b)). As the radial is approached the deviation bar gradually aligns with the selected course.

If flying manually, the crew turn the aircraft onto the selected course whilst monitoring the deviation bar; when it is centred, the radial has been intercepted and the EHSI will display 'TO' confirming that the inbound radial is being followed. The flight continues until the VOR navigation is reached, and the radial to the next navigation aid is selected. If the EHSI were still selected to the original inbound radial, the EHSI would display 'FROM'. The lateral deviation bar therefore shows if the aircraft is flying on the selected radial, or if it is to the left or right of the radial.

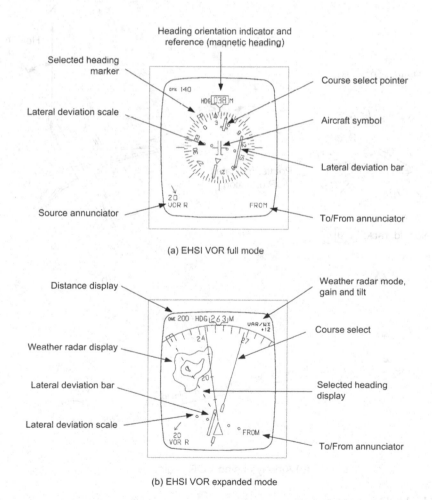

(a) EHSI VOR full mode

(b) EHSI VOR expanded mode

Figure 10.10   VOR electronic displays

Key point

Two intersecting VOR radials can be used to define unique locations known as reporting points; these are used for air traffic control purposes.

Key point

Doppler VOR was introduced to overcome siting problems found with conventional VOR. The two systems operate on different principles; however, the airborne equipment is the same.

10.3 Operational aspects of VOR

Radials from any given VOR navigation aid are the basis of **airways**; system accuracy is typically within one degree. These are the standard routes flown by aircraft when flying on instruments. When two VOR radials intersect, they provide a unique navigation position fix base (theta–theta). The accuracy of this fix is greatest when the radials intersect at right angles. Figure 10.11 illustrates how navigation charts are built up on a network of VOR radials; the accuracy of VOR radials is generally very good (1 degree). In Figure 10.11(d), the navigation aid located near Brussels (abbreviated BUB) transmits on

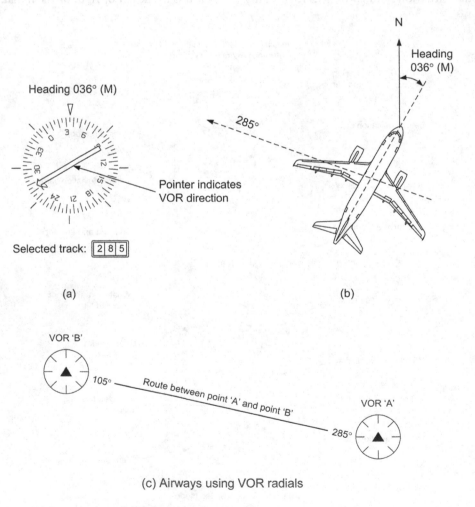

(c) Airways using VOR radials

Figure 10.11   VOR radials and airways

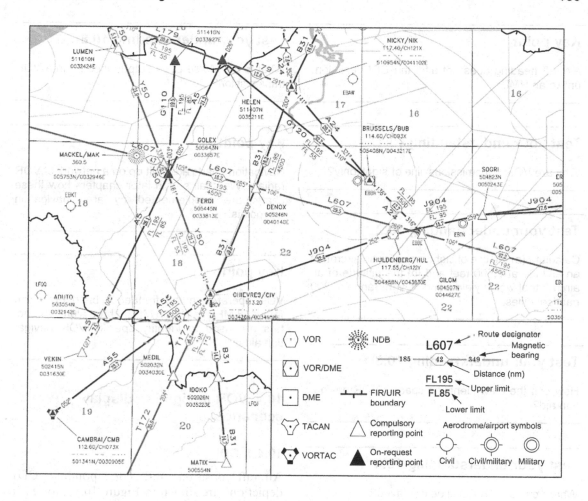

(d) Airway network over Belgium

Figure 10.11 (Continued)

114.6 MHz. Three radials can be seen projected from this navigation aid on 136°, 310° and 321°. These radials are used to define airways A24 and G120. Note the Morse code output and latitude/longitude for the navigation aid.

The intersecting radials from navigation aids are used to define reporting points for en route navigation. These reporting points are given five-letter identification codes associated with their geographic location. For example, the reporting point HELEN is defined by airways G5 and A120. The flexibility of VOR is greatly increased when co-located distance measuring equipment (DME) is used, thereby providing rho–theta fixes from a single navigation aid. There are examples of VOR-only navigation aids, e.g. Perth in Scotland (identification code PTH, frequency 110.40 MHz). The majority of VOR navigation aids are paired with DME; this system is described in the next chapter.

Key point

'Q Codes' are used in aviation to provide a succinct and accurate protocol for radio communications, initially by way of Morse code. For navigation, two assigned codes are:

- QDR - Magnetic bearing of a ground station to the aircraft radial
- QDM - Magnetic bearing of an aircraft to the radial

Key point

Aircraft heading does not alter the CDI depiction, unlike an RMI.

Test your understanding 10.1

Why are VOR transmissions 'line of sight' only?

Test your understanding 10.2

Calculate (a) the line of sight range for an aircraft at an altitude at 7,500 feet and (b) the altitude of an aircraft that would yield a line of sight range of 200 nautical miles.

Test your understanding 10.3

How can the crew identify a specific VOR navigation aid?

Test your understanding 10.4

Where can a VOR radial be displayed?

Test your understanding 10.5

What is the Morse code output used for in a VOR transmission?

Test your understanding 10.6

Why does an RMI have two VOR pointers?

Test your understanding 10.7

What is the difference in aircraft equipment between conventional and Doppler VOR?

Test your understanding 10.8

What is the Morse code output used for in a VOR transmission?

Key point

Navigation charts are built up on a network of VOR radials. We shall see in later chapters how these charts are supplemented by area navigation waypoints.

Key point

VOR operates in the frequency band extending from 108 MHz to 117.95 MHz. Three alpha characters are used to identify specific VOR navigation aids.

10.4 VOR navigation display scenarios

10.4.1 CDI

Aircraft position and corresponding CDI depictions are shown in Figure 10.12; for each scenario, note the aircraft heading, VOR radial and TO/FROM display. Flying directly towards the VOR station requires compensation for the crosswind, as shown in Figure 10.12(a), to avoid the 'homing effect'. In this example, the aircraft tracks the VOR radial of 140 degrees, with TO in the TO/FROM window. When intercepting an inbound track, as shown in Figure 10.12(b), the aircraft is steered on a heading of 360 degrees, onto the 120 degree radial. Intercepting an outbound radial of 260 degrees, Figure 10.12(c), displays FROM in the TO/FROM window.

10.4.2 OBS

Aircraft position, and corresponding OBS indicator depictions, can also be illustrated via a quadrant, as shown in Figure 10.13. In this example, the pilot has selected a 030 degrees radial (indicated at the top of the display). In the

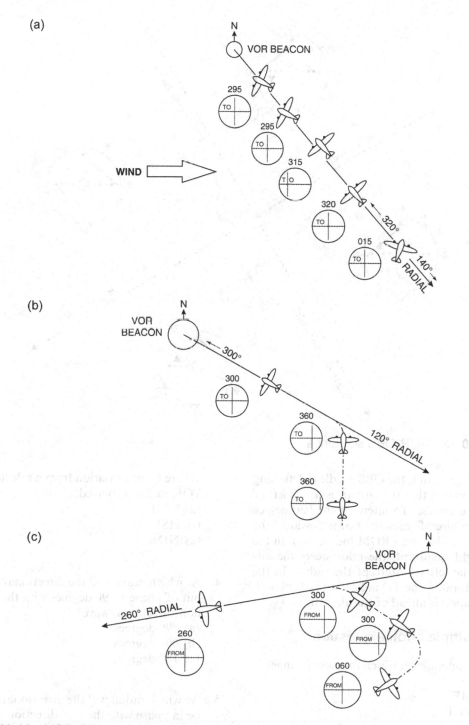

Figure 10.12 VOR-CDI scenarios

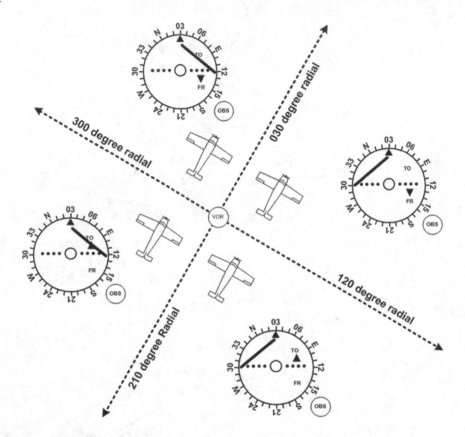

Figure 10.13   VOR-OBS scenarios

upper left quadrant, the OBS needle is deflecting right, indicating that the aircraft is to the left of the desired course. To intercept the 030 degree radial, the aircraft must be flown 'towards' the needle (note also the FROM indication). In the upper right quadrant, the pilot steers the aircraft to the left to intercept the radial. In the lower quadrants, the TO indication is displayed, i.e. the station is ahead of the aircraft.

10.5 Multiple choice questions

1. VOR operates in which frequency range?
   (a) LF
   (b) MF
   (c) VHF.

2. VOR signals are transmitted as what type of wave?
   (a) sky wave
   (b) ground wave
   (c) space wave.

3. Where is the deviation from a selected VOR radial displayed?
   (a) RMI
   (b) HSI
   (c) NDB.

4. At which radial will the directional wave be out of phase by 90 degrees with the non-directional wave?
   (a) 090 degrees
   (b) 000 degrees
   (c) 180 degrees.

5. At which radial will the directional signal be in phase with the non-directional signal?
   (a) 090 degrees
   (b) 000 degrees
   (c) 180 degrees.

6. VOR navigation aids are identified by how many alpha characters?
   (a) two
   (b) three
   (c) five.

7. VOR radials are referenced to:
   (a) non-directional signals from the navigation aid
   (b) magnetic north
   (c) true north.

8. The RMI has two pointers coloured red and green; these are used to indicate:
   (a) the bearing of two separately tuned VOR stations
   (b) directional (red) and non-directional transmissions (green)
   (c) the radials of two separately tuned VOR stations.

9. Morse code tones are used to specify the VOR:
   (a) identification
   (b) frequency
   (c) radial.

10. The intersection of two VOR radials provides what type of position fix?
    (a) rho–rho
    (b) theta–theta
    (c) rho–theta.

11. An aircraft is flying on a heading of 090 degrees to intercept the selected VOR radial of 180 degrees; the HSI will display that the aircraft is:
    (a) right of the selected course
    (b) left of the selected course
    (c) on the selected course.

12. The DVOR navigation aid has an omnidirectional transmitter located in the:
    (a) centre
    (b) outer antenna array
    (c) direction of magnetic north.

13. When flying overhead a VOR navigation aid, the reliability of directional signals:
    (a) decreases
    (b) increases
    (c) stays the same.

14. Reporting points using VOR navigation aids are defined by the:
    (a) identification codes
    (b) intersection of two radials
    (c) navigation aid frequencies.

15. With increasing altitude, the range of a VOR transmission will be:
    (a) increased
    (b) decreased
    (c) the same.

16. Referring to Figure 10.14, the instrument shown is called the:
    (a) omni-bearing selector (OBS)
    (b) radio magnetic indicator (RMI)
    (c) course deviation indicator (CDI).

Figure 10.14 See Question 16

Chapter 11

Distance measuring equipment

The previous two chapters have been concerned with obtaining directional information for the purposes of airborne navigation. In this chapter we will look at a system that provides the crew with the distance to a navigation aid. Distance measuring equipment (DME) is a short/medium-range navigation system, often used in conjunction with other systems, e.g. VOR and ILS, to provide accurate navigation fixes. DME is based on secondary radar principles and operates in the L-band of radar. Before looking at what the system does and how it operates in detail, we need to take at look at some basic radar principles.

11.1 Radar principles

The word **radar** is derived from <u>ra</u>dio <u>d</u>etection <u>a</u>nd <u>r</u>anging; the initial use of radar was to locate aircraft and display their range and bearing on a monitor (either ground based or in another aircraft). This type of radar is termed **primary radar**: energy is radiated via a rotating radar antenna to illuminate a 'target'; this target could be an aircraft, the ground or a cloud. Some of this energy is reflected back from the target and is collected in the same antenna, see Figure 11.1. The strength of the returned energy is measured and used to determine the range of the target. A rotating antenna provides the directional information such that the target can be displayed on a screen.

Primary radar has its disadvantages, one of which is that the amount of energy being transmitted is very large compared with the amount of energy reflected from the target. An alternative method is **secondary radar** that transmits a specific low-energy signal (the interrogation) to a known target. This signal is analysed and a new (or secondary) reply signal, i.e. not a reflected signal, is sent back to the origin, see Figure 11.2. Secondary radar was developed

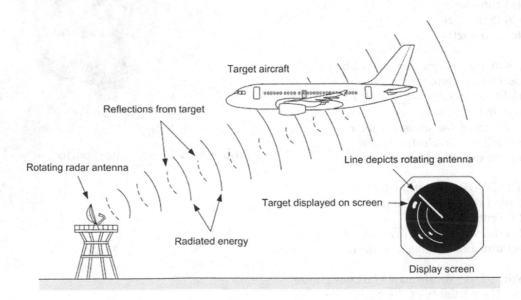

Figure 11.1 Primary radar

DOI: 10.1201/9781003411932-11

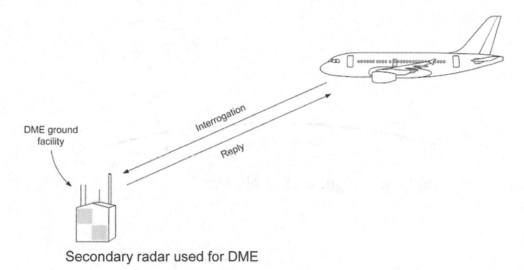

Figure 11.2   Secondary radar used for DME.

during the Second World War to differentiate between friendly aircraft and ships: Identification Friend or Foe (IFF). The principles of secondary radar now have a number of applications including distance measuring equipment (DME).

11.2 DME overview

The DME navigation aid contains a transponder (receiver and transmitter) contained within a single navigation aid. The aircraft equipment radiates energy pulses to the DME navigation aid; secondary signals are then transmitted back to the aircraft. An on-board **interrogator** measures the time taken for the signals to be transmitted and received at the aircraft. Since we know the speed of radio wave propagation, the interrogator can calculate the distance to the DME navigation aid. DME navigation aids can either be self-contained ground stations, or co-located with a VOR navigation aid.

Since the system is 'line of sight', the altitude of the aircraft will have a direct relationship with the range that the system can be used, see Figure 11.3(a). Using DME navigation aids imposes a limit on the working range that can be obtained. The maximum line of sight (LOS) distance between an aircraft and the ground station is given by the relationship:

$$d = 1.1\sqrt{h}$$

where d is the distance in nautical miles, and h is the altitude in feet above ground level (assumed to be flat terrain). The theoretical LOS range for altitudes up to 20,000 feet is given in Table 11.1.

Referring to Figure 11.3(b) it can be seen that the actual distance being measured by the interrogator is the **'slant' range**, i.e. not the true distance (horizontal range) over the ground. The effects of slant range in relation to the horizontal range are greatest at high altitudes and/or when the aircraft is close to the navigation aid. Taking this to the limit, when the aircraft is flying over a DME navigation aid, it would actually be measuring the aircraft's altitude!

Test your understanding 11.1

What is the difference between primary and secondary radar?

Test your understanding 11.2

Distinguish between slant range and horizontal range.

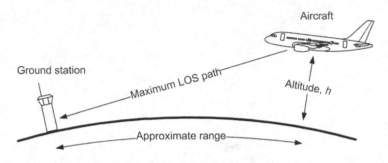

(a) Line of sight versus altitude

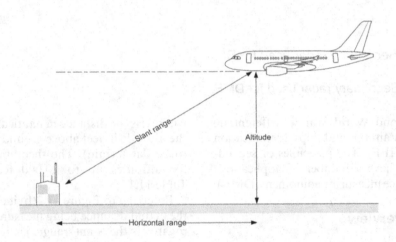

(b) DME slant range

Figure 11.3   DME range terminology. (a) Line of sight versus altitude. (b) DME slant range

Table 11.1   Theoretical LOS range

Altitude (feet)	Range (nm)
100	10
1,000	32
5,000	70
10,000	100
20,000	141

11.3 DME operation

The signals transmitted by the interrogator are a pair of pulses, each of 3.5 μs duration, modulated on the DME navigation aid frequency. The interrogator generates a pulse-pair repetition rate between 5 and 150 pulse-pairs per second.

At the DME navigation aid, the transponder receives these pulses and, after a 50 ms time delay, transmits a new pair of pulses at a frequency 63 MHz above or below the interrogator's frequency. The aircraft's interrogator receives the pulses and matches the time interval between the transmitted pair of pulses. This ensures that other aircraft interrogating the same DME navigation aid at the same time only process their own pulses.

By measuring the elapsed time between transmitting and receiving (and taking into account the 50 ms time delay) the interrogator calculates the distance to the navigation aid. DME is a line of sight system with a maximum range of approximately 200 nm; this equates to approximately 2400 ms elapsed time taken for a pair of pulses to be transmitted and received, taking into account the 50 ms time delay in the ground station. System accuracy is typically 0.5 nm, or 3% of the calculated distance, whichever is the greater.

The DME ground station responds to the interrogations of 50–100 aircraft; these all send their interrogations at the same DME station frequency. The aircraft transceiver has to recognize its own replies from replies from all other aircraft using the DME station. For this purpose, the transceiver interrogates with its own 'jitter' (or random set of pulses); it then looks for replies with a correlating, constant time difference with respect to its transmission.

Test your understanding 11.3

What is the typical accuracy and maximum range of a DME system?

Key point

The varying interval between pulse-pairs ensures that the DME interrogator recognises its own signals and rejects other signals.

Key point

There are 252 DME channels between 960 and 1215 MHz. Channels are numbered from 1 to 126; each channel number is sub-divided into channels designated 'X' and 'Y'. These are separated by varying the pulse separation time: 2 μs for X channels, and 30 μs for Y channels.

Key point

DME is based on secondary radar; it operates in the L-band between 962 MHz and 1215 MHz (UHF) with channel spacing at 1 MHz.

11.4 Equipment overview

Commercial transport aircraft are usually fitted with two independent DME systems, comprising antennas and interrogators.

The DME antennas are L-band blades, located on the underside of the aircraft fuselage, see Figure 11.4(a); note that the antenna is dual purpose in that it is used for both transmitting and receiving.

The interrogators are located in the equipment bays (Figure 11.4(b)) and provide three main functions: transmitting, receiving and calculation of distance to the selected navigation aid. Transmission is in the range 1025 to 1150 MHz; receiving is in the range 962 to 1215 MHz; channel spacing is 1 MIIz. The interrogator operates in several modes:

- Standby
- Search
- Track
- Scan
- Memory
- Fault
- Self-test.

When the system is first powered up, it enters the standby mode; transmissions are inhibited, the receiver and audio are operative; the DME display is four dashes to indicate **no computed data** (NCD). The receiver monitors pulse-pairs received from any local ground stations. If sufficient pulse-pairs are counted, the interrogator enters the search mode. The transmitter now transmits pulse-pairs and monitors any returns; synchronous pulse-pairs are converted from time into distance and the system enters the track mode. Distance to the navigation aid will now be displayed on the DME indicator (see Figure 11.5). The scan mode has two submodes: directed scanning for multiple navigation aid tuning; up to five stations can be scanned in accordance with a predetermined area navigation auto-tuning programme (described in more detail in Chapter 16). Alternatively, free scanning occurs for any DME navigation aids within range. If pulse-pairs from any navigation aids are not received after a short period of time (two seconds typical), the interrogator goes into memory mode whereby distance is calculated from the most recently received pulse-pairs. Memory mode expires after a short period of time, typically ten seconds, or until pulse-pairs are received again. If the system detects any fault conditions, the distance display is blanked

(a) Location of DME antennas

DME (right)

DME (left)

(b) Location of DME transceiver

Figure 11.4 DME equipment. (a) Location of DME antennas. (b) Location of DME transceiver

out. Self-test causes the system to run through a predetermined sequence causing the indicators to read: blank, dashes (NCD) and 0.0 nm.

Key point

VOR and DME systems operate on different frequencies. When they are co-located, the DME frequency is automatically selected when the pilot tunes into the VOR frequency.

DME outputs can be displayed in a variety of ways, see Figure 11.5. These displays include dedicated readouts, electronic flight instrument systems (EFIS), combined panels/transceivers (for general aviation) and radio distance magnetic indicators (RDMI). When selecting a co-located VOR-DME navigation aid, the crew only needs to tune into the VOR frequency; the DME frequency is automatically selected.

(a) Self-contained DME displays

(b) DME panel/transceiver for general aviation

(c) Radio distance magnetic indicator (RDMI)

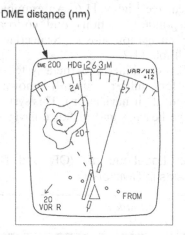

(d) Electronic instrument—DME display

Figure 11.5 Various types of DME display. (a) Self-contained DME displays. (b) DME panel/transceiver for general aviation. (c) Radio distance magnetic indicator (RDMI). (d) Electronic instrument—DME display

Key point

When no computed data (NCD) is available this condition is displayed as four dashes.

Test your understanding 11.4

List and describe four modes in which a DME interrogator can operate.

11.5 En route navigation using radio navigation aids

Basic en route navigation guidance for commercial aircraft can be readily accomplished using colocated VOR and DME systems, thereby providing rho–theta fixes from a single navigation aid. The DME frequency is paired with the VOR frequency; this means that only the VOR frequency needs to be tuned, and the DME frequency is automatically tuned as a result.

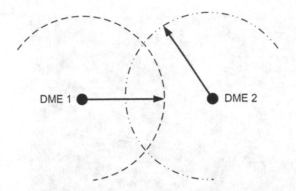

Figure 11.6 Ambiguous DME–DME position

Alternatively, rho–rho fixes can be established from a pair of DME navigation aids. Note that this produces an ambiguous fix unless another DME is used, see Figure 11.6. An example of DME transponder locations and co-located VOR-DME navigation aids in Switzerland is provided in Table 11.2.

In the US, a combined rho–theta system was introduced for military aircraft known as **TACAN** (<u>ta</u>ctical <u>a</u>ir <u>n</u>avigation). This system is a short-range bearing and distance navigation

Table 11.2 Locations of VOR and DME navigation aids in Switzerland

Name	Identification code	Type
Corvatsch	CVA	DME
Fribourg	FRI	VOR-DME
Geneva Cointrin	GVA	VOR-DME
Grenchen	GRE	VOR-DME
Hochwald	HOC	VOR-DME
Kloten	KLO	VOR-DME
La Praz	LAP	DME
Montana	MOT	VOR-DME
Passeiry	PAS	VOR-DME
Sion	SIO	VOR-DME
St. Prex	SPR	VOR-DME
Trasadingen	TRA	VOR-DME
Willisau	WIL	VOR-DME
Zurich East	ZUE	VOR-DME

Figure 11.7 TACAN navigation aid

aid operating in the 962–1215 MHz band. TACAN navigation aids (see Figure 11.7) are often co-located with VOR navigation aids; these are identified on navigation charts as '**VORTAC**'.

The TACAN navigation aid is essentially a DME transponder (using the same pulse-pair and frequency principles as the standard DME) to which directional information has been added; both operate in the same UHF band. An important feature of TACAN is that both distance and bearing are transmitted on the same frequency; this offers the potential for equipment economies. Furthermore, because the system operates at a higher frequency than VOR, the antennas and associated hardware can be made smaller. This has the advantage for military use since the TACAN equipment can be readily transported and operated from ships or other mobile platforms.

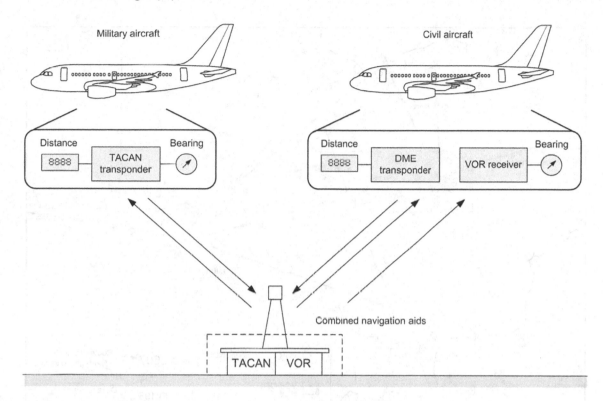

Figure 11.8 VORTAC navigation aid and associated aircraft functions

When co-located with a VOR navigation aid, military and commercial aircraft can share the VORTAC facility. Referring to Figure 11.8, military aircraft obtain their distance and bearing information from the TACAN part of the VORTAC; commercial aircraft obtain their distance information from the TACAN and bearing information from the VOR part of the TACAN. Reporting points (shown as triangles) based on DME navigation aids, e.g. the VORTAC navigation aid located at Cambrai (CMB), northern France, are illustrated in Figure 11.9. The intersecting radials from navigation aids are used to define reporting points for en route navigation. These **reporting points** are given five-letter identification codes associated with their geographic location. For example, the reporting point 'HELEN' (at the top of the chart) is defined by a distance and bearing from the Brussels VOR/DME navigation aid.

TACAN frequencies are specified as channels that are allocated to specific frequencies, e.g. Raleigh–Durham VORTAC in North Carolina, USA, operates on channel 119X. This corresponds to a:

- VOR frequency of 117.2 MHz
- DME interrogation frequency of 1143 MHz
- DME reply frequency of 1206 MHz
- pulse code of 12 ms.

Note that since DME, VOR and VORTAC navigation aids have to be located on land, the airways' network does not provide a great deal of coverage beyond coastal regions.

Referring to Figure 11.10, a combination of VOR, DME and VORTAC stations (see Figure 11.11) located in a number of European countries provides a certain amount of navigation guidance in the North Atlantic, Norwegian Sea and North Sea. This diagram assumes an LOS range of approximately 200 nm. The gaps in this radio navigation network can be overcome by the use of alternative navigation systems including: inertial and satellite navigation systems (INS), these are all described elsewhere in this book.

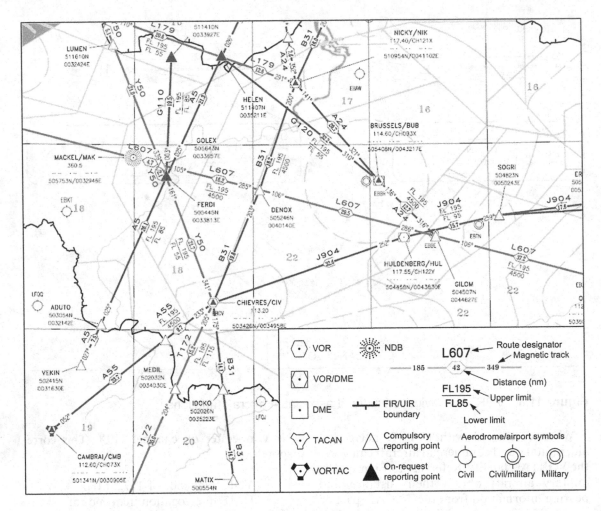

Figure 11.9   Reporting points defined by VOR–DME

Test your understanding 11.5

Explain what is meant by frequency pairing.

Test your understanding 11.6

Describe two ways in which DME distance information is displayed.

Test your understanding 11.7

DME ground stations could be responding to numerous aircraft; how does the airborne DME system recognise its own signals and reject signals intended for other aircraft?

Test your understanding 11.8

What information does an RDMI provide the crew?

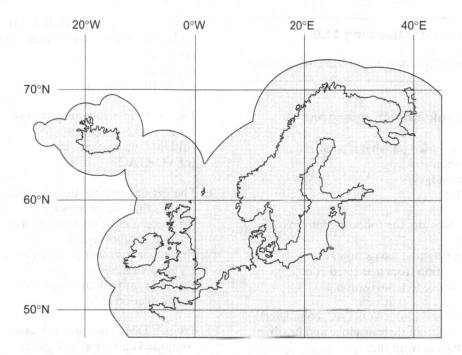

Figure 11.10 Approximate line of sight navigation coverage for northern Europe

Figure 11.11 Typical VOR–DME navigation aid

Test your understanding 11.9

What type of information does a VORTAC provide?

11.6 Multiple choice questions

1. DME is based on what type of radar?
   (a) primary
   (b) secondary
   (c) VHF.

2. DME provides the following information
   to the crew:
   (a) bearing to a navigation aid
   (b) deviation from a selected course
   (c) distance to a navigation aid.

3. When tuned into a VORTAC, commercial
   aircraft obtain their distance and bearing
   information from the:
   (a) TACAN and VOR
   (b) DME and VOR
   (c) DME and TACAN.

4. DME signals are transmitted:
   (a) by line of sight
   (b) as ground waves
   (c) as sky waves.

5. An RDMI provides the following
   information:
   (a) distance and bearing to a
       navigation aid
   (b) deviation from a selected course
   (c) the frequency of the selected
       navigation aid.

6. Slant range errors are greatest when the
   aircraft is flying at:
   (a) high altitudes and close to the
       navigation aid
   (b) high altitudes and far from the
       navigation aid
   (c) low altitudes and far from the
       navigation aid.

7. To select a co-located VOR-DME
   navigation aid, the crew tunes into the:
   (a) DME frequency
   (b) VOR frequency
   (c) NDB frequency.

8. The DME interrogator is part of the:
   (a) airborne equipment
   (b) DME navigation aid
   (c) VORTAC.

9. The varying interval between pulse-pairs
   ensures that the interrogator:
   (a) recognises its own pulse-pairs and
       rejects other signals
   (b) recognises other pulse-pairs and rejects
       its own signal
   (c) tunes into a VOR station and DME
       navigation aid.

10. When a DME indicator is receiving no
    computed data, it will display:
    (a) dashes
    (b) zeros
    (c) eights.

11. Using a collocated VOR–DME navigation
    aid produces what type of position fix?
    (a) rho-rho
    (b) rho-theta
    (c) theta-theta.

12. Distance and bearing signals from a
    TACAN navigation aid are transmitted on:
    (a) HF
    (b) UHF
    (c) VHF.

13. Using two DME navigation aids provides
    how many calculated positions?
    (a) two
    (b) one
    (c) three.

14. DME operates in which frequency band?
    (a) UHF
    (b) VHF
    (c) LF/MF.

Figure 11.12   See Question 15

Figure 11.14   See Question 17

15. The instrument shown in Figure 11.12 is
    called the:
    (a) RMI
    (b) RDMI
    (c) CDI.

16. Referring to Figure 11.13, the display is
    providing:
    (a) maximum distance
    (b) minimum distance
    (c) no computed data.

17. Referring to Figure 11.14, the installation
    on the right is a DME:
    (a) transponder
    (b) transmitter
    (c) receiver.

Figure 11.13   See Question 16

Chapter 12

Instrument landing system

The instrument landing system (ILS) is used for precision approaches and landings. The system uses a combination of VHF and UHF radio waves and has been in operation since 1946. In this chapter we will look at ILS principles and hardware in detail, concluding with how the ILS combines with the automatic flight control system (AFCS) to provide fully automatic approach and landing. Although the ILS is being gradually replaced by satellite-based augmentation systems (SBAS), described elsewhere in this book, ILS is in widespread use around the world.

12.1 ILS overview

The instrument landing system is used for the final approach and is based on directional beams propagated from two transmitters at the airfield, see Figure 12.1. One transmitter (the **glide slope**) provides guidance in the vertical plane and has a range of approximately 10 nm. The second transmitter (the **localizer**) guides the aircraft in the horizontal plane. In addition to the directional beams, two or three marker beacons are located at key points on the extended runway centreline defined by the localizer, see Figure 12.4.

12.2 ILS ground equipment

12.2.1 Localizer transmitter

The localizer transmits in the VHF frequency range, 108–112 MHz in 0.5 MHz increments. Note that this is the same frequency range as used by the VOR system (see Chapter 10). Although the two systems are completely independent and work on totally different principles, they often share the same receiver. The two systems are differentiated by their frequency allocations within this range. ILS frequencies are allocated to the odd tenths of each 0.5 MHz increment, e.g. 109.10 MHz, 109.15 MHz, 109.30 MHz, etc. VOR frequencies are allocated to the even tenths

of each 0.5 MHz increment, e.g. 109.20 MHz, 109.40 MHz, 109.60 MHz, etc. Table 12.1 provides an illustration of how these frequencies are allocated within the 109 MHz range. This pattern applies from 108 to 111.95 MHz.

The localizer antenna is located at the far end of the runway and transmits two lobes to the left and right of the runway centreline modulated at 90 Hz and 150 Hz respectively. On the extended runway centreline, see Figure 12.2, the combined depth of modulation is equal. Either side of the centreline will produce a **difference in depth of modulation** (DDM); this difference is directly proportional to the deviation on either side of the extended centreline of the runway. The localizer also transmits a two- or three-letter Morse code identifier that the crew can hear on their audio panels.

Table 12.1 Allocation of ILS and VOR frequencies

ILS frequency (MHz)	VOR frequency (MHz)
	109.00
109.10	
109.15	
	109.20
109.30	
109.35	
	109.40
109.50	
109.55	
	109.60
109.70	
109.75	
	109.80
109.90	
109.95	

DOI: 10.1201/9781003411932-12

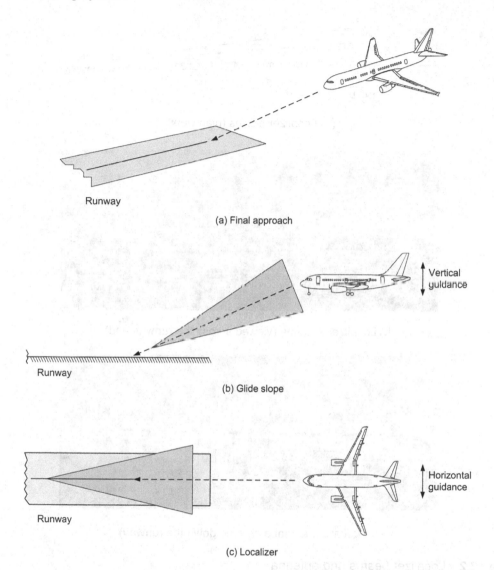

Runway

(a) Final approach

Vertical
guidance

Runway

(b) Glide slope

Horizontal
guidance

Runway

(c) Localizer

Figure 12.1 ILS overview

Key point

The instrument landing system is based on directional beams propagated from two transmitters at the airfield: localizer and glide slope.

Test your understanding 12.1

What frequency bands do the localizer and glide slope use?

12.2.2 Glide slope antenna

The glide slope antenna transmits in the UHF frequency band, 328.6 to 335 MHz at 150 kHz spacing. Upper and lower lobes are modulated at 90 HZ and 150 Hz respectively. When viewed from the side, see Figure 12.3, the two lobes overlap and produce an approach path inclined at a fixed angle between 2.5 and 3.5 degrees. Glide slope frequency is automatically selected when the crew tunes the localizer frequency.

(a) Localizer beams (plan view)

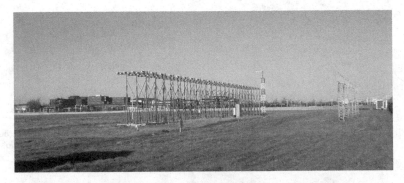

(b) Localizer antenna (viewed across the runway end)

(c) Localizer antenna (viewed down the runway)

Figure 12.2   Localizer beams and antenna.

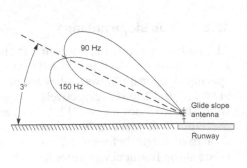

(a) Glide slope beams

(b) Glide slope antenna

Figure 12.3   Glide slope beams and antenna.

12.2.3 Marker beacons

Two or three beacons are sited on the extended runway centreline at precise distances; these are specified in the approach charts for specific runways. These beacons operate at 75 MHz and radiate approximately 3–4 W of power. The beacons provide visual and audible cues to the crew to confirm their progress on the ILS, see Figure 12.4. The **outer marker** is located between four and seven miles from the runway threshold; it transmits Morse code dashes at a tone frequency of 400 Hz and illuminates a blue light (or cyan 'OM' icon for electronic displays) when the aircraft passes over the beacon.

The outer marker provides the approximate point at which an aircraft on the localizer will intercept the glide slope. Some airfields use non-directional beacons (NDBs) in conjunction with (or in place of) the outer marker. These are referred to as **locator beacons** (compass locator in the USA).

The middle marker is located approximately 3500 feet from the runway threshold. When passing over the middle marker, the crew receive an alternating Morse code of dots/dashes modulated at 1300 Hz, and a corresponding amber light (or yellow 'MM' icon for electronic displays) is illuminated. The middle marker coincides with the aircraft being 200 feet above the runway touchdown point.

Runways that are used for low visibility approach and landings (see later in this chapter)

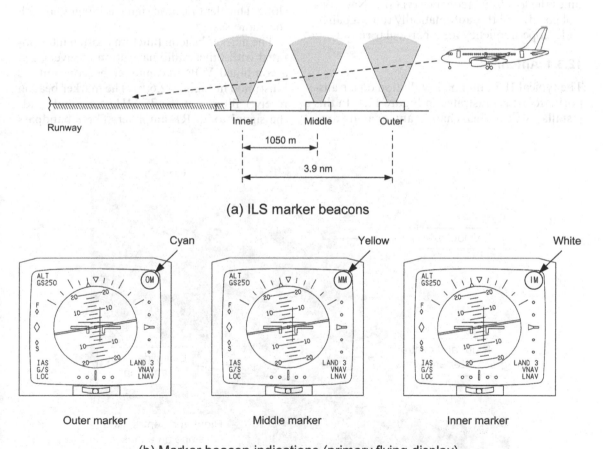

(a) ILS marker beacons

(b) Marker beacon indications (primary flying display)

Figure 12.4 ILS marker beacon system.

have a third **inner marker**. When passing over the inner marker, the crew receive Morse code dots modulated at 3000 Hz on the audio system, and a corresponding white light (or 'IM' icon for electronic displays) is illuminated. The marker beacon system is currently being phased out with the introduction of DME and GPS approaches.

12.3 ILS airborne equipment

The airborne equipment comprises localizer and glide slope antennas, ILS receiver, marker receiver and flight deck controls and displays. Most aircraft are fitted with two or three independent ILS systems (typically named left, centre and right). The localizer and glide slope frequencies are in different wave bands; the crew tunes the localizer frequency (via the 'Nav' control panel) and this automatically tunes a paired glide slope frequency for a particular runway.

12.3.1 Antennas

The typical ILS antenna installation on a transport aircraft is illustrated in Figure 12.5. In this installation, two dual channel antennas are used for localizer and two dual channel antennas for the glide slope. One channel from each of the antennas is not used; the received signals are fed to the corresponding ILS receiver.

12.3.2 Receivers

ILS receivers are often combined with other radio navigation functions, e.g. VHF omni directional range (VOR); these are located in the avionic equipment bay, see Figure 12.6. ILS receivers are based on the super-heterodyne principle with remote tuning from the control panel. The signal received from the localizer antenna is modulated with 90 and 150 Hz tones for left/right deviation; a 1020 Hz tone contains the navigation aid identification in Morse code. Filters in the ILS receiver separate out the 90 and 150 Hz tones for both localizer and glide slope. The identification signal is integrated with the audio system.

The marker beacon function is often incorporated with other radio navigation receivers, e.g. a combined VOR and marker beacon unit, as illustrated in Figure 12.6(b). The marker beacon receiver filters out the 75 MHz tone and sends the signal to an RF amplifier. Three bandpass

Figure 12.5   ILS antennas—glide slope and localizer.

(a) ILS receivers

(b) Combined VOR/marker beacon receiver

Figure 12.6   VHF/navigation receivers.

filters are then employed at 400 Hz, 1300 Hz and 3000 Hz to identify the specific marker beacon. The resulting signals are sent to an audio amplifier and then integrated into the audio system. Discrete outputs drive the visual warning lights (or PFD icons).

12.3.3 Controls and displays

A control panel is typically located on the centre pedestal on larger aircraft, see Figure 12.7(a); this is used to select the runway heading and ILS frequency. Alternatively, on General Aviation (GA) aircraft, it can be a combined navigation/controller and display, see Figure 12.7(b). Outputs from the marker receiver are sent to three indicator lights (or PFD icons) and the crew's audio system, as described in the previous section.

The method of selecting and displaying the selected runway heading depends on the type of

(a) ILS control panel (centre pedestal)

(b) GA navigation and communication units

Figure 12.7   ILS control panels.

avionic fit. Electromechanical instruments include the omni-bearing selector (OBS), course deviation indicator (CDI), horizontal situation indicator (HSI) and electronic horizontal situation indicator (EHSI), see Figure 12.8. These instruments vary slightly in their design and layout; however they have some common features. The omni-bearing selector is used to rotate the course card. This card is calibrated from 0 to 360° and indicates the selected runway heading. Each dot on the scale represents a 2° deviation from the selected runway heading. A second

pointer displays glide slope deviation; flags are used to indicate when the:

- localizer and/or glide slope signals are
- beyond reception range
- pilot has not selected an ILS frequency
- ILS system is turned off, or is inoperative.

Note that these indicators can also be used with the VOR navigation system; refer to Chapter 10 for a detailed description of this system.

(a) Omni-bearing indicator (OBI)

(b) Course deviation indicator (CDI)

(c) Horizontal situation indicator (HSI)

Figure 12.8   ILS displays.

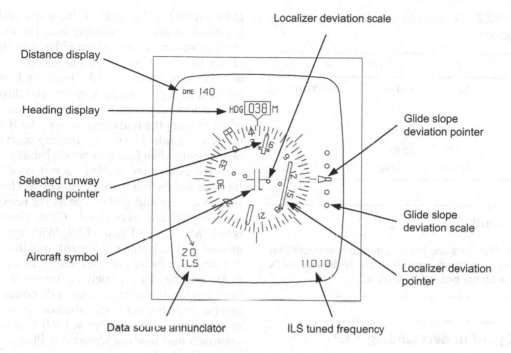

Localizer deviation scale

Distance display

Heading display

Glide slope deviation pointer

Selected runway heading pointer

DME 140

HDG 038 M

Glide slope deviation scale

Aircraft symbol

20
ILS

110.10

Localizer deviation pointer

Data source annunciator

ILS tuned frequency

(d) Electronic display of ILS

Figure 12.8 (Continued)

A pointer moves left/right over a deviation scale to display lateral guidance information. The glide slope deviation pointer moves up/down over a scale to indicate vertical deviation. The strength of the 90 Hz and 150 Hz tones is summed to confirm the presence of the localizer and glide slope transmissions; this summed output is displayed in the form of a 'flag'. If either of the two transmissions is not present, the warning flag is displayed.

Key point

ILS frequencies are selected by tuning the localizer, which automatically selects the glide slope.

Key point

The marker beacon system is being phased out and replaced by GPS/DME approaches.

Key point

The ILS glide slope is inclined at a fixed angle between 2.5 and 3.5 degrees from the ground.

12.4 Low range radio altimeter

The low range radio altimeter (LRRA) described in Chapter 8 is an integral part of the ILS; most aircraft are fitted with two or more independent systems. The LRRA incorporates an adjustable altitude bug, controlled by a minimum descent altitude (MDA) control, see Figure 12.12, that creates a visual or aural warning to the pilot when the aircraft reaches the selected altitude. Typically, the pilot will abort the approach if the runway is not visible when the decision height is reached, see Table 12.2 for typical approach and landing categories.

Table 12.2 Automatic approach and landing categories

Category	DH	RVR (min)	RVR (max)
1	200'	550 m	1000 m
2	100'	300 m	–
3A	< 100'	200 m	–
3B	< 50'	75 m	–
3C	None	< 75 m	–

Key point

Radio altimeters are the only airborne sensors that can make a direct measurement of the aircraft's altitude above ground level (AGL).

Test your understanding 12.2

Where are the runway's localizer and glide slope antennas located?

Test your understanding 12.3

What type of radar system is used for the low range radio altimeter (LRRA)?

Test your understanding 12.4

(a) When flying overhead ILS marker beacons, what indications are provided to the crew?
(b) What is the preferred sequence to capture the localizer and glide slope?
(c) What are the decision heights for Category 1, 2 and 3 landings?

12.5 ILS approach

The normal procedure is to capture the localizer first and then the glide slope. The crew select the ILS frequency on the navigation control panel as described above. Runway heading also needs to be sent to the ILS receiver; the way of achieving this depends on the avionic fit of the aircraft. Desired runway heading can either be selected on a CDI/HSI or via a remote selector located on a separate control panel. Deviation from the localizer and glide slope is monitored throughout the approach together with confirmation of position from the marker beacons. The ILS can be used to guide the crew on the approach using instruments when flying in good visibility.

In the event that visibility is not good, then the approach is flown using the AFCS. The crew select localizer and glide slope as the respective roll and pitch modes on the AFCS mode control panel (MCP), see Figure 12.9. With approved ground and airborne equipment, qualified crew can continue the approach through to an automatic landing. To complete an automatic landing (autoland) the pitch and roll modes need precise measurement of altitude above the ground; this is provided by the LRRA. A typical approach and landing scenario is illustrated in Figure 12.10.

An alternative ILS approach is the localizer back-beam, or back-course (BC). All localizer antennas actually transmit in two directions; with primary (forward-course) and secondary (back-course) lobes, see Figure 12.11. Instead of having a separate ILS for each runway direction, back-course approaches utilise the same localizer beam. For back course approaches, the localizer display s are reversed; this is identified on the navigation display with a 'BC' flag. Note that most back course approaches do not have glide slopes.

12.6 Autoland

The development of airborne and ground equipment, together with crew training, led to trials being carried out on the effectiveness and reliability of fully automatic landings using the ILS. In 1947, the Blind Landing Experimental Unit (BLEU) was established within the UK's Royal Aircraft Establishment. The world's first fully automatic landing was achieved in 1950. Equipment and procedures were further developed, leading to the world's first automatic landing in a passenger carrying aircraft (the HS121 Trident) in July 1965.

Figure 12.9 AFCS mode control panel (LOC/GS modes).

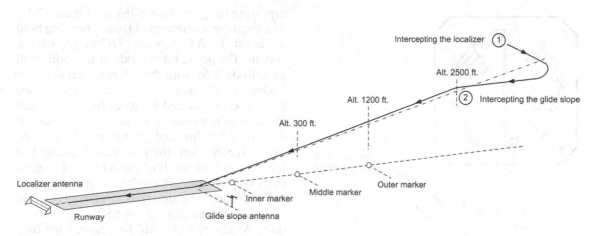

Figure 12.10 Automatic approach and landing.

Automatic approach and landings are categorised by the certifying authorities as a function of ground equipment, airborne equipment and crew training. The categories are quoted in terms of decision height (DH) and runway visual range (RVR). These categories are summarised in Table 12.2; JAR OPS provides further details and notes. Category 3 figures depend on aircraft type and airfield equipment,

e.g. quality of ILS signals and runway lighting (centreline, edges, taxi ways, etc.).

An operator has to have approval from the regulatory authorities before being permitted to operate their aircraft with automatic Category 2 and 3 approach and landings. This applies in particular to Category 3 decision heights.

Automatic approaches are usually made by first capturing the localizer (LOC) and then

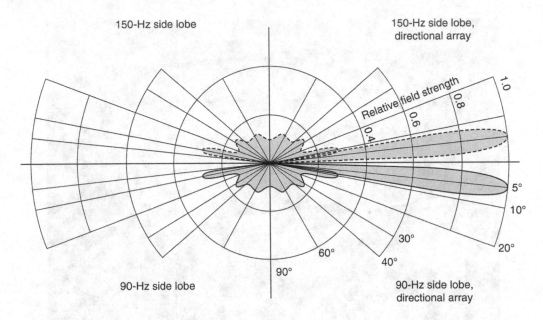

Figure 12.11   ILS back course.

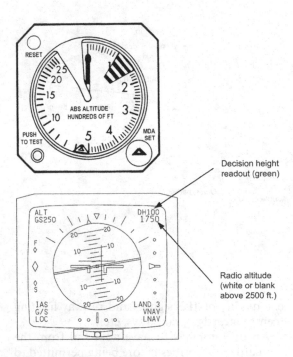

Figure 12.12   Radio altitude and decision height setting and displays

capturing the glide slope (GS), see Figure 12.10. The localizer is intercepted from a heading hold mode on the AFCS, with LOC armed on the system. The active pitch mode at this point will be altitude hold, with the GS mode armed.

Once established on the localizer, the glide slope is captured and becomes the active pitch mode. The approach continues with deviations from the centreline and glide slope being sensed by the ILS receiver; these deviations are sent to roll and pitch channels of the AFCS, with sensitivity of pitch and roll modes being modified by radio altitude. The auto throttle controls desired airspeed. Depending on aircraft type, two or three AFCS channels will be engaged for fully automatic landings, thus providing levels of redundancy in the event of channel disconnects.

Although the glide slope antenna is located adjacent to the touchdown point on the runway, it departs from the straight-line guidance path below 100 feet. The approach continues with radio altitude/descent rate being the predominant control input into the pitch channel. At approximately 50 feet, the throttles are retarded and the aircraft descent rate and airspeed are reduced by the '**flare**' mode, i.e. a gradual nose-up attitude that is maintained until touchdown. The final pitch manoeuvre is to put the nose of the aircraft onto the runway. Lateral guidance is still provided by the localizer at this point until such time as the crew take control of the aircraft.

12.7 Operational aspects of the ILS

ILS remain installed throughout the world and are the basis of automatic approach and landing for many aircraft types. Limitations of ILS are the single approach paths from the glide slope and localizer; this can be a problem for airfields located in mountainous regions. Furthermore, any vehicle or aircraft approaching or crossing the runway can cause a disturbance to the localizer beam, which could be interpreted by the airborne equipment as an unreliable signal. This often causes an AFCS channel to disconnect, with the possibility of a missed approach. The local terrain can also have an effect on ILS performance, e.g. multi-path errors can be caused by reflections of the localizer; the three-degree glide slope angle may not be possible in mountainous regions or in cities with tall buildings. These limitations led to the development of the Microwave Landing System (MLS).

ILS approaches, together with autoland, are described further in Aircraft Flight Instruments and Guidance Systems (AFIGS).

Key point

Positive touchdown can be achieved by the Low Range Radio Altimeter (LRRA) being programmed to 'aim' for minus 20 feet. 'Flare' mode is initiated at 50 feet (typical), followed by the auto throttle reducing to idle at 15 feet. In a typical system, when the LRRA reduces to zero feet, the weight-on-wheel switches take over, modifying the control laws to alleviate ground affect. (Autoland is covered in more detail in Aircraft Flight Instruments and Guidance Systems—refer to Preface notes.)

Key point

When the aircraft has touched down with an automatic landing, the ILS continues to provide lateral Guidance via the localizer.

12.8 Multiple choice questions

1. Frequency bands for ILS are:
   (a) localizer (UHF) and glide slope (VHF)
   (b) localizer (VHF) and glide slope (VHF)
   (c) localizer (VHF) and glide slope (UHF).

2. Localizer transmitters are located:
   (a) at the threshold of the runway, adjacent to the touchdown point
   (b) at the stop end of the runway, on the centreline
   (c) at three locations on the extended centreline of the runway.

3. The LRRA provides:
   (a) deviation from the runway centreline
   (b) deviation from the glide path angle
   (c) altitude in feet above the ground.

4. Some airfields use NDBs in conjunction with (or in place of) the:
   (a) localizer
   (b) glide slope
   (c) outer marker.

5. When viewed from the antenna, the localizer is characterised by two lobes modulated:
   (a) 90 Hz to the right, 150 Hz to the left of the centreline
   (b) 150 Hz to the right, 90 Hz to the left of the centreline
   (c) equally either side of the centreline.

6. ILS frequencies are selected by tuning:
   (a) the glide slope which automatically selects the localizer
   (b) the localizer which automatically selects the glide slope
   (c) the localizer and glide slope frequencies independently.

7. The ILS glide slope is inclined at a fixed angle between:
   (a) 2.5 and 3.5 degrees
   (b) zero and 2.5 degrees
   (c) 2.5 degrees and above.

8. The glide slope is characterised by two
   lobes modulated:
   (a) 90 Hz above, 150 Hz below the glide
       slope angle
   (b) 150 Hz above, 90 Hz below the glide
       slope angle
   (c) equally either side of the glide
       slope angle.

9. Marker beacons transmit on which
   frequency?
   (a) 75 MHz
   (b) 1300 Hz
   (c) 400 Hz.

10. With three marker beacons installed in an
    ILS system, they will be encountered along
    the approach as:
    (a) outer, middle, inner
    (b) inner, middle, outer
    (c) outer, inner, middle.

11. Marker beacon outputs are given by:
    (a) coloured lights and Morse code tones
    (b) deviations from the runway centreline
    (c) deviations from the glide slope.

12. The decision height and runway visual
    range for a Category 2 automatic
    approach are:
    (a) 100 ft. and 300 m respectively
    (b) 200 ft. and 550 m respectively
    (c) less than 100 ft. and 200 m respectively.

13. The outer marker is displayed on the
    primary flying display as a coloured icon
    that is:
    (a) yellow
    (b) white
    (c) cyan.

14. When the aircraft has touched down with
    an automatic landing, the ILS continues to
    provide:
    (a) lateral control via the localizer
    (b) lateral control via the glide slope
    (c) vertical control via the LRRA.

Chapter 13

Doppler navigation

Doppler navigation is a self-contained dead reckoning system, i.e. it requires no external inputs or references from ground stations. Ground speed and drift can be determined using a fundamental scientific principle called Doppler shift. Doppler navigation systems were developed in the mid-1940s and introduced in the mid-1950s as a primary navigation system with many features, including continuous calculations of ground speed and drift. Being self-contained, the system can be used for long distance navigation over oceans and undeveloped areas of the globe. Doppler sensors are also used in other airborne applications, including weather radar and missile warning systems. In this chapter, we will review some basic scientific principles, look at Doppler navigation as a stand-alone system, and then review some of the other Doppler applications. The use of Doppler navigation systems in civil aircraft applications is mainly with helicopters.

13.1 The Doppler effect

The 'Doppler effect' is named after Christian Doppler (1803–53), an Austrian mathematician and physicist. His hypothesis was that the frequency of a wave apparently changes as its source moves closer to, or further away from, an observer. This principle was initially proven to occur with sound; it was subsequently found to occur with any wave type including electromagnetic energy. An excellent example of the Doppler effect can be observed when aircraft or fast trains (or racing cars) pass by an observer.

To illustrate this principle, consider Figure 13.1, an observer located at a certain distance from a sound source that is emitting a fixed-frequency tone. As the approaches the observer, the number of cycles 'received' by the observer is the fixed tone, plus the additional cycles received as a function of the speed.

This will have the effect of increasing the tone (above the fixed frequency) as heard by the observer. At the instant when the is adjacent to the observer, the true fixed-frequency will be heard. When the travels away from the observer, fewer cycles per second will be received and the tone will be below the fixed-frequency as heard by the observer. The difference in tone is known as the **Doppler shift**; this principle is used in Doppler navigation systems. Doppler shift is, for practical purposes, directly proportional to the relative speed of movement between the source and observer. The relationship between the difference in frequencies and velocity can be expressed as:

$$F_{D} = \frac{vf}{c}$$

where F_{D} = frequency difference, v = aircraft velocity, f = frequency of transmission and c = speed of electromagnetic propagation (3×10^8 meters/second).

13.2 Doppler navigation principles

Doppler navigation systems in aircraft have a focused beam of electromagnetic energy transmitted ahead of the aircraft at a fixed angle (theta, θ), as shown in Figure 13.2. This beam is scattered in all directions when it arrives at the surface of the earth. Some of the energy is received back at the aircraft. By measuring the difference in frequency between the transmitted and received signals, the aircraft's **ground speed** can be calculated. The signal-to-noise ratio of the received signal is a function of a number of factors including:

- aircraft range to the terrain
- backscattering features of the terrain
- atmospheric conditions, i.e. attenuation and absorption of radar energy
- radar equipment.

DOI: 10.1201/9781003411932-13

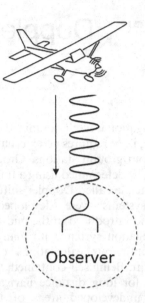

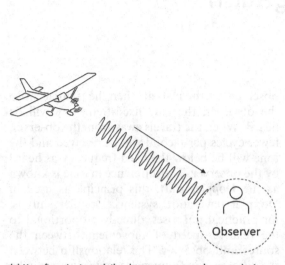

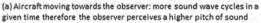

(a) Aircraft moving towards the observer: more sound wave cycles in a given time therefore the observer perceives a higher pitch of sound

(b) Aircraft nearest to the observer with the exact pitch of sound

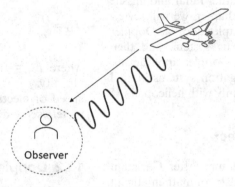

(c) Aircraft moving away from the observer: less sound wave cycles in a given time therefore the observer perceives a lower pitch of sound

Figure 13.1   The Doppler effect

Note that the aircraft in Figure 13.2 is flying straight and level. If the aircraft were pitched up or down, this would change the angle of the beam with respect to the aircraft and the surface; this will change the value of Doppler shift for a given ground speed. This situation can be overcome in one of two ways; the transmitter and receiver can be mounted on a stabilised platform or (more usually) two beams can be transmitted from the aircraft (forward and aft), as shown in Figure 13.3. By comparing the Doppler shift of both beams, a true value of ground speed can be derived. The relationship between the difference in frequencies and velocity in an aircraft can be expressed as:

$$F_D = \frac{2\cos\theta \times vf}{c}$$

where F_D = frequency difference, θ = the angle between the beam and aircraft, v = aircraft velocity, f = frequency of transmission and c = speed of electromagnetic propagation (3×10^8 meters/second).

Note that a factor of two is needed in the expression since both the transmitter and

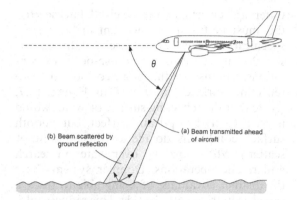

Figure 13.2 Doppler navigation principles

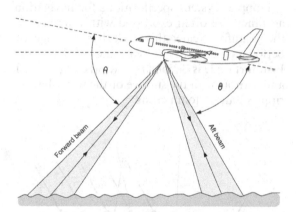

Figure 13.3 Compensation for aircraft pitch angle

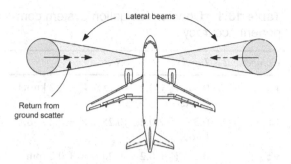

Figure 13.4 Measuring drift by Doppler shift

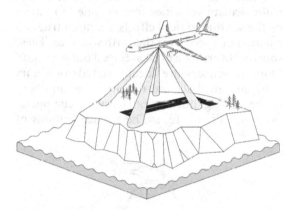

Figure 13.5 Measuring ground speed and drift using directional beams

receiver are moving with respect to the earth's surface. It can be seen from this expression that aircraft altitude is not a factor in the basic Doppler calculation. Modern Doppler systems operate up to 15,000 feet (rotary wing) and 50,000 feet (fixed wing).

Having measured velocity along the track of the aircraft, we now need to calculate **drift**. This can be achieved by directing a beam at right angles to the direction of travel, see Figure 13.4. Calculation of drift is achieved by utilising the same principles as described above. In practical installations, several directional beams are used, see Figure 13.5.

The calculation of ground speed and drift provides 'raw navigation' information. By combining these two values with directional information from a gyro-magnetic compass system, we have the basis of a complete self-contained navigation system. By integrating the velocity calculations, the system can derive the distance travelled (along track) and cross track deviations. The Doppler system has a resolution of approximately 20–30 Hz (frequency difference) per knot of speed. Note that, in addition to ground speed and drift information, Doppler velocity sensors can also detect **vertical displacement** from a given point. Errors accumulate as a function of distance travelled; typical Doppler navigation system accuracy can be expressed in knots (Vt) as follows:

$$V_t = \sqrt{V_x^2 + V_y^2 + V_z^2}$$

The individual components of velocity along the x, y and z axes (ground speed, drift and vertical components) have accuracies given in Table 13.1 for both sea (Beaufort scale of 1) and land conditions.

Table 13.1 Doppler navigation system component accuracy

Component	Land	Sea
Vx	0.3% V_t + 0.2 knots	0.25% V_t + 0.2 knots
Vy	0.3% V_t + 0.2 knots	0.25% V_t + 0.2 knots
Vz	0.2% V_t + 0.2 fpm	0.20% V_t + 0.2 fpm

When flying over oceans, the Doppler system will calculate velocities that include movement of the sea due to tidal effects, i.e. not a true calculation of speed over the earth's surface. These short-term errors will be averaged out over time. Doppler sensors are ideally suited for rotary wing aircraft that need to hover over an object in the sea, e.g. during search and rescue operations, see Figure 13.6. The surface features of

water are critical to the received **backscatter**; this must be taken into account in the system specification. The 'worst case' conditions for signal-to-noise ratios are with smooth sea conditions; to illustrate this point consider the two reflecting surfaces illustrated in Figure 13.7. (Note that the reflecting surface of water would never actually be optically perfect, but smooth surface conditions do reduce the amount of scatter.) When hovering over water in search and rescue operations, Doppler systems have the distinct advantage of being able to track a vessel as it drifts with the tide. This reduces pilot workload, particularly if the Doppler system is coupled to an automatic control system.

Doppler system specifications for navigation accuracy are often expressed with reference to the Beaufort scale; this scale has a range of between 1 and 12. A sea state of 1 on the Beaufort scale is defined by a wind of between 1 and 3 knots with the surface of the water having ripples, but no foam crests.

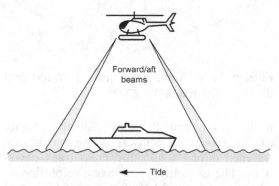

(a) Aircraft tracks object using forward/aft beams

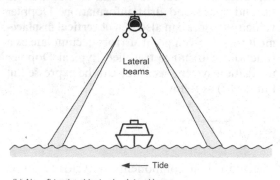

(b) Aircraft tracks object using lateral beams

Figure 13.6 Using Doppler during hover

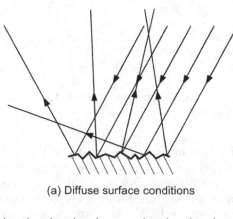

(a) Diffuse surface conditions

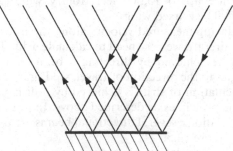

(b) Smooth surface conditions

Figure 13.7 Surface reflections

Key point

Christian Doppler's hypothesis was that the frequency of a wave apparently changes as its source moves closer to, or farther away from, an observer.

Key point

Doppler sensors are ideally suited for rotary wing aircraft that need to hover over an object in the sea, e.g. during search and rescue operations.

Test your understanding 13.1

Explain Christian Doppler's hypothesis for the sound waves

Test your understanding 13.2

What effect does increasing the frequency of a transmitted Doppler beam have on sensitivity of the frequecy shift?

13.3 Airborne equipment overview

Doppler navigation systems use directional beams to derive ground speed and drift as previously described; these beams are arranged in a number of ways, as illustrated in Figure 13.8. The fore and aft beams are referred to as a 'Janus' configuration (after the Roman god of openings and beginnings, Janus, who could face in two directions at the same time)

Three beams can be arranged in the form of the Greek letter lambda (λ). The four-beam arrangement is an X configuration; only three beams are actually required; the fourth provides a level of monitoring and redundancy. In the four-beam arrangement, the fore and aft signals are transmitted in alternative pairs.

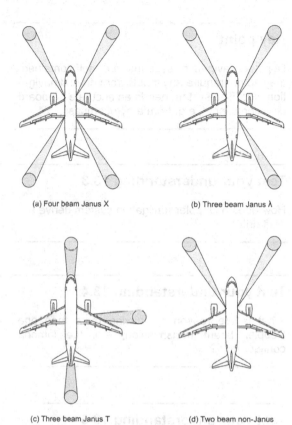

(a) Four beam Janus X   (b) Three beam Janus λ

(c) Three beam Janus T   (d) Two beam non-Janus

Figure 13.8   Doppler beam arrangements

Referring to the relationship:

$$F_{D} = \frac{2\cos\theta \times vf}{c}$$

it can be seen that the sensitivity of Doppler velocity calculations increases with the transmitted frequency; this means that a smaller antenna can be used.

The frequencies allocated to Doppler navigation systems are within the SHF range, specifically 13.25–13.4 GHz; some Doppler systems operate within the 8.75–8.85 GHz range.

Key point

By measuring the difference in frequency between the transmitted and received signals, the aircraft's velocity in three axes can be calculated using the Doppler shift principle.

Key point

Doppler navigation systems are self-contained; they do not require any inputs from ground navigation aids. The system needs an accurate on-board directional input, e.g. from a gyrocompass.

Test your understanding 13.3

How does a Doppler navigation system derive aircraft drift?

Test your understanding 13.4

What is the reason for having more than one Doppler beam transmission, e.g. the lambda configuration?

Test your understanding 13.5

What effect does the sea state have on the back scattering of a Doppler beam?

The basic Doppler system comprises an antenna, transmitter and receiver. The antenna can be fixed to the airframe, thereby needing corrections for pitch attitude (achieved via the Janus configuration). Alternatively the antenna could be slaved to the aircraft's attitude reference system. The antenna produces a very narrow conical- or pencil-shaped beam.

The Doppler navigation system has been superseded for commercial airline use by inertial and satellite navigation systems. Rotary wing aircraft, however, use Doppler sensors to provide automatic approach and stabilisation during hover manoeuvres; in this case the display would provide vertical displacement above/below the selected hover altitude and lateral/longitudinal deviation from the selected hover position.

13.4 Typical Doppler installations

Doppler principles can either be used in self-contained navigation systems or as stand-alone velocity sensors. The standalone velocity sensor is in the form of a radar transmitter–receiver, item 1. This sensor has a resolution of less than 0.1 knots and can be interfaced with other avionic systems and displays using data bus techniques. With increasing digital processing capability, the Doppler velocity sensor can be integrated with other navigation sensors to provide filtered navigation calculations. This subject is addressed in Chapter 14.

Typical self-contained Doppler navigation systems comprise the radar transmitter–receiver, signal processor, control display unit and steering indicator. A typical navigation system transmits at 13.325 GHz, using frequency modulation/continuous wave signals at a low radiated power of 20 mW.

Digital signal processing is used for continuous spectrum analysis of signal returns; this leads to enhanced tracking precision accuracy and optimises signal acquisition over marginal terrain conditions (sand, snow and calm sea conditions). Doppler systems compensate for attitude changes as described earlier; these manoeuvres can be aggressive for certain helicopter operations. Changes of up to 60 degrees per second can be accommodated for pitch and roll excursions; the system can accommodate rate changes of up to 100 degrees per second in azimuth.

13.5 Doppler summary

In summary, Doppler navigation has a number of advantages:

- Velocity and position outputs from the system are provided on a continuous basis
- It requires no ground navigation aids, i.e. it is self-contained and autonomous
- Velocity outputs are very accurate
- Navigation is possible over any part of the globe, including oceans and polar regions

- The system is largely unaffected by weather (although certain rainfall conditions can affect the radar returns)
- The system does not require any preflight alignment.

The disadvantages of Doppler navigation are:

- It is dependent upon a directional reference,
- e.g. a gyro-magnetic compass
- It requires a vertical reference to
- compensate for aircraft attitude
- Position calculations degrade with distance travelled
- Short-term velocity calculations can be inaccurate, e.g. when flying over the tidal waters, the calculated aircraft velocity will be in error depending on the tide's direction and speed. (This effect will average out over longer distances and can actually be used to an advantage for rotary wing aircraft)
- Military users have to be aware that the radar transmission is effectively giving away the location of the aircraft.

13.6 Other Doppler applications

In addition to self-contained navigation, the Doppler shift principle is also employed in several other aerospace systems. Missile warning systems and military radar applications include pulse-Doppler radar target acquisition and tracking; the Doppler principle is employed to reduce clutter from ground returns and atmospheric conditions.

In Chapter 10 reference was made to siting errors of conventional VOR ground stations. Many VOR stations now employ the Doppler principle to overcome these errors; these are referred to as Doppler VOR (DVOR) navigation aids; more details are provided in Chapter 10. Enhanced weather radar systems for commercial aircraft have the additional functionality of being able to detect turbulence and predict wind shear (see Chapter 18).

Key point

Doppler navigation system accuracy can be expressed in knots; errors accumulate as a function of distance travelled.

Test your understanding 13.6

When crossing over a coastal area, from land towards the sea, what effect will tidal flow have on the Doppler system's calculated ground speed and drift?

13.7 Multiple choice questions

1. Doppler navigation systems operate in which frequency range?
   (a) SHF
   (b) VHF
   (c) UHF.

2. When moving towards a sound source, what effect will Doppler shift have on the pitch of the sound as heard by an observer?
   (a) no effect
   (b) increased pitch
   (c) decreased pitch.

3. What effect does increasing the frequency of a transmitted Doppler beam have on sensitivity of the frequency shift?
   (a) decreased
   (b) no effect
   (c) increased.

4. Raw Doppler calculations include:
   (a) pitch and roll
   (b) directional information
   (c) ground speed and drift.

5. Velocity and position outputs from a Doppler navigation system are provided:
   (a) only when the aircraft is moving
   (b) on a continuous basis
   (c) only in straight and level flight.

6. The backscattering features of the terrain affect the Doppler navigation system's:
   (a) accuracy
   (b) signal-to-noise ratio
   (c) coverage.

7. Integrating Doppler ground speed calculations will provide:
   (a) distance travelled
   (b) drift angle
   (c) directional information.

8. Drift can be measured by directing a beam:
   (a) at right angles to the direction of travel
   (b) in line with the direction of travel
   (c) directly below the aircraft.

9. Doppler position calculations degrade with:
   (a) time
   (b) attitude changes
   (c) distance travelled.

10. When hovering directly over an object in the sea with a six-knot tide, the Doppler system will indicate:
    (a) six knots drift in the opposite direction of the tide
    (b) six knots drift in the direction of the tide
    (c) zero drift.

11. When hovering over water, the 'worst case' conditions for signal-to-noise ratios are with:
    (a) smooth sea conditions
    (b) rough sea conditions
    (c) tidal drift.

12. Doppler system beams in the lambda arrangement have beams directed in the following way:
    (a) forward and to each side of the aircraft
    (b) forward, aft and to one side of the aircraft
    (c) forward, aft and to each side of the aircraft.

13. Backscattering of a Doppler beam from the surface of water is:
    (a) low from a rough surface
    (b) low from a smooth surface
    (c) high from a smooth surface.

Chapter 14

Area navigation

Area navigation (RNAV) is a means of combining, or filtering, inputs from one or more navigation sensors and defining positions that are not necessarily co-located with ground-based navigation aids. This facilitates aircraft navigation along any desired flight path within range of navigation aids; alternatively, a flight path can be planned with autonomous navigation equipment. Typical navigation sensor inputs to an RNAV system include VHF omni directional range (VOR); distance measuring equipment (DME); and global navigation satellite system (GNSS). This chapter describes RNAV systems using VOR and DME navigation aids to establish the basic principles of RNAV. The chapter concludes with a review of Kalman filters and how RNAV systems are specified with a 'required navigation performance' (RNP).

14.1 RNAV overview

Two basic ground navigation aids that can be used for RNAV are VOR and DME, see Figures 14.1 and 14.2. RNAV is a guidance system that uses various inputs, e.g. VOR and/or DME, to compute a position. The VOR system transmits specific bearing information, referred to as **radials**, see Figure 14.1. The pilot can select any radial from a given VOR navigation aid and fly to or from that aid.

(DME) is a short/medium-range navigation system based on secondary radar. Both VOR and DME are described in earlier chapters of this book. Conventional airways are defined by VOR and DME navigation aids, see Figure 14.3.

Since the VOR–DME systems are **line of sight (LOS)**, the altitude of the aircraft will have a direct relationship with the range that the system can be used, see Figure 14.4. Using VOR–DME navigation aids imposes a limit on the working range that can be obtained. The maximum LOS distance between an aircraft and the ground station is given by the relationship:

$$d = 1.1\sqrt{h}$$

where d is the distance in nautical miles, and h is the altitude in feet above ground level (assumed to be flat terrain). The theoretical LOS range for altitudes up to 20,000 feet is given in Table 14.1. At higher altitudes, it is possible to receive VOR signals at greater distances but with reduced signal integrity. Although the actual range also depends on transmitter power and receiver sensitivity, the above relationship provides a good approximation.

The positions defined in an RNAV system are called **waypoints**; these are geographical positions that can be created in a number of ways. RNAV systems can store many waypoints in a sequence that comprises a complete route from origin to destination. Creating waypoints that are not co-located with fixed ground aids provides a very flexible and efficient approach to flight planning. These waypoints are stored in a **navigation database** (NDB) as permanent records or entered by the pilot. Waypoints can be referenced to a fixed position derived from VOR and/or DME navigation aids, see Figure 14.5. The desired track between waypoints (Figure 14.6) is referred to as an RNAV **leg**. Each leg will have a defined direction and distance; a number of legs in sequence becomes the route. The advent of digital computers has facilitated comprehensive area navigation systems that use a combination of ground navigation aids and airborne equipment.

DOI: 10.1201/9781003411932-14

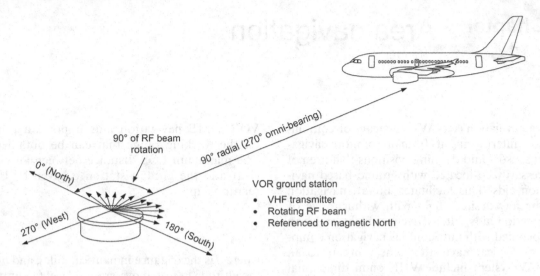

(a) VHF omni directional range (VOR) overview

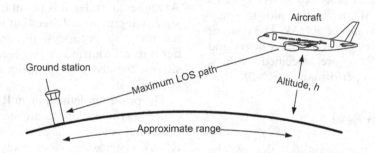

(b) VHF omni directional range—line of sight

Figure 14.1   VOR principles

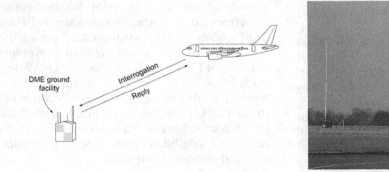

(a) Secondary radar used for DME   (b) DME transponder (right of photo)

Figure 14.2   DME principles

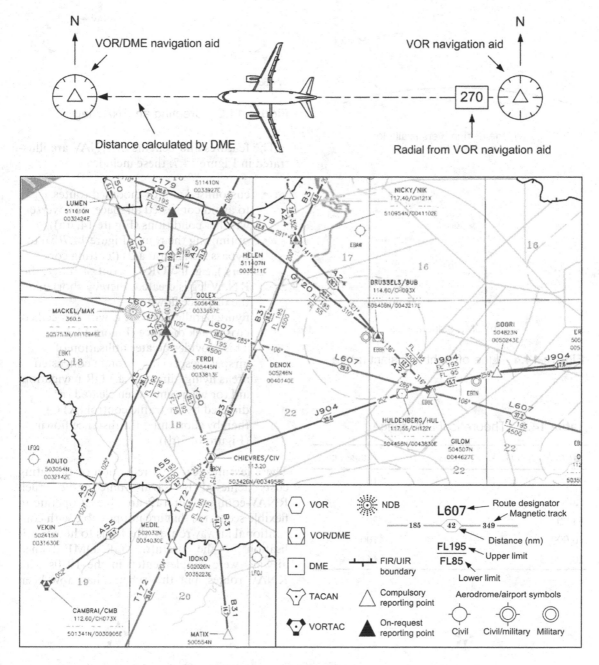

Figure 14.3 Conventional airways defined by VOR-DME

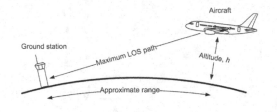

(a) Line of sight versus altitude

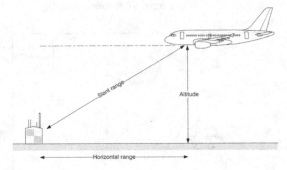

(b) DME slant range

Figure 14.4   Line of sight and slant range

Table 14.1   Theoretical LOS range

Altitude (feet)	Range (nm)
100	10
1,000	32
5,000	70
10,000	100
20,000	141

Figure 14.6   Creating an RNAV leg

The features and benefits of RNAV are illustrated in Figure 14.7; these include:

- customised and/or modified routes, e.g. to avoid congested airspace, or adverse weather conditions (Figure 14.7(a))
- optimising the route (Figure 14.7(b)) to bypass navigation aids ('cutting corners'), e.g. if VOR-C is out of range, the RNAV leg is created, thereby shortening the distance flown
- flying parallel tracks, i.e. with a specified cross track distance (Figure 14.7(c)). This provides greater utilisation of airspace, especially through congested areas flying 'direct to' a VOR navigation aid (or waypoint) when cleared or directed by air traffic control (ATC), thereby shortening the distance flown (Figure 14.7(d)).

These features lead to a reduction of operating costs achieved by saving time and/or fuel. RNAV-equipped aircraft are able to operate in flexible scenarios that are not possible with conventional airway routes; this leads to higher utilisation of the aircraft. VOR–DME-defined airways were supplemented in the 1970s with RNAV routes, but this scheme has now been

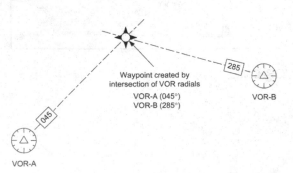

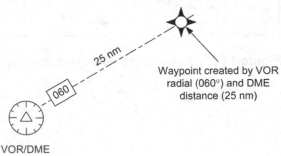

Figure 14.5   Creating waypoints from VOR and DME navigation aids

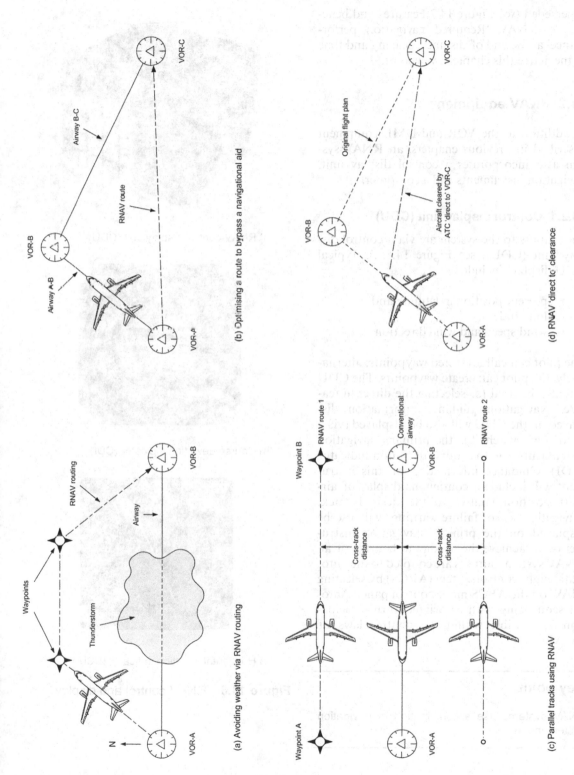

(a) Avoiding weather via RNAV routing

(b) Optimising a route to bypass a navigational aid

(c) Parallel tracks using RNAV

(d) RNAV 'direct to' clearance

Figure 14.7 Features and benefits of RNAV

superseded (see Figure 14.7 Features and bene-fits of RNAV 'Required navigation perfor-mance' at the end of distance, bearing and time to the active this chapter). waypoint.

14.2 RNAV equipment

In addition to the VOR and DME equipment described in previous chapters, an RNAV sys-tem also incorporates a control display unit, navigation instruments and a computer.

14.2.1 Control display unit (CDU)

Pilot inputs to the system are via a control dis-play unit (CDU), see Figure 14.8(a). Typical CDU displays include:

- present position in latitude and longitude
- wind speed and wind direction

The pilot can call up stored waypoints; alterna-tively, the pilot can create waypoints. The CDU can also be used for selecting the **direct to** fea-ture. Navigation guidance information dis-played on the CDU will also be displayed (via a 'Rad/Nav' switch) for the primary navigation instruments, e.g. the course deviation indicator (CDI). Guidance information on this instru-ment will include a continuous display of air-craft position relative to the desired track. Navigation sensor failure warnings will also be displayed on the primary navigation instru-ments. To achieve the maximum benefits of an RNAV system, outputs are coupled to the auto-matic flight control system (AFCS) by selecting 'NAV' on the AFCS mode control panel. Auto-leg sequencing with associated turn anticipa-tion is possible within the control laws of the AFCS.

Key point

RNAV systems use a combination of navigation system inputs.

(a) RNAV control display unit (CDU)

(b) Course deviation indicator (CDI)

(c) Horizontal situation indicator (HSI)

Figure 14.8   RNAV control and display

Key point

RNAV-equipped aircraft are able to operate in conditions and scenarios that would not have been previously possible, thereby obtaining higher utilisation of the aircraft.

14.2.2 Navigation instruments

Instruments that can be used for the display of RNAV information are the course deviation indicator (CDI) and horizontal situation indicator (HSI); see Figures 14.8(b) and (c). The course selector is set to the desired leg; a deviation pointer moves left or right of the aircraft symbol to indicate if the aircraft is to the right or left of the selected leg. (The CDI/HSI can also be used in VOR/ILS radio navigation systems; see Chapters 10 and 11.)

14.2.3 Computer

The RNAV computer is used to resolve a variety of navigation equations. In order to realise the benefits of RNAV, systems contain a **navigation database** (see below). Simple area navigation is achieved by solving geometric equations; the data required for these calculations is obtained from the relative bearings of VOR stations and/or distances from DME stations.

In navigation calculations, bearings are referred to as theta (θ) and distances as rho (ρ). The RNAV definition of a waypoint using a colocated VOR–DME navigation aid is illustrated in Figure 14.9. Accurate horizontal range can be calculated by the computer based on:

- DME slant range
- DME transponder elevation
- aircraft altitude.

This calculation provides the true range, as illustrated in Figure 14.10. Transponder elevation is obtained from the computer's navigation database. Altitude information is provided by an encoding altimeter or air data computer (ADC). Cross track deviation, either intentional or otherwise, is calculated as shown in Figure 14.11.

Computers in more sophisticated systems are also able to **auto-tune** navigation aids to provide the optimum navigation solution. The system decides on whether to use combinations of VOR–VOR (theta–theta), VOR–DME (theta–rho) or DME–DME (rho–rho). Note that when two DME navigation aids are used, there is an ambiguous position fix, see Figure 14.12; this can be resolved in a number of ways, e.g. by tuning in to a third DME navigation aid or tuning in to a VOR station. Systems use algorithms to determine which combination of navigation aids to use; this will depend on signal strength and geometry. Fourdimensional waypoints can also be defined by specifying the required time of arrival over a three-dimensional waypoint. This is discussed further in the flight management system chapter. Other aircraft sensor inputs such as initial fuel quantity, fuel flow, airspeed and time provide the means of calculating range, estimated time of arrival (ETA), endurance, etc. This data can be provided for specific waypoints or the final destination.

14.2.4 Navigation database (NDB)

The navigation database (stored within the RNAV computer's memory) contains permanent records for VOR, DME and VORTAC navigation aids. Table 14.2 illustrates the locations, identification codes and navigation aid type for a typical European country. Details that are stored in the database include specific information for each navigation aid such as:

- name
- identification code
- navigation aid type
- latitude and longitude
- elevation
- transmission frequency.

The navigation database is updated every 28 days to take into account anything that has changed with a navigation aid, e.g. frequency changes, temporary unavailability, etc. The pilot can enter new or modified details for navigation aids that might not be contained in the navigation database. Waypoints can either be entered as they appear on navigation charts, or the pilot can create them. The navigation database in more sophisticated RNAV systems will also include standard instrument departures (SIDs),

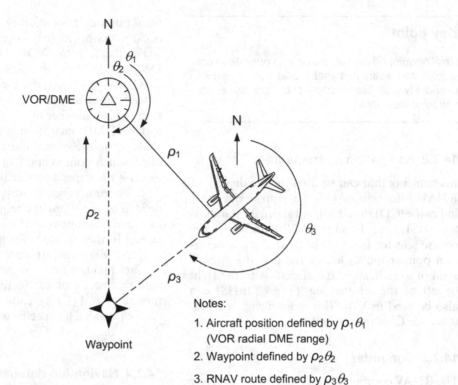

Notes:

1. Aircraft position defined by $\rho_1\theta_1$
   (VOR radial DME range)
2. Waypoint defined by $\rho_2\theta_2$
3. RNAV route defined by $\rho_3\theta_3$

(a) RNAV triangulation

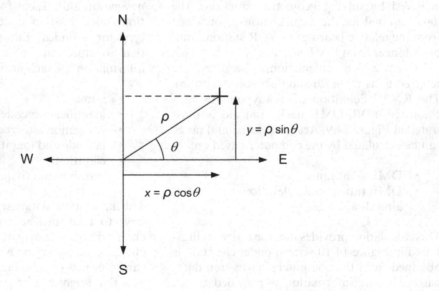

(b) RNAV calculation

Figure 14.9   RNAV geometry

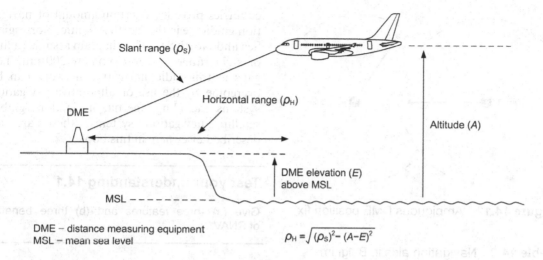

DME – distance measuring equipment
MSL – mean sea level

$$\rho_H = \sqrt{(\rho_S)^2 - (A-E)^2}$$

Figure 14.10   RNAV geometry—vertical profile

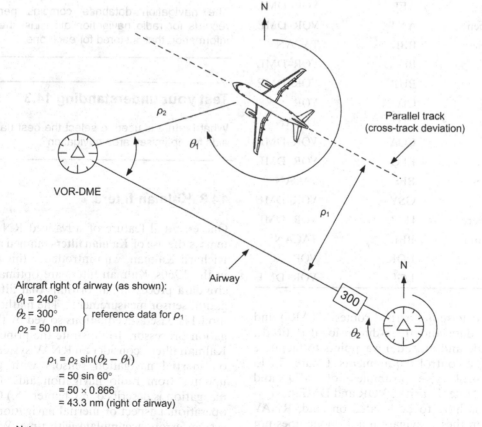

Aircraft right of airway (as shown):
$\theta_1 = 240°$
$\theta_2 = 300°$ } reference data for ρ_1
$\rho_2 = 50$ nm

$\rho_1 = \rho_2 \sin(\theta_2 - \theta_1)$
  $= 50 \sin 60°$
  $= 50 \times 0.866$
  $= 43.3$ nm (right of airway)

Note:
If the aircraft was to the left of the airway, a negative value of ρ would be calculated – this is
interpreted as being left of the airway

Figure 14.11   RNAV geometry—lateral profile

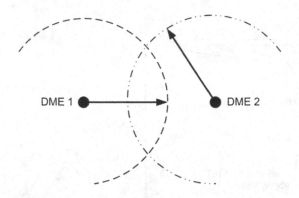

Figure 14.12   Ambiguous DME position fix

Table 14.2   Navigation aids in Belgium

Name	Identification	Type
Affligem	AFI	VOR–DME
Antwerpen	ANT	VOR–DME
Beauvechain	BBE	TACAN
Bruno	BUN	VOR–DME
Brussels	BUB	VOR–DME
Chievres	CIV	VOR
Chievres	CIV	TACAN
Costa	COA	VOR–DME
Flora	FLO	VOR–DME
Florennes	BFS	TACAN
Gosly	GSY	VOR–DME
Huldenberg	HUL	VOR–DME
Kleine Brogel	BBL	TACAN
Koksy	KOK	VORTAC
Liege	LGE	VOR–DME

standard terminal arrival routes (STARs) and runway data and three-dimensional (latitude, longitude and altitude) waypoints to facilitate air traffic control requirements. Figures 14.13 and 14.14 give examples of SID and STAR. Note that since VOR and DME navigation aids have to be located on land, RNAV based on these navigation aids alone does not extend far beyond coastal regions. Referring to Figure 14.15, a combination of radio navigation stations located in a number of European countries provides a certain amount of navigation guidance in the North Atlantic, Norwegian Sea and North Sea. This diagram assumes a line of sight range of approximately 200 nm. The gaps in this radio navigation network can be overcome by the use of alternative navigation systems including: inertial navigation, global satellite navigation systems; these are all described elsewhere in this book.

Test your understanding 14.1

Give (a) three features and (b) three benefits of RNAV.

Test your understanding 14.2

The navigation database contains permanent records for radio navigation aids. List the typical information that is stored for each one.

Test your understanding 14.3

What feature is used to select the best navigation aids for optimised area navigation?

14.3 Kalman filters

One essential feature of advanced RNAV systems is the use of **Kalman filters**, named after Dr Richard Kalman, who introduced this concept in the 1960s. Kalman filters are optimal **recursive** data processing algorithms that filter navigation sensor measurements. The mathematical model is based on equations solved by the navigation processor. To illustrate the principles of Kalman filters, consider an RNAV system based on inertial navigation sensors with periodic updates from radio navigation aids. (Inertial navigation is described in Chapter 15.) One key operational aspect of inertial navigation is that system errors accumulate with time. When the system receives a position fix from navigation aids, the inertial navigation system's errors can be corrected.

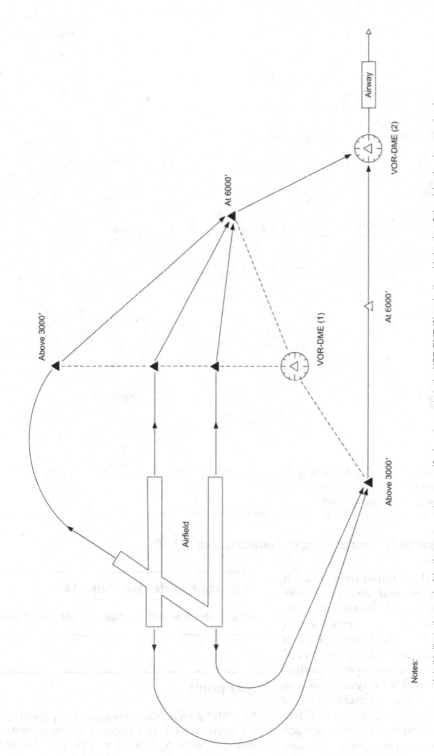

Notes:

1. In this illustration, each of the three runways has a specific departure route to the VOR-DME (2) navigation aid   the aircraft then joins the airways network
2. The SIDs are typically referenced to navigation aids, e.g. VOR-DME or marker beacons
3. There would also be published departure routes for aircraft joining airways to the south, east and north
4. Reporting points (triangles) are often specified with altitude constraints, e.g. at, below or above 3000'

Figure 14.13   Illustration of standard instrument departures (SIDs)

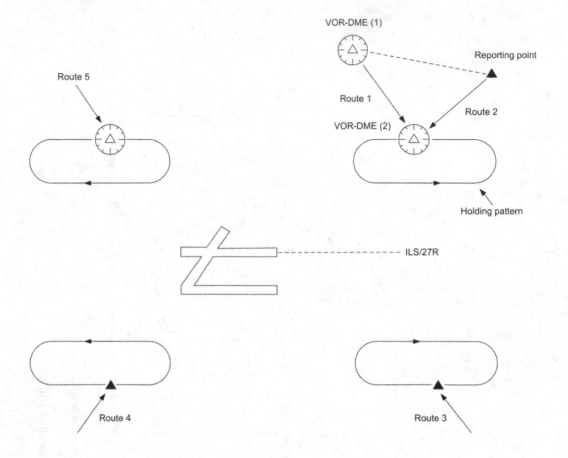

Notes:

1. In this illustration, each of the three arrival routes is associated with a navigation aid (VOR-DME) and reporting point (solid triangles)
2. Each arrival route is normally allocated a holding pattern
3. Minimum sector altitudes are published for each route
4. When cleared by ATC, the aircraft would leave the holding pattern and be given a heading to join the ILS for the active runway, e.g. 27R

Figure 14.14   Illustration of standard terminal arrival routes (STAR)

The key feature of the Kalman filter is that it can analyse these errors and determine how they might have occurred; the filters are recursive, i.e. they repeat the correction process on a succession of navigation calculations and can 'learn' about the specific error characteristics of the sensors used. The numerous types of navigation sensors employed in RNAV systems vary in their principle of operation, as described in the specific chapters of this book. Kalman filters take advantage of the dissimilar nature of each sensor type; with repeated processing of errors, complementary filtering of sensors can be achieved.

Test your understanding 14.4

What is the difference between a SID and STAR?

Key point

The RNAV navigation database is updated every 28 days to take into account anything that has changed with a navigation aid, e.g. frequency changes, temporary unavailability, etc.

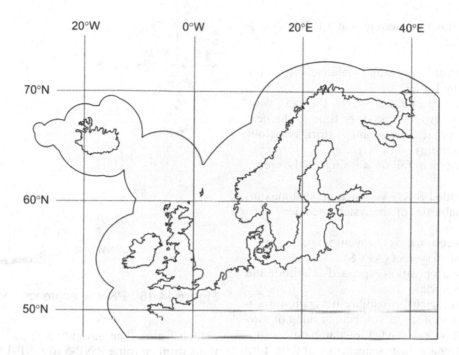

Figure 14.15 Line of sight coverage of radio navigation aids in Northern Europe

Test your understanding 14.5

Explain the purpose of a Kalman filter.

14.4 Required navigation performance (RNP)

Simple area navigation systems can use radio navigation aid inputs such as VOR and DME to provide definitions of waypoints as described in this chapter. Comprehensive area navigation systems use a variety of sensors such as satellite and inertial reference systems; these specific systems are addressed in more detail in subsequent chapters. The accuracy and reliability of area navigation systems has led to a number of navigation performance standards and procedures for the aircraft industry; these are known as **required navigation performance** (RNP). Various RNAV systems, together with their associated RNP, are evolving via individual aviation authorities. This is embraced by the generic term of **performance-based navigation (PBN)**.

Factors that contribute to overall area navigation accuracy include:

- external navigation aids
- the aircraft's navigation equipment
- (including displays)
- automatic flight control system (AFCS).

The International Civil Aviation Organization (ICAO) has defined RNAV accuracy levels covering terminal, en route, oceanic and approach flight phases with specific navigation performance values between 1 and 10 nm. These values are expressed by a number, e.g. RNP-5. This indicates that (on a statistical basis) the aircraft's area navigation system must maintain the aircraft for 95% of the flight time within 5 nm of the intended flight envelope, i.e. either side of and along the track. RNP-5 is used for basic RNAV (**B-RNAV**) in Europe. It is not specified how this navigation performance should be achieved, or what navigation equipment is to be used. RNP for terminal operations is less than 1 nm; these systems require performance monitoring and alert messages to the crew in the event of system degradation.

Typical functions required of a B-RNAV system include:

- aircraft position relative to the desired track
- distance and bearing to the next waypoint ground speed, or time to the next waypoint waypoint storage (four minimum)
- equipment failure warnings to the crew.

Recommended B-RNAV functions that maximise the capabilities of the system include:

- roll commands to an automatic flight
- control system (AFCS)
- aircraft position expressed as latitude and
- longitude
- a 'direct to' capability navigation accuracy indication automatic tuning of navigation aids navigation database
- automatic leg sequencing and/or turn anticipation.

Note that if an inertial reference system (IRS) is used as a sensor, the B-RNAV system must have the capability of automatically tuning into radio navigation aids after a maximum period of two hours; this is because an IRS-derived position will drift (see Chapter 15). If a global navigation satellite system (GNSS) is used as a sensor into the RNAV system, the GNSS must have fault detection software known as receiver autonomous integrity monitoring (RAIM); see Chapter 16. Single RNAV systems are permissible; however, the aircraft must be able to revert to conventional navigation using VOR, DME and ADF in the event of RNAV equipment failure.

Precision area navigation (P-RNAV) is the European terminal airspace application evolved from B-RNAV. The P-RNAV track keeping accuracy equates to cross track accuracy of RNP-1 (+/−1nm). P-RNAV applications are intended primarily for terminal airspace where there is higher traffic density, terrain clearance considerations etc. These considerations require more accuracy than en route B-RNAV navigation.

Referring to Figure 14.16, the RNP approach concept can then be extended to RNP0.3

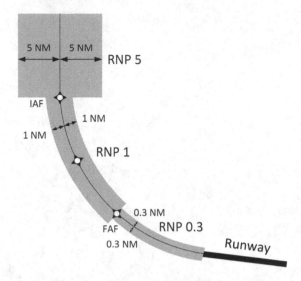

Figure 14.16   PRNAV approach concept

(+/− 0.3 nm). The navigation sequence transitions from en route RNP-5 to RNP-1 at the initial approach fix (IAF). The navigation sequence transitions from RNP-1 to at the final approach fix (FAF)

In more remote areas, e.g. isolated oceanic regions where it is impossible to locate ground navigation aids, RNP-10 applies. This allows spacing of 50 nm between aircraft in place of 100 nm. The RNAV system now needs two independent long-range systems, e.g. IRS and/or GNSS. If using IRS as a sensor, the system has to receive a position fix with a specified period, typically 6.2 hours. The GNSS has to have fault detection and exclusion (FDE) capability, a technique used to exclude erroneous or failed satellites from the navigation calculations by comparing the data from six satellites.

Key point

Waypoints can be based on existing navigation aids and defined mathematically as:

- rho–theta (using one DME and one VOR navigation aid)
- rheta–theta (using two VOR navigation aids)
- rho–rho (using two DME navigation aids).

Key point

Auto-tuning of navigation aids is used by RNAV systems to select the best navigation aids for optimised area navigation.

Key point

Required navigation performance (RNP) is the performance-based successor to area navigation (RNAV).

Test your understanding 14.6

Explain why RNAV systems using VOR–DME are generally unavailable beyond land and its immediate coastal regions.

Test your understanding 14.7

Explain what is meant by RNP and why it is needed.

Test your understanding 14.8

Explain why an RNAV database needs to be updated every 28 days.

14.5 PBN system errors

All navigation systems need to be defined in terms of performance. Ground-based (or fixed) systems, e.g. VOR and DME, both have levels of performance defined by quantified tolerances. In this case, the navigation performance relates to the transmitted signals and the ability of the aircraft to accurately utilise these signals. With (PBN) the performance requirements are specified, and the navigation system is required to meet the specified level of performance. The resulting total system error is calculated from three components:

Navigation System Error (NSE). This represents the capability of the navigation system to determine position, relative to the aircraft's actual position. NSE is dependent on the accuracy of the input sensors being used for the position solution, e.g. the accepted accuracy of VOR/DME signals, or global navigation system errors.

Flight Technical Error (FTE). This represents the ability of the aircraft guidance system to follow the computed flight path. FTE is normally evaluated during flight trials and/or in-service data. FTE values can vary for a particular aircraft type depending on the flight control method, e.g. where the autopilot is coupled compared to the manual control using flight director.

Path Definition Error (PDE). Area navigation routes are defined by segments between waypoints. The definition of each route segment is therefore dependent on the accuracy of the waypoint, and the ability of the navigation system to manage the waypoint data. Waypoints can be defined and managed very effectively with digital avionic systems, with a resulting high level of accuracy; PDE is typically minimal and is generally considered to be zero.

Total System Error (TSE) is calculated as the statistical sum of the NSE, FTE and PDE errors, and can be calculated from the root of the sum of squares:

$$TSE = \sqrt{\left(NSE^2 + FTE^2 + PDE^2 \right)}$$

14.6 Actual navigation performance (ANP)

A key feature of RNP is the ability of the aircraft's navigation system to monitor its actual navigation performance during flight. The system also needs to identify to the crew whether this performance is, or is not, being met during an operation. When ANP exceeds RNP, an alerting message is displayed to the flight crew, typically via the CDU. This indicates that the navigation system's position does not meet RNP accuracy. The concept of ANP can be applied to lateral navigation as described in this chapter, and also vertical navigation which is described in the flight management system chapter.

14.7 Multiple choice questions

1. Waypoints are defined geographically by:
   (a) latitude and longitude
   (b) VOR frequency
   (c) DME range.

2. SIDs are used during the following flight phase:
   (a) arrival
   (b) cruise
   (c) departure.

3. Accurate area navigation using DME-DME requires:
   (a) slant range
   (b) horizontal range
   (c) VOR radials.

4. Rho–theta is an expression for which area navigation solution?
   (a) DME-DME
   (b) VOR-DME
   (c) VOR-VOR.

5. Navigation legs are defined by:
   (a) speed and distance
   (b) bearing and distance
   (c) bearing and speed.

6. Specific information for each navigation aid is contained in the:
   (a) navigation database
   (b) control display unit
   (c) course deviation indicator.

7. Flying a parallel track requires a specified:
   (a) cross track deviation
   (b) bearing
   (c) distance to go.

8. A three-dimensional waypoint is defined by:
   (a) VOR-DME
   (b) latitude, longitude, altitude
   (c) rho–theta–rho.

9. Autotuning of navigation aids is used by RNAV systems to:
   (a) update the navigation database
   (b) create waypoints in the CDU
   (c) select the best navigation aids for optimised area navigation.

10. Cross track deviation is displayed on the CDU and:
    (a) RMI
    (b) DME
    (c) HSI.

11. The navigation database is normally updated:
    (a) at the beginning of each flight
    (b) every 28 days
    (c) when selected by the pilot.

12. A four-dimensional waypoint is defined by:
    (a) lateral position, altitude, and time
    (b) latitude, longitude, altitude, and speed
    (c) altitude, direction, speed and time.

13. VORTAC navigation aids comprise which two facilities:
    (a) VOR and DME
    (b) VOR and TACAN
    (c) TACAN and NDB.

14. RNP-2 requires that the aircraft:
    (a) uses a minimum of two different navigation sensor inputs
    (b) is maintained within two nautical miles of the specified flight path
    (c) is maintained within two degrees of the specified flight path.

15. An area navigation position calculated from two DME stations is referred to mathematically as:
    (a) theta-theta
    (b) rho-theta
    (c) rho-rho.

16. The feature marked X in Figure 14.17 is a:
    (a) VOR-DME
    (b) STAR
    (c) waypoint.

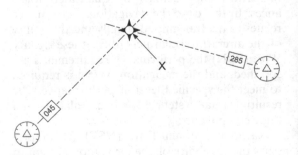

Figure 14.17 See Question 16

Chapter 15

Inertial navigation systems

Inertial navigation is an autonomous navigation system, i.e. it requires no external inputs or references from ground stations. The system was developed in the 1950s for use by the US military and subsequently the space programmes. Inertial navigation systems (INS) were introduced into commercial aircraft service during the early 1970s. The system does not receive or transmit radio frequencies. Being self-contained, the system is ideally suited for long distance navigation over oceans and undeveloped areas. As with many avionic systems, significant developments have occurred with inertial navigation systems in recent decades; INS is often integrated with other avionic units, and there are a variety of system configurations being operated. This chapter seeks to provide an introduction to the principles of inertial navigation together with some examples of typical hardware.

15.1 Inertial navigation principles

The primary sensors used in the system are **accelerometers** and **gyroscopes** (hereinafter gyro) to determine the motion of the aircraft. These sensors provide reference outputs that are processed to develop navigation data.

To illustrate the principle of inertial navigation, consider the accelerometer device illustrated in Figure 15.1; this is formed with a mass and two springs within a housing. Newton's second law of motion states that a body at rest (or in motion) tends to stay at rest (or in motion) unless acted upon by an outside force. Moving the accelerometer to the right causes a relative movement of mass to the left. If the applied force is maintained, the mass returns to the neutral position. When the accelerometer is moved to the left, or brought to rest, the relative movement of the mass is to the right. The mass continues in its existing state of rest or movement unless the applied force changes; this is the property of **inertia**. Attaching an electrical pick-up to the accelerometer creates a transducer that can measure the amount of relative movement of the mass. This relative movement is in direct proportion to the acceleration being applied to the device, expressed in m/s². If we take this electrical output and mathematically **integrate** the value, we are effectively multiplying the acceleration output by time; this can be expressed as:

$$Time \times acceleration = s \times m / s^2 = m / s = velocity$$

If we now take this velocity output and mathematically integrate the value, we are once again multiplying the output by time; this can be expressed as:

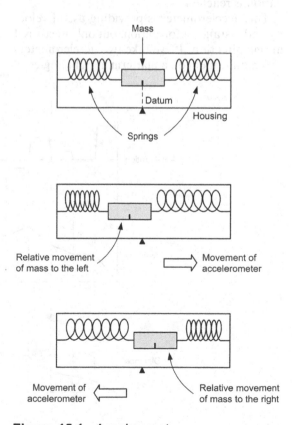

Figure 15.1 Accelerometer.

DOI: 10.1201/9781003411932-15

Time × velocity = s × m / s = m = *distance*

In summary, we started by measuring acceleration and were able to derive velocity and distance information by applying the mathematical process of integration. To illustrate this principle, consider a body accelerating at 5 m/s²; after ten seconds the velocity of the body will be 50 m/s. If this body now travels at a constant velocity of 50 m/s for ten seconds, it will have changed position by 500 m.

Referring to the profile illustrated in Figure 15.2, at t_0, the accelerometer is at rest and its output is zero. When the accelerometer is moved during the time period $t_0 - t_1$, there is a positive output from the accelerometer; this is integrated to provide velocity and distance, as illustrated in Figure 15.2. As the accelerometer reaches a steady velocity during the time period $t_1 - t_2$, the distance travelled increases. Acceleration increases and decreases during the journey (shown as positive and negative), until the destination is reached.

This accelerometer is providing useful velocity and distance information, but only measured in one direction. If we take two accelerometers and mount them on a **platform** at right angles to

each other, we can measure acceleration (and subsequently velocity and distance information) in any **lateral direction**. Thinking of an aircraft application, if we can align the platform with a known reference, e.g. **true north**, the two accelerometers are then directed N–S and W–E respectively, see Figure 15.3. We now have the means of calculating our velocity and distance travelled in any lateral direction.

An illustration of how the basic navigation calculations are performed is given in Figure 15.4. In a practical inertial navigation system, there is very little actual movement of the mass.

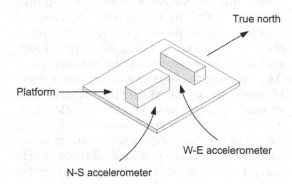

Figure 15.3   Platform NS-WE.

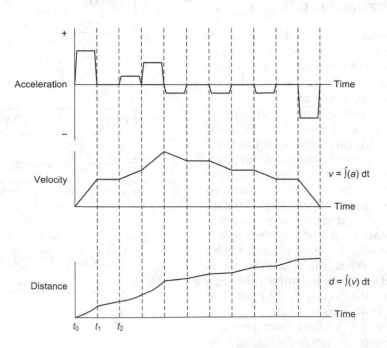

Figure 15.2   Integration profile.

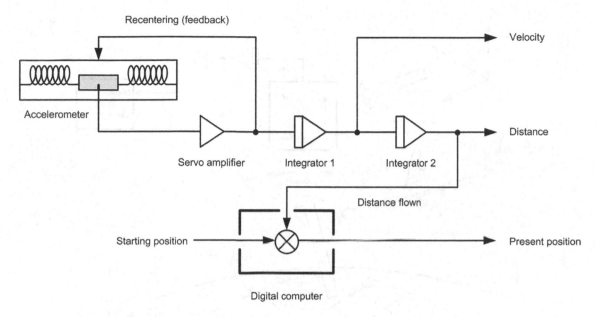

Figure 15.4 Navigation calculations (1).

The relative displacement between the mass and housing is sensed by an electrical pick-off signal. A closed loop servomechanism feedback signal (proportional to acceleration) is then amplified and used to restrain the mass in the **null position**. The amount of feedback required to maintain the null position is proportional to the sensed acceleration; this becomes the accelerometer's output signal. Calculation of basic navigation data is illustrated in Figure 15.5. By combining two accelerometer outputs in the directions N–S and W–E, we can sum the vector outputs and calculate distance and velocity in the horizontal plane. By comparing the distance travelled with the starting position (see 'Alignment process' in Section 15.4) we can calculate our present position.

Since the aircraft will be operating through a range of pitch and roll manoeuvres, it is vital that when measuring acceleration in the N–S and W–E directions, we do not measure the effects of gravity. The original inertial navigation systems maintained a physical platform such that it was always aligned with true north, and always level with respect to the earth's surface. Electromechanical gyros and torque motors mounted within gimbals in each of the

three axes achieved these requirements. The platform is aligned with true north and levelled at the beginning of the flight; this condition is maintained throughout the flight.

A by-product of aligning and levelling the platform is that attitude information is available for use by flight instruments and other systems. Modern-day commercial aircraft inertial navigation systems are equipped with **strap-down** devices including solid-state gyros and accelerometers. The alignment process and attitude compensation is now achieved in the computer's software, i.e. there is no physical platform.

15.2 System overview

The key principles of inertial navigation are based on accelerometer and gyro references together with a navigation processing function. These can be combined within a single INS and dedicated crew interface, as illustrated in Figure 15.6. Alternatively, the accelerometers and gyros are contained within an **inertial reference unit** (IRU); the processing function and crew interface is then integrated within the flight management computer system (FMCS). Within the FMCS, the flight management computer (FMC)

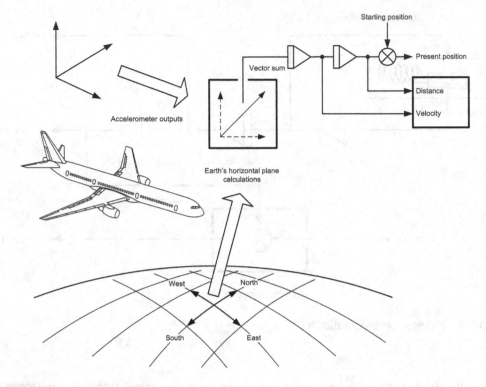

Figure 15.5 Navigation calculations (2).

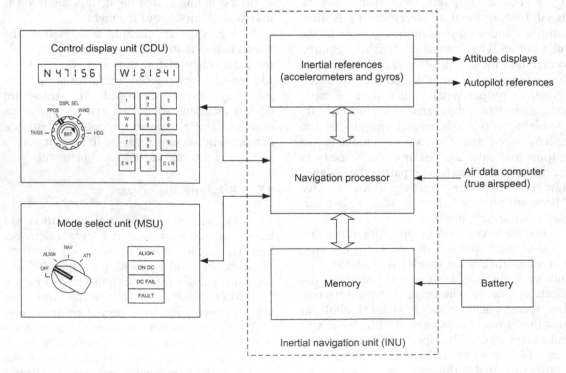

Figure 15.6 Inertial navigation system

combines area navigation and performance management into a single system (described in Chapter 16).

Key point

The primary sensors used in the inertial navigation system are accelerometers and gyros to determine the aircraft's movement. These sensors provide outputs that are processed to provide basic navigation data.

15.3 System description

The components described in this section are for devices used in typical commercial transport aircraft; note that system architecture varies

considerably with different aircraft types. A long-range aircraft will have three independent inertial reference systems, each providing its own navigation information.

The inertial navigation system can be considered to have three functions:

- References
- Processing
- Crew interface.

15.3.1 References

Accelerometers

These can be single or three axis devices; a typical single axis device is packaged in a 25 × 25 mm casing weighing 45 grams (see Figure 15.7). This contains a pendulum (**proof-mass**) that

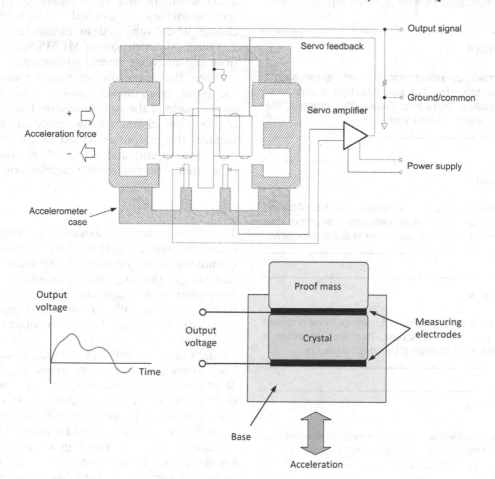

Figure 15.7 Accelerometer arrangement

senses acceleration as previously described over the range ±40 g; relative displacement between the pendulum and casing is sensed by a high gain capacitance pick-off and a pair of coils.

A closed loop servomechanism feedback signal (proportional to acceleration) is then amplified and demodulated. This feedback signal (analogue current or digital pulses) is applied to the coils to restrain the pendulum at the **null position**.

The feedback required to maintain the null position is proportional to the sensed acceleration; this becomes the accelerometer's output signal. Because of the high gain of the servomechanism electronics used, pendulum displacements are limited to microradians. An integral temperature sensor provides thermal compensation. The IRU contains three devices, measuring acceleration in the longitudinal, lateral and normal axes of the aircraft.

Key point

The inertial navigation system needs to establish a local attitude reference and direction of true north for navigation purposes. During this process, the aircraft should not be moved.

Key point

Synthesised magnetic variation can be obtained from inertial navigation systems, meaning that remote sensing compass systems are not required.

Key point

By comparing the position outputs of three onboard inertial navigation systems, this also provides a means of error checking between systems.

Key point

Errors in the inertial navigation system are random and build up as a function of time; this applies even if the aircraft is stationary.

Test your understanding 15.1

The output from an accelerometer goes through two stages of integration; what does each of these integration stages produce?

Developments in micro-electromechanical systems (MEMS) technology has led to silicon accelerometers that are more reliable and can be manufactured onto an integrated circuit. MEMS is the integration of mechanical elements, sensors and electronics on a common silicon substrate through micro-fabrication technology.

A typical MEMs accelerometer measures the deformation of a semiconductor material when forces act upon it; the principles are illustrated by the schematic diagram in Figure 15.7(b). The semiconductor's electrical characteristics change when subjected to mechanical stress. The physical dimensions of MEMS devices vary from one micron to several millimetres.

When the accelerometer senses movement, the proof mass moves, causing tension and compression of the semiconductor. The changes of electrical characteristics create an output voltage; this output is sensed by an external measuring circuit and amplified, producing a signal that is proportional to acceleration.

Gyros

The original inertial navigation systems used electromechanical gyros; these were subsequently replaced by a more reliable and accurate technology: the **ring laser gyro** (RLG). Ring laser gyros use interference of a laser beam within an optic path, or ring, to detect rotational displacement. An IRU contains three such devices (see Figure 15.8) for measuring changes in pitch, roll and azimuth. (Note that laser gyros are not actually gyroscopes in the strict sense of the word—they are in fact sensors of angular rate of rotation about an axis.) Two laser beams are transmitted in opposite directions (contrarotating) around a cavity within a triangular block of **cervit glass**; mirrors are located in two of the corners. The cervit glass (ceramic) material is very hard and has an

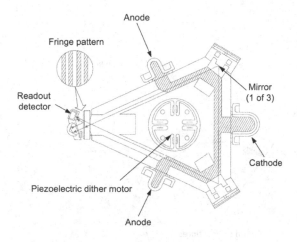

Figure 15.8 Ring laser gyro arrangement

ultra-low thermal expansion coefficient. The two laser beams travel the same distance, but in opposite directions; with a stationary RLG, they arrive at the detector at the same time.

The principles of the laser gyro are based on the **Sagnac effect**, named after the French physicist Georges Sagnac (1869–1926). This phenomenon results from interference caused by rotation. Interferometry is the science and technique of superposing (interfering) two or more waves, which creates a resultant wave different from the two input waves; this technique is used to detect the differences between input waves. In the aircraft RLG application, when the aircraft attitude changes, the RLG rotates; the laser beam in one path now travels a greater distance than the beam in the other path; this changes its phase at the detector with respect to the other beam. The angular position, i.e. direction and rate of the RLG, is measured by the **phase difference** of the two beams. This phase difference appears as a fringe pattern caused by the interference of the two wave patterns. The fringe pattern is in the form of light pulses that can be directly translated into a digital signal. Operating ranges of typical RLGs are 1000° per second in pitch, roll and azimuth. In theory, the RLG has no moving parts; in practice there is a device required to overcome a phenomena called **lock-in**. This occurs when the frequency difference between the two beams is low (typically 1000 Hz) and the two beams merge their frequencies. The solution is to mechanically oscillate the RLG to minimise the amount of time in this lock-in region.

Ring laser gyros are very expensive to manufacture; they require very high quality glass, cavities machined to close tolerances and precision mirrors. There are also life issues associated with the technology.

A variation of this laser gyro technology is the fibre optic gyroscope (FOG), where the transmission paths are through coiled fibre optic cables packaged into a canister arrangement to sense pitch, roll and yaw, see Figure 15.9. The fibre optic gyro also uses the interference of light through several kilometers of coiled fibre optic cable to detect angular rotation. Two light beams travel along the fibre in opposite directions and produce a phase shift due to the Sagnac effect. Three fibre optic gyroscope sensors are arranged for each of the aircraft's three axis of movement, pitch, roll and yaw. Fibre optic gyros have a life expectancy in excess of 3.5 million hours.

15.3.2 Inertial signal processing

The acceleration and angular rate outputs from the IRU are transmitted via a data bus to the navigation processor, in the flight management computer (FMC). Aside from navigation purposes, the IRU outputs are also supplied to other systems, e.g. the primary flying display and weather radar for attitude reference. Acceleration is measured as a **linear function** in each of the three aircraft axes; normal, lateral and longitudinal. Attitude is measured as an angular rate in pitch, roll and yaw.

These outputs are resolved and combined with air data inputs to provide navigation data, e.g. latitude, longitude, true heading, distance to the next waypoint, ground speed, wind speed and wind direction. The processor simultaneously performs these navigation calculations using outputs from all three accelerometers and angular rate sensors; in addition to these calculations, the processor has to compensate for three physical effects of the earth.

- Gravity
- Rotation
- Geometry

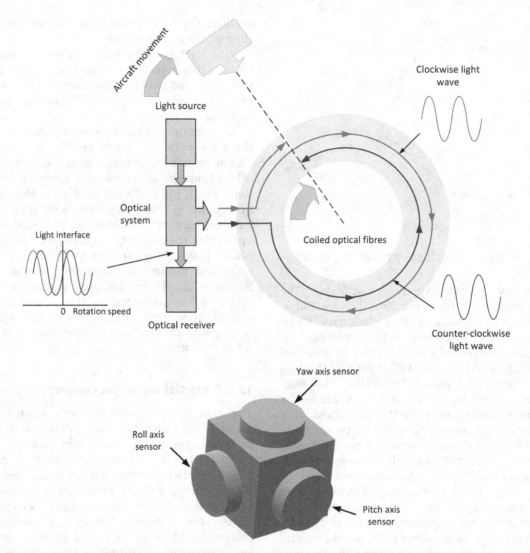

Figure 15.9   (a) Fibre optic gyroscope schematic; (b) Three axis fibre optic gyroscope

Effects of gravity

The navigation processor needs to determine the relationship between the aircraft attitude and surface of the earth such that the accelerometers only measure aircraft motion, not **gravity**. Outputs from each of the laser gyros are angular rates of rotation about an axis. These outputs are integrated, i.e. multiplied by time, to provide measurements of pitch, roll and heading. To illustrate this principle, a yaw rate of 4.5 degrees/second over a ten-second period equates to a heading change of 45°. If the aircraft's heading is now a constant 090°, and pitch/roll

rates are zero, the only acceleration measured is along the longitudinal axis of the aircraft. With a constant velocity of 500 knots, after one hour the navigation processor calculates that the aircraft has changed position by 500 nm in an easterly direction.

Now consider the same scenario, but with the aircraft climbing with a 10 degrees nose-up attitude. The processor needs to separate out vertical and lateral accelerations caused by gravity and motion of the aircraft respectively. The accelerometer is needed to measure motion parallel to the earth's surface, but if the aircraft is pitching or rolling, it will not be able to distinguish between gravity and aircraft acceleration. The component of gravity has to be separated out of the measured acceleration.

Effects of the earth's rotation

The processor now has to take into account the effect of the **earth's rotation**. To illustrate this effect, consider a platform device with an accelerometer and gyro as shown in Figure 15.10(a). In this schematic illustration, the gyro is used to

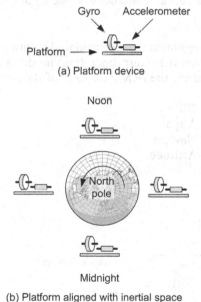

(a) Platform device

(b) Platform aligned with inertial space

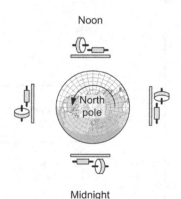

(c) Platform aligned with the earth's surface

Figure 15.10 Effect of gyro and accelerometer alignment relative to the earth's rotation

maintain the platform in a stable position. As the earth rotates, the platform maintains its position with respect to inertial space (Figure 15.10(b)); however, it is moving relative to the earth's surface. The platform has to be aligned with the earth's surface (Figure 15.10(c)) so that we can use it practically for navigation purposes.

With a strap-down system, each laser gyro will measure the angular rate of rotation of the aircraft about an axis. Since each laser gyro is fixed in position within the IRU, with the aircraft on the ground, it will also measure the rotation of the earth in an easterly direction. This motion includes (a) the rotation of the earth about its own axis and (b) the orbit around the sun:

- 360° over 24 hours = 15°/hour
- 360° over 365 days = 0.04°/hour.

This **earth rate** of up to 15.04 degrees per hour depends on latitude. Earth rate is a component that is subtracted from any measurement of aircraft angular rate sensed in an easterly direction.

Effects of the earth's geometry

The final consideration for the processor to address is the **spherical geometry** of the earth. As the aircraft travels around the earth in straight and level flight (parallel to the surface) it actually describes an arc. The pitch laser gyro senses this as an angular rate with respect to **inertial space**. When integrated, this rate output is converted into a change of pitch attitude. Clearly no pitch change has actually occurred due to this **transport rate**; the processor needs to subtract this component from the pitch laser gyro measurement. To calculate transport rate, the distance travelled (described by an arc) and angle subtended from the earth's centre are divided by time. This relationship can be developed to relate lateral (tangential) velocity and angular rate. The navigation processor calculates transport rate from lateral velocity divided by an estimate of the earth's radius plus the aircraft altitude. Transport rate is then subtracted from any gyro output using a process known as **Schuler tuning** (after the Austrian physicist Max Schuler, who solved the problem of accelerations due to the effect of ship manoeuvres on pendulum-based gyro-magnetic compass systems). Schuler tuning is achieved by feeding back aircraft rate terms such that the system is always aligned to the local vertical as the aircraft travels over the spherical earth.

15.3.3 Crew interface

A self-contained inertial navigation system (INS) is illustrated in Figure 15.6. With the integration of avionic systems, the navigation function is now undertaken by the flight management system (FMS), with the inertial system serving as an inertial reference sensor (IRS) as illustrated in Figure 15.11. (Flight management systems are described in the next chapter.) The IRS still retains a dedicated interface panel, the inertial system display unit or inertial reference mode panel depending on the aircraft type. Crew interface with the IRU is via the FMS control display unit (Figure 15.12) and inertial reference mode panel (IRMP) on the overhead panel, as shown in Figure 15.13.

The IRMP is used to initiate the alignment process before selection of the navigation mode.

Figure 15.12　FMS control display unit (CDU).

(NB Alignment must be achieved before moving the aircraft.) Four operating modes can be selected via the IRMP for each of the systems:

- Off
- Align
- Navigate
- Attitude.

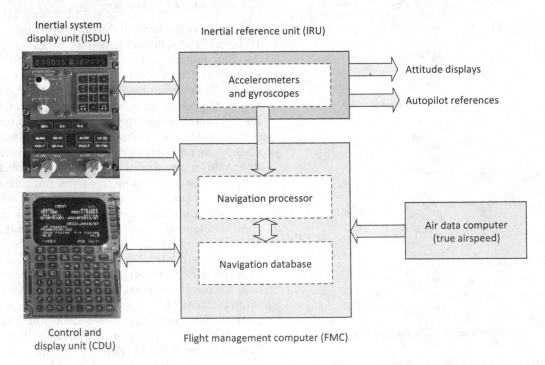

Figure 15.11　Inertial reference system (IRS)

Figure 15.13 Inertial reference mode panel (IRMP)

When the system is selected from off to align, the initialisation process is started. Present position is entered via the CDU; this is checked for accuracy within the IRU. When the system is aligned (see Section 15.4), the navigation mode can be selected. In the event of navigation computer failure, the IRU can be selected to provide attitude references only for the flight instruments. Four annunciators are provided for each of the systems. These provide the inertial reference system (IRS) status and fault indications, as illustrated in Table 15.1.

Table 15.1 IRS status and fault indications

Caption	Colour	Purpose
Align	White	IRU is in the align mode, initial attitude mode or powering down
On DC	Amber	IRU has switched to backup battery power
DC fail	Amber	DC power failure to the IRU
Fault	Amber	Built-in test has detected a failure, or certain alignment problems have occurred

The IRMP also has two alphanumeric displays; the data being displayed depends on what has been selected by a rotary switch. This displayed information is illustrated in Table 15.2. (A second rotary switch selects which system information is being displayed, e.g. left, centre or right.)

The IRMP normally displays data that has been entered via the control display unit (Figure 15.14); however, the IRMP's alphanumeric keyboard can also be used to enter data including latitude, longitude and magnetic heading.

Test your understanding 15.2

What is the difference between an RLG and an FOG?

Table 15.2 Inertial reference mode panel (IRMP) displays

Switchposition	Left display	Right display
TK/GS	Track angle	Groundspeed
PPOS	Latitude	Longitude
Wind	Wind angle	Wind speed
HDG	True heading	Blank

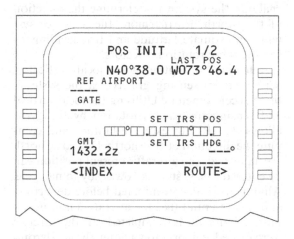

Figure 15.14 CDU position information

15.4 Alignment process

A fundamental requirement of inertial navigation is the initial alignment process; this is required to determine a **local vertical** and direction of **true north**. Alignment must be carried out with the aircraft on the ground and stationary. (Note that in certain cases, e.g. on the flight deck of an aircraft carrier, alignment has to be accomplished whilst the aircraft is moving. In this case, an external reference is required, e.g. the carrier's own inertial navigation system.) Systems using an electromechanical gimballed platform need this time for the gyros to level the platform with respect to the local vertical and align the platform in the direction of true north. For strap-down systems, there is no platform as such; however, the system still needs to establish a local attitude reference and direction of true north for navigation purposes.

For illustration purposes, a strap-down IRS using ring laser gyros (RLG) is described. With the aircraft's longitudinal axis lined up exactly with true north (Figure 15.15(a)), the roll RLG will sense an angular rate corresponding to the earth's rotation; the pitch RLG output (as a function of the earth's rotation) will be zero. If the aircraft's longitudinal axis were lined up exactly to the east (Figure 15.15(b)), the pitch RLG would correspond to the earth's rotation and the roll RLG (as a function of the earth's rotation) would be zero. Any aircraft position other than these two examples will provide **N–S and W–E components** of the earth's rotation, enabling the system to determine the direction of true north. Furthermore, the reference system can estimate latitude and true heading by sensing these rotational vectors.

Referring to Figure 15.16, local vertical is computed by sensing gravity via the system's three accelerometers. Utilising the local vertical and sensing the earth's rotation by the gyros allows the IRU to estimate latitude and compute the direction of true north. Once true north is established, the aircraft's present position can be entered; the system is now ready to navigate. Alignment is always initiated before departure, and it is essential that the aircraft is not moved until alignment is completed. If the aircraft were moved, e.g. by a towing tug, the accelerometers would measure this, thereby corrupting the sensing of a local vertical. A warning (flashing

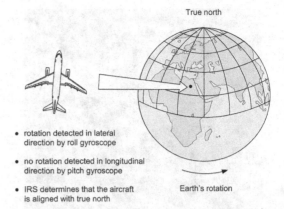

- rotation detected in lateral direction by roll gyroscope
- no rotation detected in longitudinal direction by pitch gyroscope
- IRS determines that the aircraft is aligned with true north

(a) Aircraft's longitudinal axis aligned with true north

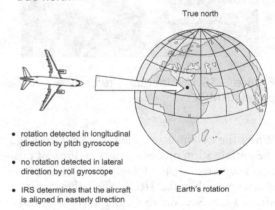

- rotation detected in longitudinal direction by pitch gyroscope
- no rotation detected in lateral direction by roll gyroscope
- IRS determines that the aircraft is aligned in easterly direction

(b) Aircraft's longitudinal axis aligned due east

Figure 15.15 Inertial system alignment

'align' light) is provided during the alignment mode to indicate if:

- present position has not been entered
- there is a significant difference between the position entered and the last known position
- the aircraft has been moved.

If any of these events occur, the entire alignment process would have to be started again, thereby causing a delay. The mode takes between 5 to 10 minutes to complete, depending on the system and latitude.

At the equator, earth rate is a maximum and the direction of true north can be determined relatively quickly. This process takes longer up

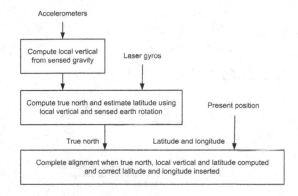

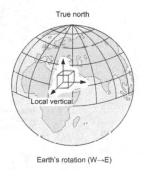

Figure 15.16 Computation of local vertical, true north and aircraft position by the navigation processor.

to latitudes of 70 degrees, above which system accuracy and performance are degraded.

Once aligned, the inertial navigation computer is always referenced to true north. It is therefore possible to establish the variation of the earth's magnetic field by reference to a look-up table in the computer's memory.

The reader will be aware from Chapter 8 that **magnetic variation** is the difference between true north and magnetic north; this variation depends on where the observer is on the earth's surface. Magnetic variation also changes over the passage of time and so the computer's memory must be updated on a periodic basis.

The **synthesised** magnetic variation that can be obtained from inertial navigation systems means that remote reading compass systems are not required, thereby saving weight and system installation costs.

By entering the origin airport's position prior to departure, and then calculating distance travelled, as shown in Figure 15.5, the navigation processor calculates the aircraft's present position and desired track to the destination at any given time. Waypoints can be entered into the memory for a given route, and the direction to that waypoint will be calculated and displayed. Additional information that can be supplied by the inertial navigation system is provided in Table 15.3 and illustrated in Figure 15.17.

Table 15.3 Navigation terminology

Term	Abbreviation	Description
Cross track distance	XTK	Shortest distance between the present position and desired track
Desired track angle	DSRTK	Angle between north and the intended flight path of the aircraft
Distance	DIS	Great circle distance to the next waypoint or destination
Drift angle	DA	Angle between the aircraft's heading and ground track
Ground track angle	TK	Angle between north and the flight path of the aircraft
Heading	HDG	Horizontal angle measured clockwise between the aircraft's centreline (longitudinal axis) and a specified reference
Present position	POS	Latitude and longitude of the aircraft's position
Track angle error	TKE	Angle between the actual track and desired track (equates to the desired track angle minus the ground track angle)
Wind direction	WD	Angle between north and the wind vector
True airspeed	TAS	Measured in knots
Wind speed	WS	Measured in knots
Ground speed	GS	Measured in knots

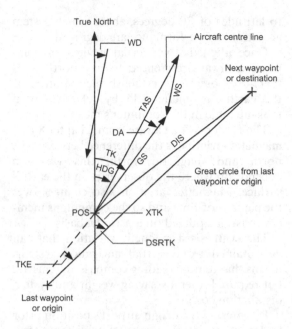

Figure 15.17 Navigation terminology.

Test your understanding 15.3

How does an inertial navigation system derive the magnetic variation?

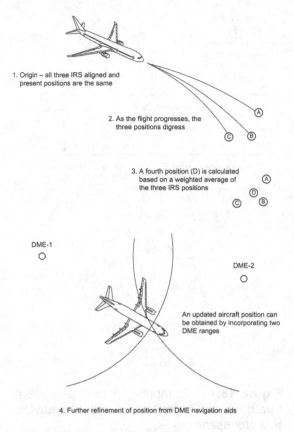

Figure 15.18 Position drift.

15.5 Inertial navigation accuracy

The accuracy of an inertial navigation system depends on a number of factors including the precision of the accelerometers and gyros and the accuracy of alignment with respect to true north and local vertical. Errors in the system will build up as a function of time; typical errors of between one and two nautical miles per hour should be allowed for; however, smaller errors can be achieved.

These errors are random and start to accumulate from the moment that the navigation mode starts. When three systems (A, B and C) are aligned at the origin, their present positions are identical. As the flight progresses, see Figure 15.18, the three positions digress. By combining the present positions of three on-board systems it is possible to derive an optimised position (D) within the triangle of positions.

By incorporating other navigation sensor outputs into the navigation computer, e.g. two

DME navigation aids, or global navigation sensors (see Section 15.7), it is possible to develop an updated and accurate position calculation. Furthermore, by comparing the position outputs of three on-board systems, we are also providing a means of error checking between systems. For example, if one system's position differs from the other two by a predetermined amount the crew can be alerted to this, and they might decide to deselect the system.

15.6 Inertial navigation summary

Inertial navigation has a number of advantages and disadvantages compared with other systems. The disadvantages include:

* the position calculation degrades with time (even if the aircraft is not moving) the equipment is expensive

- initial alignment is essential (this process is degraded at high latitudes, above 70 degrees)
- if the alignment process is interrupted, it has to be repeated, leading to potential delays.

The advantages of inertial navigation include:

- instantaneous velocity and position outputs
- autonomous operation, i.e. it does not rely on ground-based navigation aids
- passive operation, i.e. it does not radiate signals and cannot be jammed
- the system can be used on a global basis and is unaffected by the weather.

15.7 System integration

There have been several references in this chapter to stand-alone inertial navigation systems and those integrated with the flight management system. Many inertial systems are also integrated with global positioning systems and air data computers. An example of such a system is the global navigation air data inertial reference unit (GNADIRU). This provides a powerful and accurate (RNP 0.1) navigation system and overcomes the problem of accumulated position errors. The system integrates inertial and global navigation satellite system (GNSS) measurements to provide highly accurate aircraft position with the high navigation integrity. Inertial sensing is based on state-of-the-art fibre optic gyros and micro-electromechanical systems silicon accelerometers. The system also provides air data information such as altitude, airspeed, angle of attack and other air data parameters. (GNSS principles are described in Chapter 16.)

Other examples of system integration include the attitude heading reference system (AHRS) into a single line replaceable unit. (LRU). AHRS consist of laser or MEMS gyroscopes, accelerometers and magnetometers on all three axes. The AHRS will also incorporate integrated processing that calculates attitude and heading solutions. AHRS can also be integrated with air data computers to form an air data, attitude and heading reference system (ADAHRS). More

information on these systems is given in another book in the series, Aircraft flight instruments and guidance systems.

Test your understanding 15.4

What are the sources of error in an inertial system?

Test your understanding 15.5

List (a) three advantages and (b) three disadvantages of inertial navigation systems compared with other systems used for aircraft navigation.

15.8 Multiple choice questions

1. What output is produced from an accelerometer after the first integration process?
   (a) Acceleration
   (b) Velocity
   (c) Distance.

2. If the applied force on an accelerometer is maintained, the mass:
   (a) stays in the same position
   (b) moves in the direction of the force
   (c) returns to the neutral position.

3. During the 'align' mode, local vertical is sensed by:
   (a) accelerometers
   (b) gyros
   (c) the earth's rotation.

4. Establishing the orientation of true north is achieved by sensing:
   (a) local vertical
   (b) the earth's rotation
   (c) the earth's magnetic field.

5. Inertial navigation system errors are a factor of:
   (a) the aircraft's velocity
   (b) how long the system has been in the 'align' mode
   (c) how long the system has been in the 'navigation' mode.

6. Align mode is selected by the crew on the:
   (a) mode select unit
   (b) control display unit
   (c) inertial navigation unit.

7. During flight, with zero output from the accelerometers, the aircraft's ground speed and distance travelled are:
   (a) constant ground speed, increasing distance travelled
   (b) increasing ground speed, increasing distance travelled
   (c) decreasing ground speed, increasing distance travelled.

8. Magnetic north can be derived by an inertial reference system through:
   (a) knowledge of the present position and local magnetic variation
   (b) remote sensing of the earth's magnetic field
   (c) the earth's rotation.

9. Once aligned, the inertial navigation computer is always referenced to:
   (a) magnetic north
   (b) true north
   (c) latitude and longitude.

10. Errors in an inertial navigation system are:
    (a) random and build up as a function of time
    (b) fixed and irrespective of time
    (c) random and irrespective of time.

11. Magnetic variation depends on:
    (a) the location of magnetic north
    (b) the navigation computer's memory
    (c) where the observer is on the earth's surface.

12. Alignment of the inertial navigation system is possible:
    (a) at any time in flight
    (b) at any time on the ground
    (c) only when the aircraft is on the ground and stationary.

13. In 'attitude' mode, the inertial navigation system provides:
    (a) pitch and roll information
    (b) present position
    (c) drift angle.

14. The angle between the actual track and desired track is called:
    (a) track angle error
    (b) heading
    (c) drift angle.

15. During the alignment mode, a flashing align light indicates:
    (a) the system is ready to navigate
    (b) the aircraft was moved during align mode
    (c) the present position entered agrees with the last known position.

16. Referring to Figure 15.19, drift angle is defined as the:
    (a) difference between heading and ground track
    (b) angle between north and intended flight path
    (c) angle between north and the aircraft's flight path.

17. Referring to Figure 15.19, the shortest distance between the present position and desired track is:
    (a) desired track angle
    (b) ground track angle
    (c) cross track distance

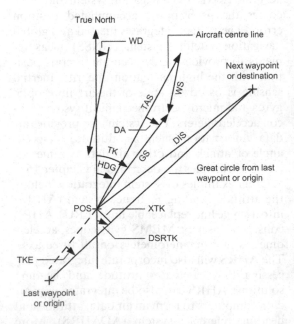

Figure 15.19   See Questions 16 and 17.

Chapter 16

Global navigation satellite systems

Global navigation satellite system (GNSS) is a generic reference for any navigation system based on satellites; the system in widespread use today is the United States' global positioning system (GPS). Other types of GNSS include the Russian global navigation satellite system GLONASS and European system Galileo. For the purposes of explaining the principles and operation of GNSS, in this chapter we will refer to GPS. The chapter concludes with a review of augmentation systems used to increase GPS accuracy, availability and integrity for aircraft navigation.

16.1 GPS overview

The US GPS was initiated in 1973 and referred to as Navstar (**navigation satellite with timing and ranging**). The system was developed for use by the US military; the first satellite was launched in 1978, and the full constellation was in place and operating by 1994.

GPS is now widely available for use by many applications including aircraft navigation. The system comprises a space segment, user segment and control segment. Twenty-four satellites (the **space segment**) in orbit around the earth send data via radio links that allow aircraft receivers (the **user segment**) to calculate precise position, altitude, time and speed on a 24-hour, worldwide, all-weather basis. The principles of satellite navigation are based on radio wave propagation, precision timing and knowledge of each satellite's position above the earth; this is all monitored and controlled by a network of stations (the **control segment**).

16.2 Principles of wave propagation

The reader will have witnessed the effect of sound wave propagation by observing lightning

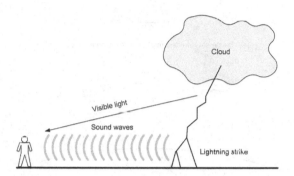

Figure 16.1 Delay in sound waves versus visible light

and thunder during an electrical storm. If the storm is some distance away, there is a time delay between seeing the lightning flash and then hearing the thunder, see Figure 16.1. This delay is caused by the difference in time taken for the light and sound to travel from the lightning to the observer. The same principle applies when an electromagnetic wave is transmitted, except that the wave is propagated at the speed of light, 3×10^8 m/s (in a vacuum).

16.3 Satellite navigation principles

This property of wave propagation can be exploited for satellite navigation purposes. In the first instance, we need to know the exact position of a satellite in orbit above the earth. When this satellite transmits a radio wave to an observer on the earth's surface, the **time delay** between when the radio signal was transmitted and received provides the means of calculating the spherical **range** between the satellite and observer. (Note that the term range is used here when defining the distance from a target object.)

Consider an observer located at a point somewhere on the earth's surface receiving radio waves from a satellite (Figure 16.2). The range

DOI: 10.1201/9781003411932-16

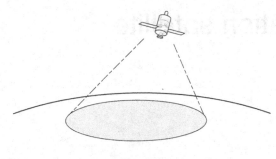

(a) Single satellite describes a circle on the earth's surface

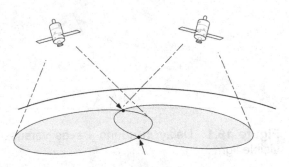

(b) Two satellites define two unique positions

Figure 16.2 Satellite ranging to determine position

between the satellite and observer can be determined by the principle described above; however, this same range can occur at any position described by a circle around the globe. We can reduce this ambiguity through basic geometry by taking range measurements from a second satellite; this will now identify one of two positions on the earth's surface. By using a third satellite, we can remove all ambiguity and define our unique **two-dimensional position** on the earth's surface. Furthermore, a fourth satellite can be used to determine a three-dimensional position, i.e. latitude, longitude and **altitude**. Accuracy of the system depends on having good visibility of these satellites to provide angular measurements. Once the user's position has been calculated, the GPS receiver can derive other useful navigation information, e.g. track, ground speed and drift angle.

16.4 GPS segments

Referring to Figure 16.3, the GPS comprises three segments: space, ground and user. (Note

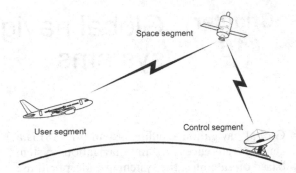

Figure 16.3 Global positioning system segments

that the GPS is being updated and modernized with next-generation satellites, alternative radio frequencies and higher specifications. The reader is encouraged to refer to recognised websites and relevant aircraft documentation for new developments.)

16.4.1 Space segment

There are typically 32 satellites in orbit; 24 are operational, and the others are used as backups. Each satellite is approximately 17 feet across (see Figure 16.4) and weighs approximately 2000 lb. The satellites are in orbit 10,900 nm (approximately 20,200 km) above the earth; this orbit provides optimum ground coverage with the least number of satellites. Each satellite is installed with four **atomic clocks** that are extremely accurate, typically maintaining accuracy within three nanoseconds (3×10^{-9} seconds) per day. (Four clocks are installed for backup purposes in the event of failure.) The satellites are powered by the sun's energy via solar panels; nickel cadmium batteries provide electrical power backup. Each satellite orbits the earth twice per day at an inclination angle of 55° with respect to the equatorial plane; there are six defined orbits, each containing four satellites. Figure 16.5 provides an illustration of these orbital patterns. The net result of this orbital pattern is that a minimum of five satellites should be in view to a receiver located almost anywhere on the earth's surface. Satellites have a finite operational life, typically five to ten years.

Satellites also download **almanac** data; this is a set of orbital parameters statuses for all satellites in the constellation. The receiver uses

Figure 16.4 Typical navigation satellite

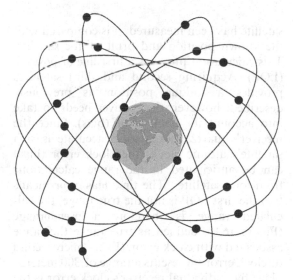

Figure 16.5 GPS space segment—six orbits, each with four satellites

almanac data during initial acquisition of satellite signals. **Ephemeris** data is also downlinked by each satellite; this data contains current satellite position and timing information.

16.4.2 Control segment

The control segment comprises one master control station (MCS) located at Schriever (formerly Falcon) Air Force Base in Colorado Springs, USA; five monitoring stations (located in Colorado Springs, Hawaii, Kwajalein, Diego Garcia and Ascension Island); and three ground

antennas (located on Ascension Island, Diego Garcia and Kwajalein). The locations of the monitoring stations provide ground visibility for each satellite.

Although each satellite's clock is very accurate, the relative timing between satellites gradually drifts over time. The individual clocks are monitored and synchronised mathematically relative to Coordinated Universal Time (UTC) by the master station. (UTC is the basis for the worldwide system of time.)

Key point

Three satellites are required to define a unique two-dimensional position on the earth's surface. A fourth satellite can be used to determine an aircraft's altitude.

Key point

The principles of satellite navigation are based on radio wave propagation, precision timing and knowledge of each satellite's position above the earth.

Test your understanding 16.1

How many satellites need to be in view to be able to calculate a two-dimensional position fix?

Each of the monitoring stations tracks all satellites in view; ranging data and satellite health information is collected on a continuous basis. This data is processed at the MCS to establish precise satellite orbits and to update each satellite with its ephemeris (orbital) data. Updated data is transmitted to each of the satellites via one of the ground antennas.

16.4.3 User segment

GPS installed on an aircraft comprises two receivers and two antennas located in a forward position on the top of the fuselage, see Figure 16.6. Antennas are typically flat devices, $7 \times 5 \times 0.75''$, with a single coaxial connector. Satellites

Figure 16.6 Location of GPS antennas

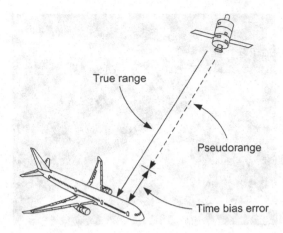

Figure 16.8 Illustration of pseudorange

that are less than 5° from the horizon are rejected as an inherent feature of the antenna's design. Other design features include the ability to reject signals that are reflected, e.g. from the sea, by rejecting incorrectly polarised signals. The antennas receive signals directly from whichever GPS satellites are visible, i.e. within line of sight.

GPS receivers are often incorporated into multi-mode receivers (MMR) along with other radio navigation systems. In this chapter we shall refer to this item simply as the 'receiver', remembering that different aircraft types will have different configurations of equipment. The receiver contains RF filters, a quartz clock (to reduce equipment costs versus atomic clocks) and a processor.

The receiver and satellite generate identical pulse coded signals at precisely the same time (Figure 16.7); these signals are compared in the receiver to provide the basis of time delay (Δt) measurements. When the time delay from the

satellite has been measured, it is compared with the known position and orbit of the satellite. This calculation provides a first **line of position** (LOP). Acquiring second and third satellites provides a unique position as previously described; however, the receiver needs to take into account its clock error (bias). Since the receiver's quartz clock is not as accurate as each satellite's atomic clock, the clock error (bias) can be anticipated in the range calculations from four satellites. The **time bias error** means that the first LOP is not the true range; the calculated range is therefore a **pseudorange** (Figure 16.8), defined as true range the range associated with clock error. (Every microsecond of clock error represents a range of 300 meters.) Since the individual receiver's clock error is the same with respect to any satellite, using four satellites defines a precise and unique position, as illustrated in Figure 16.9. Note that, since the satellites are in the order of 11,000 nm from the receiver and are all in different orbits, we need to know the exact position of each satellite via its ephemeris data (transmitted as part of the message code).

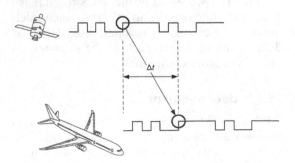

Figure 16.7 Pulse-coded signals

Test your understanding 16.2

How many GPS satellites are there and how are they arranged into orbits?

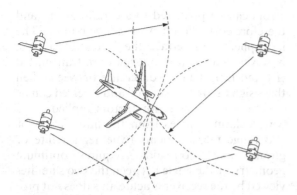

Figure 16.9 Pseudorange and position fixing with four satellites

Test your understanding 16.3

What is the purpose of the control segment?

Test your understanding 16.4

What is the difference between ephemeris and almanac data?

16.5 GPS signals

Each satellite transmits low power (20–50 watt) signals on two carrier frequencies: L1 (1575.42 MHz) and L2 (1227.60 MHz). Two carrier waves are transmitted so that the effects of refraction through the ionosphere can be compared between the two signals and corrections applied. These carrier frequencies are modulated with complex digital codes that appear like random electrical noise; these are called **pseudorandom** codes, and they are a fundamental part of a GPS. There are three sets of data to be modulated on the L1 and L2 carrier waves:

- Course acquisition (C/A) code
- Precise (or protected) P-code
- Navigation/system data.

The **coarse acquisition** (C/A) code is a pseudorandom string of digital data used primarily by commercial GPS receivers to determine the range of the transmitting satellite. The C/A code modulates the carrier wave at 1.023 MHz and repeats every 1 ms. The P-code (not available to civilian users) is modulated on both the L1 and L2 carriers at a frequency of 10.23 MHz. The P-code can be further encrypted as a Y-code to provide a high level of security for military users.

Data is exchanged between each satellite and the monitoring stations via uplink and downlink frequencies in the S-band (2227.5 and 1783.74 MHz respectively).

Key point

Pseudorandom noise (PRN) codes allow all satellites to transmit at the same frequency. Each satellite uses a unique PRN code; these carry the navigation message and are used for acquisition, tracking and ranging.

Key point

The European satellite-based augmentation systems (SBAS) are based on EGNOS and complement existing global navigation satellite systems (GNSS). SBAS improves accuracy, integrity, continuity and availability. This is achieved by measurements from reference stations; errors are then transferred to a computing centre, which calculates differential corrections and integrity messages. These are broadcast via geostationary satellites as an augmentation or overlay of the original GNSS message.

16.6 GNSS Operation

GNSS has various levels of operation depending on how many satellites are in view. Three satellites provide a two-dimensional position fix; four satellites or more are desirable for optimum navigation performance. The receiver seeks out atleast four satellites by monitoring their signal transmissions; this acquisition process takes about 15–45 seconds. To speed up the navigation process, the receiver can obtain an initial position fix from the inertial reference system; this allows the receiver to search for satellites that should be in view. In the event of poor satellite coverage for defined periods (typically less than 30 seconds), the system uses other navigation sensor inputs to enter into a dead reckoning mode. For prolonged periods of poor satellite reception, the system re-enters the acquisition mode.

16.6.1 Selective availability

Selective availability (SA) was a feature of GPS that intentionally introduced errors (typically 10 metres horizontally, and 30 metres vertically) into the publicly available L1 signals. This was a political strategy at the time that denied any advantage for hostile forces acting against the USA. The highest GPS accuracy was available (in an encrypted form) for the US military, its allies and US government users. During the 1990s, a number of political factors were mounting in the USA including a shortage of military standard GPS units during the 1990s Gulf War; the widespread availability of civilian products; the FAA's long-term desire to replace ground based radio navigation aids with GPS. This led to a presidential decision in 2000 allowing all users access to the L1 signal without the intentional errors.

16.6.2 GNSS vulnerability

GNSS signals are relatively weak at the receiver antenna, so are vulnerable to interference. Although conventional radio navigation aids can also be disrupted by interference, GNSS typically serves more aircraft simultaneously. The interference may affect wide geographic areas, in particular atmospheric effects. Navigation errors can arise from poor satellite visibility or less-than-optimum geometry from the satellites that are visible. Accuracy of ephemeris data (i.e. each satellite's positional information) is fundamental to the accuracy of the system. There are external effects that will affect the GNSS signal, introduce errors and subsequently affect accuracy. Multi-path ranging errors can be caused by reflections of the GNSS signals from mountains and tall buildings. Atmospheric conditions in the ionosphere and troposphere will affect GNSS signals; these

errors can be predicted to a certain extent and therefore correction factors can be built in. The ionosphere will refract the satellites' signals; however, since two frequencies are transmitted (L1 and L2), the time difference between when these signals are transmitted and received can be compared, and correction factors applied.

Calculating ranges from the intersection of two range measurements (whether satellite or ground navigation aids) requires optimum geometry. If the angle between the two satellites viewed by the receiver is acute, this does not provide an accurate position fix. In satellite navigation, this is referred to as geometric dilution of precision (GDOP). The closer two satellites are, when viewed from the aircraft, the greater is the GDOP. This dilution of precision (DOP) can be broken down into specific components:

- PDOP (position DOP based on geometry only)
- HDOP (horizontal contribution to PDOP)
- VDOP (vertical contribution to DOP)
- TDOP (range equivalent of clock bias)

Almanac data within the receiver, together with ephemeris data from the satellite, is used to assist the receiver in acquiring specific satellites for optimum geometry.

The aforementioned errors are all unintentional. There is, however, the ongoing concern of intentional interference known as 'spoofing', i.e. the deliberate attempt to disrupt GNSS signals. Aviation authorities are constantly testing the quality of GNSS signals and working on ways to mitigate such threats.

GNSS sensors are used by primary navigation systems, e.g., area navigation (RNAV) and flight management systems (FMS), as described in this book. Loss of the GNSS sensor can be mitigated by using other navigation systems, e.g., distance measuring equipment (DME); see automatic DME tuning in the FMS chapter.

16.6.3 GNSS integration

GNSS equipment can be used in isolation or with other avionic systems to provide differing levels of operation. For example, in general aviation (GA), GNSS receivers are often integrated

Figure 16.10 Integrated GPS, navigation and communication panel

with ILS–VOR and VHF communication systems, see Figure 16.10. This example is a self-contained, panel-mounted device. Graphics are used to provide a multifunction type display; text is displayed on the screen for selected frequencies, distances, bearings, etc.

16.7 GNSS evolution

16.7.1 European GNSS

Galileo is the European system, intended to be compatible with, but more advanced than, GPS or GLONASS. The system is based on 24 operational satellites plus 6 in-orbit spares, in a higher (23,000 km) orbit; the satellites form three orbital planes, each comprising 10 satellites. With this higher orbit is an increased time to circle the earth: 14 hours. The ground system comprises two control centres, five monitoring and control stations and five uplink stations. Initial services will be made available by the end of 2016, with system completion scheduled for 2020. As a further feature, Galileo is providing a global Search and Rescue (SAR) function based on the operational Cospas-Sarsat system (refer to Chapter 7 for details). The European Geostationary Navigations Overlay Service (EGNOS) is the first phase of Galileo. EGNOS utilises a network of ground stations and three geostationary satellites to provide increased

accuracy, integrity and reliability of any global satellite navigation system.

16.7.2 Russian GNSS

The Russian global navigation satellite system (GLONASS) features 24 satellites (21 active, 3 spare) orbiting at a lower altitude of 19,100 km in three orbital planes. The Russian defence organization owns the system, and civilian usage is managed by the Russian Space Agency. At the time of writing there is limited take-up of GLONASS outside of Russia for civilian applications compared with the worldwide acceptance and usage of GPS. Several satellites have exceeded their design life, thereby reducing system capability; these are being replaced on a progressive basis.

10.7.3 Other GNSS

Japan, China and India their own satellite navigation systems; other nations are either joining or forming partnerships. Some novel ideas include optimising orbits such that the satellite(s) remain visible over certain areas of the globe for longer periods to obtain maximum usage. Estimates vary, but it is conceivable that over 100 navigation satellites could be in orbit over the next 20 years. Since the original global navigation systems were established, there remains a political debate about the deployment of additional systems. This debate is fuelled by a number of factors, e.g. national security aspects (remember that GPS was originally established as a US military asset). From the commercial aviation perspective, if there are numerous systems in place, there is the added 'complication' of what equipment will be approved in each and every nation. The political and economic aspects of these factors are beyond the scope of this book; however, the reader is encouraged to monitor events through the press and other media.

16.8 GNSS augmentation

GNSS constellations require augmentation systems to meet ICAO performance requirements for en route, terminal, approach and landings.

A combination of on-board avionics and (where available) external references are implemented via:

- Aircraft-based augmentation system (ABAS)
- Satellite-based augmentation system (SBAS)
- Ground-based augmentation system (GBAS).

ABAS is the on-board avionics implementation that processes GNSS signals to achieve the accuracy and integrity required to support en route, terminal and non-precision approach (NPA) operations. SBAS uses a network of ground reference stations and geostationary earth orbit (GEO) satellites to augment en route navigation and approaches with vertical guidance. GBAS uses airport monitoring stations to process signals from GNSS constellations and broadcast corrections and approach path data to support precision approach and landing operations. GBAS also has the potential to support surface movement operations.

SBAS in geographic regions has the potential to support seamless guidance where their service areas overlap:

- USA Wide Area Augmentation System (WAAS)
- European Geostationary Navigation Overlay Service (EGNOS)
- Japan – MTSAT Satellite Augmentation System (MSAS)
- India – GPS and GEO Augmented Navigation System (GAGAN)
- Russia System for Differential Correction and Monitoring (SDCM).

16.8.1 Error detection

The GNSS receiver can be installed with error detection software known as receiver autonomous integrity monitoring (RAIM). Monitoring is achieved by comparing the range estimates made from five satellites. RAIM is a GNSS receiver function that performs a consistency check on all tracked satellites. This ensures that the available satellite geometry allows the receiver to calculate a position within a specified protection limit:

- Oceanic 4 nm
- En route 2 nm
- Terminal 1 nm
- Non-precision approaches 0.3 nm.

During oceanic, en route and terminal phases of flight, RAIM is available nearly 100% of the time. The prediction function also indicates whether RAIM is available at a specified date and time. Computations predict satellite co verage within ±15 minutes of the specified arrival date and time. Because of the tighter protection limit on approaches, there may be times when RAIM is not available. RAIM prediction must be initiated manually if there is concern over SBAS/WAAS coverage at the destination or some other reason that compromises navigation precision. If RAIM is not predicted to be available for the final approach course, the approach does not become active. If RAIM is not available when crossing the final approach fix (FAF), the missed approach procedure must be flown.

Key point

GNSS/FMS equipment incorporates a RAIM prediction calculator with fields for destination, arrival time and arrival date; these will produce a RAIM status value for the destination location.

RAIM requires redundant satellite range measurements to detect faulty signals and alert the pilot. If the GPS displays the 'RAIM UNAVAIL' alert during an approach and inside the final approach fix, the pilot should immediately abandon the approach and follow the non-GPS missed approach procedures.

Barometric aiding ('baro-aiding') is an integrity augmentation process whereby the GPS uses a non-satellite input source (e.g. the aircraft's pitot-static system) to provide vertical reference, thereby reducing the number of required satellites from five to four. Baro-aiding requires four satellites and a barometric altimeter input to detect an integrity anomaly. Barometric vertical navigation ('baro-VNAV') uses barometric altitude information from the aircraft's pitot-static system and air data computer to compute vertical guidance for the pilot. The specified vertical path is typically computed between two waypoints or an angle from a single waypoint.

Key point

The current altimeter setting may need to be manually entered into the receiver as described in the operating manual.

Key point

When using baro-VNAV guidance, pilots should check for any published temperature limitations on the approach chart which may result in approach restrictions.

In addition to RAIM, failed satellite(s) can be excluded from the GPS range estimates by comparing the data from six satellites. This technique is called fault detection and exclusion (FDE); the GPS continues to provide guidance. Aircraft typically carry dual systems; operators perform pre-flight predictions to ensure that there will be enough satellites in view to support the planned flight. This provides operators with a cost-effective alternative to inertial navigation systems in oceanic and remote airspace.

Key point

RAIM algorithms require a minimum of five visible satellites in order to perform fault detection, i.e. detecting an unacceptably large position error for a given mode of flight. FDE uses a minimum of six satellites, not only to detect a faulty satellite but also to exclude it from the navigation solution so that the navigation function can continue without interruption.

16.8.2 Augmented approaches

Localizer performance with vertical guidance (LPV) approaches in Europe utilise the satellite-based augmentation systems (SBAS)*, based on EGNOS, for improved accuracy, integrity, continuity and availability. This is achieved by measurements from reference stations; errors are then transferred to a computing centre, which calculates differential corrections and integrity messages. These are broadcast via geostationary satellites as an augmentation or overlay of the original GNSS message. LPV provides lateral and vertical guidance to provide an approach very similar to a Category I instrument landing system (ILS). As with ILS, LPV has vertical guidance and is flown to a Decision Altitude (DA).

*broadly equivalent to the US wide area augmentation system (WAAS)

16.8.3 Steep approach

Airports surrounded by mountains and/or built up areas cannot always use the standard three degrees glide slope, as described in Chapter 12 (ILS). Certain airports are approved for 'steep approach' capabilities using between 4.5 and 5.5 degrees pseudo glide slope, based on GPS coordinates.

16.8.4 Point in space

Point-in-space (PinS) procedures are being developed and introduced for helicopters. PinS is an approach procedure that includes both an instrument and a visual segment.

Approaches are flown from the final approach fix (FAF) up to the 'point in space'; the PinS also serves as the missed approach point (MAPt). The ICAO definition of MAPt is, "That point in an instrument approach procedure at or before which the prescribed missed approach procedure must be initiated in order to ensure that the minimum obstacle clearance is not infringed". From this point, the pilot can elect to fly the final segment, e.g. to a helipad on the roof of a hospital, or initiate a missed approach, see Figure 16.11. The main benefit of a PinS approach is the flexibility of selecting the MAPt.

16.9 GNSS – The future

The long-term intention of the aviation community is to rationalise the air traffic management through increased use of GNSS; this will be realised with the various augmentation systems discussed in this chapter and no doubt the additional GNSS constellations. There are programmes in place to eventually replace ADF

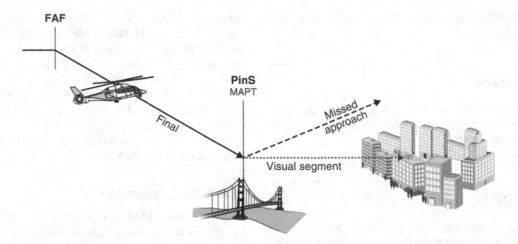

Figure 16.11 Point-in-space (PinS) approach

and VOR navigation aids. DME navigation aids will be retained for a longer period to optimise the existing network in support of PBN. Automatic approach and landing with GNSS will use to mitigate for satellite outage and/or disruption. With increased dependence on GNSS, the impact of any disruption is significant. The reader is encouraged to monitor the industry press for developments on this subject.

16.10 Multiple choice questions

1. Ephemeris data refers to the satellite's:
   (a) orbital position
   (b) current status
   (c) frequency of radio transmission.

2. GPS accuracy and integrity for en route operation can be increased by:
   (a) local area augmentation system (LAAS)
   (b) wide area augmentation system (WAAS)
   (c) differential GPS (DGPS).

3. The GPS orbital pattern is such that a minimum of how many satellites should be in view to a receiver?
   (a) five
   (b) four
   (c) three.

4. Selective availability is a feature of GPS that:
   (a) applies correction factors to known causes of error
   (b) intentionally introduces errors
   (c) determines which users can receive signals.

5. In the event of poor satellite coverage, the system:
   (a) automatically selects other navigation sensors and enters into a dead reckoning mode
   (b) continues using the same satellites
   (c) automatically selects other satellites.

6. To speed up the satellite acquisition process, the aircraft receiver can obtain an initial position fix from the:
   (a) flight management system
   (b) internal clock
   (c) inertial reference system.

7. The deliberate attempt to disrupt GPS signals is known as:
   (a) spoofing
   (b) selective availability
   (c) satellite acquisition.

8. Fault detection is achieved by comparing the position calculations made from how many satellites?

(a) five
(b) four
(c) six.

9. The GPS navigation concept is based upon calculating satellite:
   (a) speed
   (b) altitude
   (c) range.

10. Multipath reflections of GPS signals are caused by:
    (a) mountains and tall buildings
    (b) atmospheric conditions
    (c) poor satellite visibility.

11. The local area augmentation system (LAAS) provides integrity messages to the aircraft via:
    (a) geostationary satellites
    (b) VHF datalinks
    (c) the GPS satellites.

12. How many GPS satellites need to be in view to be able to define a unique two-dimensional position on the earth's surface?
    (a) two
    (b) three
    (c) one.

13. Failed satellite(s) can be excluded from the navigation calculations by comparing the data from how many satellites?
    (a) four
    (b) five
    (c) six.

14. During prolonged periods of poor satellite reception, the aircraft receiver:
    (a) enters into a dead reckoning mode
    (b) re-enters the acquisition mode
    (c) rejects all satellite signals.

15. In the diagram shown in Figure 16.12, which feature represents the control segment?
    (a) A
    (b) B
    (c) C.

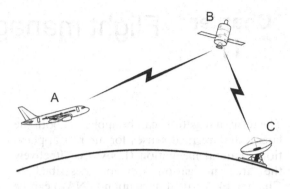

Figure 16.12 See Question 15

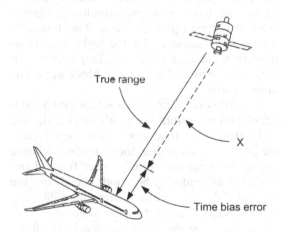

Figure 16.13 See Question 16

16. In the diagram shown in Figure 16.13, X represents the:
    (a) actual range
    (b) pseudo range
    (c) distance error.

17. GPS satellites occupy orbits at a typical altitude of:
    (a) 20,000 km
    (b) 120,000 km
    (c) 200,000 km.

18. GPS transmissions are in which radar band?
    (a) C-band
    (b) L-band
    (c) X-band.

Chapter 17 — Flight management systems

The term 'navigation' can be applied in both the lateral and vertical senses for aircraft applications. Lateral navigation (LNAV) is effectively the area navigation function described in Chapter 14. Vertical navigation (VNAV) can be used to supplement or replace approach and landings using radio navigation aids; it can also be used to optimise the performance of the aircraft to reduce operating costs. This has been traditionally achieved by the flight crew (particularly the flight engineer) making reference to data contained within charts, tables and performance manuals.

Aircraft performance data is based on a number of factors including aircraft weight, altitude and outside air temperature. Since these factors are constantly changing, the task of calculating optimum engine thrust limits, aircraft speed and altitude has gradually been automated with the advent of digital systems. During the 1980s, lateral and vertical navigation functions were combined into a single system known as the flight management system (FMS). Various tasks previously performed by the crew can now be automated with the intention of reducing crew workload. In this chapter we will review the principles of FMS and explore some the key features and benefits.

17.1 FMS overview

The flight management system (FMS) combines area navigation and performance management into a single system. The two primary components of the system are the flight management computer (FMC) and control display unit (CDU). Primary aircraft interfaces with the FMC are the inertial reference system and automatic flight control system, including the autothrottle. Flight management systems were introduced at a time of rising operating costs; the contributing factors to these costs include fuel and time. The cost of fuel is self-evident; the

cost of time includes aircraft utilisation, e.g. if the aircraft is being leased on a cost per flying hour basis. Reducing aircraft speed will decrease fuel burn, but this leads to a longer flight time and increased 'cost of time'. Flying faster will reduce the cost of time but increase fuel burn.

Four-dimensional navigation (4D navigation) is possible with flight management systems. The aircraft's latitude, longitude, altitude and arrival time requirements can be planned, calculated and subsequently predicted on an ongoing basis. Each airline will have its own financial model in terms of fuel and time costs; the FMS can be customised accordingly and expressed as a **cost-index**; this is entered into system within the range 0–100 to represent the extremes of minimum fuel through to minimum time. In order to perform the key functions of area navigation and performance management, the system interfaces with many other systems on the aircraft.

Flight management systems were the first examples of integrated multi-mode avionics. On transport category aircraft, the FMS integrates many systems including radio navigation systems, inertial navigation systems, global positioning systems, and centralised maintenance monitoring.

17.2 Flight management computer system (FMCS)

The two primary components of the system are the FMC and CDU; these are a subset of the FMS referred to as the flight management computer system (FMCS).

17.2.1 Flight management computer

The FMC contains an operational program, navigation database and performance database. We have already come across the navigation database (NDB) in Chapter 14. The FMC's

DOI: 10.1201/9781003411932-17

Table 17.1 Navigation database

Content	Details
Radio navigation aids	VOR, DME, VORTAC, ADF identification codes, frequencies, locations, elevations
Waypoints	Names and locations, pre-planned within company routes
Airports and runways	Locations, ILS frequencies, runway identifiers, lengths
Standard instrument departures (SIDs)	Published departure procedures including altitude restrictions
Standard terminal arrival routes (STARs)	Published arrival procedures including altitude restrictions
En route airways	Navigation aid references, bearings, distance between navigation aids
Holding patterns	Fix point, inbound course, turn direction
Company routes	A combination of all the above, as specified by the airline

navigation database (see Table 17.1) is a comprehensive version of what has already been discussed in area navigation systems.

Emerging avionic technology allows, inter alia, updating the navigation database (NDB) remotely via cloud connectivity. This could be typically achieved via an electronic flight bag (EFB).

The **performance database** (PDB) contains a detailed model of the aircraft's aerodynamic characteristics. This includes the aircraft's speed and altitude capabilities together with operating limits for both normal operation and abnormal conditions, e.g. engine failure. Engine parameters are also stored in the PDB, these include fuel flow and thrust models for the type of engine installed on the aircraft. Note that aircraft can be certified to fly with more than one engine type; these are all stored in the PDB.

An important feature of the FMC are the **program pins**, also known as "straps". Rather than

producing many different FMC software configurations for each aircraft type and each engine combination, one FMC part number can be installed with software covering a number of aircraft and engine types in the PDB. The FMC (like most avionic computers) is installed in the equipment rack and connects to the wiring looms via pins/sockets at the rear of the computer.

Program pins, see Figure 17.8 are used to select various software options within the computer; these connections are to ground, 28 V DC power supply or not connected. Logic circuits inside the computer are thereby set into predetermined configurations depending on how the program pins are configured. For example, a program pin could be connected to ground for one engine type, and set to 28 V DC for another engine type. When the FMC is installed, it effectively recognises which engine type is installed and the relevant engine software is used. The same FMC installed on another aircraft with different engine type will recognise this via the program pin(s) and utilise the relevant engine software.

Certain functions are fixed and cannot be changed, e.g. the aircraft type/model. Other program pins are airline options; examples of these options are the use of metric or imperial units, e.g. Centigrade or Fahrenheit, pounds (×1000) or kilograms (×1000).

17.2.2 Control display unit

The CDU is the primary interface between the crew and FMC. It is designed such that data entry and displays are in the language used by ATC. The location of a CDU on a typical transport aircraft is shown in Figure 17.1. The CDU comprises a variety of features, referring to Figure 17.2 these include the:

- data display area
- line-select keys (LSK)
- function and mode keys
- alpha numeric key pad
- warning annunciators.

The display area is arranged in the form of chapters and pages of a book. When the system is first powered up, the CDU displays the IDENT page, see Figure 17.3.

Figure 17.1 Location of FMCS control and display unit

The 'IDENT' page contains basic information as stored in the FMC including aircraft model, engine types, etc. Other pages are accessed from this page on a menu basis using the line-select keys, or directly from one of the function or mode select keys.

17.3 System initialisation

Before the system can be used for lateral and vertical navigation, the FMC needs some basic initialisation data. Certain information required by the system has to be entered by the crew; other information is stored as a default and can

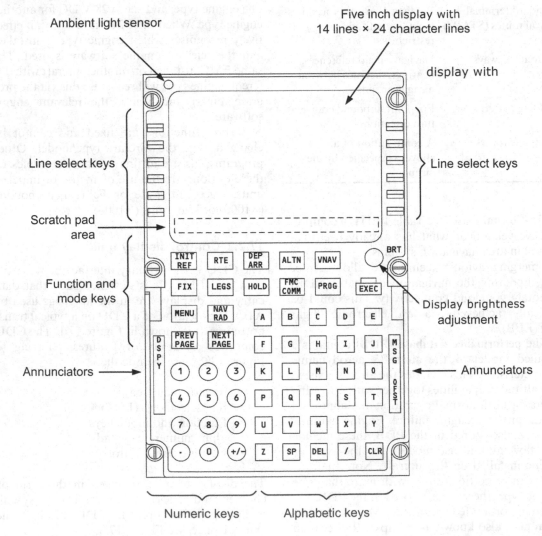

Figure 17.2 Typical FMCS CDU features

Figure 17.3 Typical CDU keyboard and display variations

be overwritten by the crew. To simplify the process, To simplify the process, box prompts indicate where essential information required by the system, e.g., present position and gross weight. Dash prompts indicate where additional information can be entered that will optimise the system, e.g., wind forecasts and outside air temperature. There are a number of ways that individual pages can be accessed, and there is a variety of information on each page. The description below illustrates an initialisation procedure, starting with position initialisation through to performance initialisation, following a logical process. During this initialisation description, we will be making reference to **fields**; these are specific areas on the CDU screen where data is either displayed and/or entered. In the following text, each **line-select key** (LSK) will be referred to by its location: left/right of the display and 1–6 from top to bottom.

17.3.1 Position initialisation

When the system is powered up, the IDENT page is displayed, see Figure 17.4 (more details about this page are provided after this section). Pressing LSK-6R, identified POS INIT for

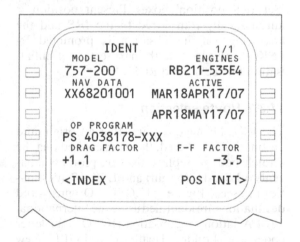

Figure 17.4 IDENT page-POS INIT prompt

position initialisation, displays the POS INIT page, see Figure 17.5. The information needed at this point (indicated by box prompts) is present position for inertial reference system (IRS) alignment. Position can be entered in a number of ways, but let's assume at this point that we want to load present position by manually keying in latitude and longitude. Using the alphanumeric keys, latitude and longitude are entered

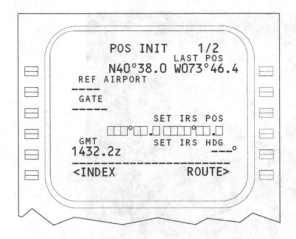

```
        POS INIT      1/2
                LAST POS
          N40°38.0 W073°46.4
   ⊏  REF AIRPORT                ⊐
      ----
   ⊏  GATE                       ⊐
      -----
                SET IRS POS
   ⊏  □□□°□□.□ □□□□□°□□.□        ⊐
      GMT       SET IRS HDG
   ⊏  1432.2z            ---°    ⊐
      ---------------------------
   ⊏  <INDEX           ROUTE>    ⊐
```

Figure 17.5 Position initialisation (POS INIT) page

via the key pad, entries appear in the bottom of the display (referred to as the **scratch pad**). When this data is confirmed in the scratch pad, LSK-4R (adjacent to the position boxes) is pressed and the scratch pad data replaces the 'Set IRS position' boxes. Present position is automatically transferred to the IRS and the next stage of initialisation is prompted by LSK-6R; this leads to the next section of initialisation for the desired ROUTE.

17.3.2 Route selection

The ROUTE page requires that an origin and destination be entered; these are entered (via the scratch pad) to replace the box prompts adjacent to LSK-1L (origin) and LSK-1R (destination), see Figure 17.6(a). Origins and destinations are identified using the International Civil Aviation Organisation (ICAO) four-letter codes, e.g. London Heathrow is EGLL, New York Kennedy international airport is coded KJFK. This system is used in preference to the International Air Transport Association (IATA) three-letter codes, e.g. LHR and JFK, since some of these three-letter codes are duplicated for some airfields. Note that most airlines have predetermined company routes, these are stored in the navigation database and can be entered (as a code) via LSK-2L. There may be more than one route between the origin and destination; when the company route code is entered

into an appropriate field, this will automatically enter the origin and destination together with all en route waypoints. Specific departure details, e.g. runway and initial departure fix, can also be contained within the company route as illustrated in Figure 17.6(b). Once the route is activated (LSK-6R), the bottom right field changes to PERF INIT for **performance initialisation**.

17.3.3 Performance initialisation

The system requires gross weight (GW) or zero fuel weight (ZFW), reserve fuel, cost index and cruise altitude. Required entries are indicated as before with box prompts, see Figure 17.7. Note that since the total fuel onboard (52.3 tonnes in this example) is known by the FMC (via an input from the fuel quantity system) entering ZFW will automatically calculate GW and vice versa. Cost index can be entered manually, or it may be contained within the company route. All other entries on the page are optional; entry of data in these fields will enhance system performance. Once the performance initialisation details are confirmed, the system is ready for operation. Further refinement of the flight profile can be made by entering other details, e.g. take-off settings, standard instrument departures, wind forecasts, etc.

17.4 FMCS operation

The flight management computer system (FMCS) calculates key performance data and makes predictions for optimum operation of the aircraft based on the cost-index. We have already reviewed the system initialisation process, and this will have given the reader an appreciation of how data is entered and displayed. The detailed operation of a flight management system is beyond the scope of this book; however, the key features and benefits of the system will be reviewed via some typical CDU pages. Note that these are described in general terms; aircraft types vary and updated systems are introduced on a periodic basis. CDU pages can be accessed at any time as required by the crew; some pages can be accessed via the line-select keys as described in section 17.3; some pages are accessed

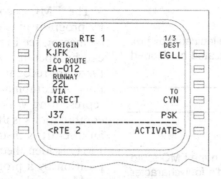

(a) Route page – origin and destination entered

Runway, first waypoint (CYN)
and airway (J37) to second
waypoint (PSK)

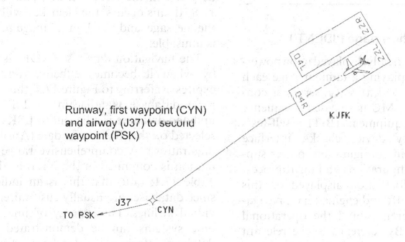

(b) Route page – departure details

Figure 17.6   Route (RTE) page details

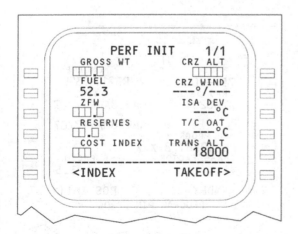

Figure   17.7 Performance   initialisation
(PERF INIT) page

via function/mode keys. The observant reader
may have already noticed that in the top right of
each CDU page is an indication of how many
sub-pages are available per selected function.

Key point

The FMS comprises the following subsystems:
FMCS, AFCS and IRS.

Key point

The page automatically displayed upon FMC
power-up is the identification page; this confirms
that the FMC has passed a sequence of self-tests.

Key point

Required entries into the CDU are indicated by box prompts; optional entries are indicated by dashed line prompts.

Key point

To define the destination airport on the FMC route page requires entry of the airfield's four-character identifier.

17.4.1 Identification page ('IDENT')

This page is automatically displayed upon power-up; aside from displaying a familiar page each time the system is used, this also serves as confirmation that the FMC has passed a sequence of built-in test equipment (BITE) self-tests including: memory device checks, interface checks, program pin configuration, power supplies, software configurations and microprocessor operation. Information displayed on this page includes aircraft and engine types, navigation database references and the operational program number. By reference to the relevant aircraft documentation, one FMC part number could be fitted to a number of different aircraft types. Each aircraft type will have different aerodynamic characteristics and these differences will be stored in the FMC's memory.

The FMC recognises specific aircraft types by program pins contained within the aircraft connector, see Figure 17.8. Given aircraft types can be operated with different engine models; these are recognised by using specific program pins. A combination of different connections representing various aircraft types and engine combinations is illustrated in Table 17.2. When an aircraft is installed with a dual FMS, each FMC is identified as being in the left or right position by a program pin. If there is a wiring problem to these program pins, a message will be given in the CDU scratch pad, see Figure 17.8. If this occurs, the clear key will not erase the message, and the flight management system is unusable.

The navigation database (NDB) is identified by when it becomes effective, and when it expires. Referring to Figure 17.4, the active (current) database is adjacent to LSK-2R. The updated database is adjacent to LSK-3R; this is selected on the changeover date (April 18 in this illustration). A comprehensive range of information is contained in the NDB as detailed in Table 17.1; note that this is an indicative list since databases are usually customised for individual airlines. The synergy of integrated avionic systems can be demonstrated by FMC database information also being displayed on the EHSI (Figure 17.9 is displaying a number of airports contained in the database).

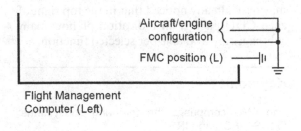

Flight Management
Computer (Left)

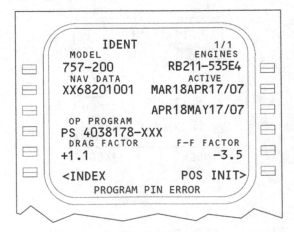

Figure 17.8   Program pin schematic

Table 17.2

Aircraft/engine configuration	Connector pins		
	0 = Pin connected to ground		
	1 = Pin connected to 28 VDC		
	A	B	C
Aircraft type 1, engine type 1	0	0	0
Aircraft type 1, engine type 2	0	0	1
Aircraft type 2, engine type 1	0	1	0
Aircraft type 2, engine type 2	0	1	1
Aircraft type 3, engine type 1	1	0	0
Aircraft type 3, engine type 2	1	0	1
Aircraft type 4, engine type 1	1	1	0
Aircraft type 4, engine type 2	1	1	1

17.4.2 Progress page

There are many pages available to the crew for managing and modifying data required by the system depending on circumstances. One of the pages used to monitor key flight information is the progress page, see Figure 17.10. By describing the information on the progress pages, the reader will gain an appreciation of the features and benefits of the flight management computer system.

The progress page can be accessed via the PROG key on the CDU. There are no entries

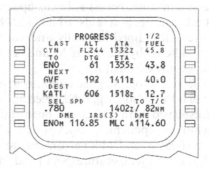

(a) Progress page (page 1/2)

(b) Progress page (page 2/2)

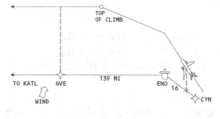

(c) Flight profile

Figure 17.10 FMCS progress pages and flight profile

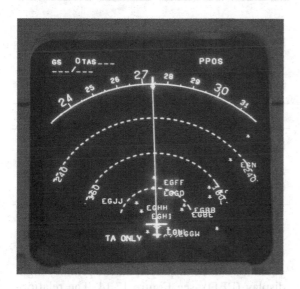

Figure 17.9 EHSI displaying nearby airports

required on this page; it is used for information only. The top line of the page displays details for the previous waypoint (CYN) in the active flight plan; name, altitude, actual time of arrival and fuel. The next three lines display details for the active waypoint (ENO), next waypoint (GVE) and final destination (KATL). Details include: distance to go (DTG), estimated time of arrival (ETA) and predicted fuel. The fifth line gives selected speed, predicted time and distance to an altitude change point, e.g. top of climb (T/C) as illustrated in Figure 17.10. The last line of the page is providing navigation source information. In this case, the FMC selected inertial reference system (number 3) is being updated by two DME navigation aids ENO and MLC; these are being tuned manually and automatically as indicted by the letters M and A next to the navigation aid identifier.

The second progress page contains a variety of useful information, e.g. wind speed and direction (displayed with associated head, tail and crosswind components), cross track (XTK) error, vertical track (VTK) error, true airspeed (TAS), static air temperature (SAT) and various fuel quantity indications.

17.4.3 Legs page

Figure 17.11 provides an illustration of how en route lateral and vertical profiles are integrated within the FMC. In this example, the aircraft is flying towards waypoint CYN on a track of 312°. There is an altitude constraint of 6000 feet over CYN, climb speed is 250 knots.

Altitude constraints are associated with waypoints and are characterised in several ways, see Figure 17.11(c) for a typical profile from the top of descent (T/D):

- "At" the altitude
- "At or above" the altitude
- "At or below" the altitude
- Between two given altitudes, i.e., a window

This combined lateral and vertical profile is depicted by the tracks, distances, speeds and altitudes for each waypoint. This level of detail also applies for **standard instrument departures** (SIDs) and **standard terminal arrival procedures** (STARs), see Figures 17.12 and 17.13.

17.4.4 Other CDU pages

A detailed review of every page available on the CDU is beyond the scope of this book; however, a summary of typical pages is provided in Table 17.3. Note that this table is provided for illustration purposes. Aircraft types vary together with the type and model of FMC installed.

Test your understanding 17.1

(a) What is the meaning of four-dimensional navigation?
(b) How can you confirm that the FMC has passed its power-up test?
(c) What is the significance of box and dash prompts on the CDU?

Test your understanding 17.2

What is the purpose of program pins?

Test your understanding 17.3

Where would you confirm details of each of the following: navigation database, operational program, aircraft and engine type?

Key point

The highlighted waypoint on the progress page is the active waypoint.

Key point

Alerting messages require attention from the crew before guided flight can be continued.

17.4.5 LNAV and VNAV guidance

During flight, the FMC's lateral and vertical guidance can be displayed on the primary flight display (PFD), see Figure 17.14. The relationship between ANP, RNP and deviations from the desired lateral and vertical flight paths need

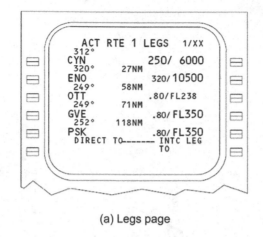

(a) Legs page

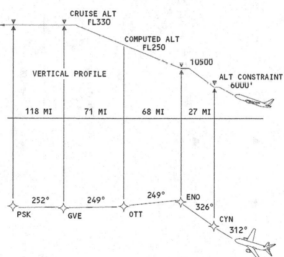

(b) Lateral and vertical profiles

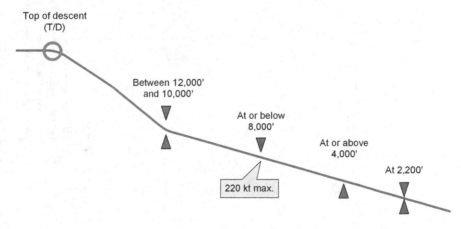

Figure 17.11   Legs page and associated flight profiles

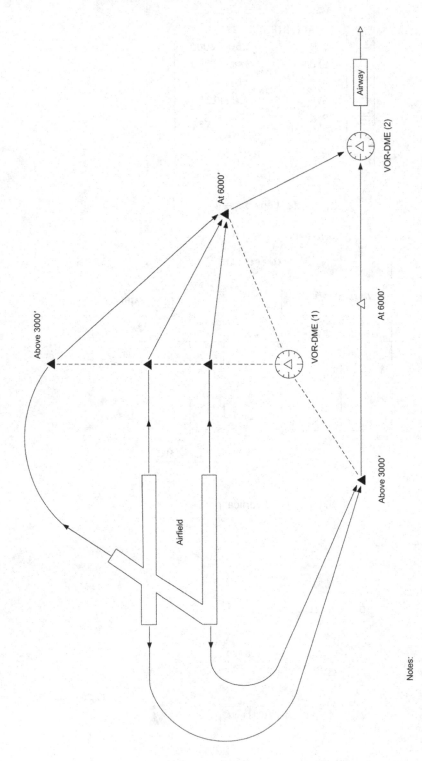

Notes:

1. In this illustration, each of the three runways has a specific departure route to the VOR-DME (2) navigation aid; the aircraft then joins the airways network
2. The SIDs are typically referenced to navigation aids, e.g. VOR-DME or marker beacons
3. There would also be published departure routes for aircraft joining airways to the south, east and north
4. Reporting points (triangles) are often specified with altitude constraints, e.g. at, below or above 3000'

Figure 17.12   Standard instrument departure (SID)

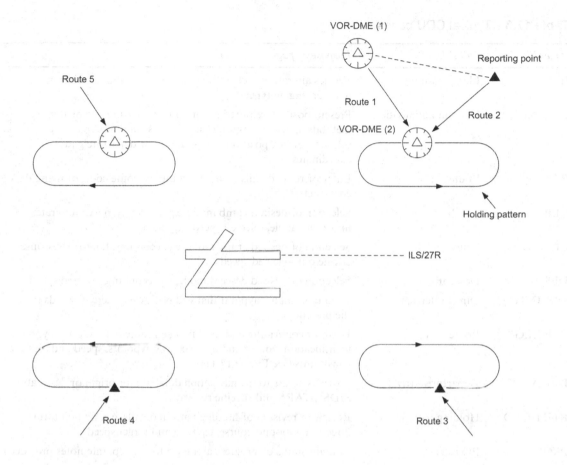

Notes:

1. In this illustration, each of the three arrival routes is associated with a navigation aid (VOR-DME) and reporting point (solid triangles)
2. Each arrival route is normally allocated a holding pattern
3. Minimum sector altitudes are published for each route
4. When cleared by ATC, the aircraft would leave the holding pattern and be given a heading to join the ILS for the active runway, e.g. 27R

Figure 17.13 Standard terminal arrival routing (STAR)

to be displayed to the flight crew. The typical presentation is via LNAV and VNAV deviation scales and pointers, similar to what has already been described in previous chapters for radio navigation systems, e.g., glide slope and localiser.

Key features of this typical display:

- FMC SPD = FMC calculated speed
- LNAV = Lateral guidance
- VNAV PTH = Vertical guidance
- ANP = Actual navigation performance
- RNP = Required navigation performance

NB The actual displays depend on modes selected by the crew, and will vary depending on aircraft type, and equipment installed. RNP progress can also be displayed on the FMS CDU.

17.5 General Aviation FMS

General Aviation FMS include touch screen navigators, see Figure 17.14. Each provides SBAS/LPV precision navigation and are designed to meet the accuracy and integrity requirements for ADS-B as part of the Next

Table 17.3 Typical CDU pages

Page title	Full title	Purpose of page
IDENT	Identification	Verifies aircraft model, active database, operational program number, engine type(s)
POS INIT	Position initialisation	Present position required by entering data using one of three methods: latitude/longitude coordinates via the keypad, line selection of last position, line selection of departure gate coordinates
RTE	Route	Entry of route details, either by company route code, or manual construction
CLB	Climb	Selection of desired climb mode, e.g. economy, maximum rate, maximum angle, selected speed, engine out
CRZ	Cruise	Selection of desired cruise mode, e.g. economy, long-range cruise, engine out, selected speed
DES	Descent	Selection of desired descent mode, e.g. economy, selected speed
DIR INTC	Direct intercept	Used to select a waypoint that will be flown directly towards from the present position
RTE LEGS	Route legs	Used for confirming and modifying en route details, e.g. waypoint identification, course and distance to waypoints, speed and altitude constraints (see Figure 17.11)
DEP ARR	Departure/Arrival	Provides access to the navigation database for origin or destination SIDS, STARS and specific runways
RTE HOLD	Holding pattern	Review or revision of holding pattern details, e.g. fix point, turn direction, inbound course, leg time and target speed
PROG	Progress	In-flight status of progress along route (see separate notes provided)
N1 LIMIT	N1 limit	The N1 limit is automatically selected and controlled by the FMC. This page provides a range of manually selected N1 limit options including go-around, maximum continuous, climb and cruise
FIX INFO	Fix	Used to create fix points on the current flight leg from known waypoints using radials and distances from the waypoint

Generation airspace initiative (refer to Chapter 19 for more on these subjects)

In addition to the dedicated knobs and buttons that many pilots prefer for frequent pilot actions, the IFD Hybrid Touch capability allows pilots to perform virtually all of those same functions via the touch screen interface. Additionally, the IFD range offers Multi-Touch functionalities such as pinch-zoom, map panning, and graphical flight plan editing. This can be done with the Multi-Touch screen, or knobs and buttons depending on the pilot's phase of flight. This provides ease-of-use, especially during single-pilot IFR operations.

Graphical flight planning capability allows for easy editing of the flight plan with the touch of the screen. The convenient 'rubber banding' feature, Figure 17.15, allows the pilot to "stretch" any leg in the flight plan, e.g. to deviate for weather or to accommodate an amendment from ATC.

17.6 Four-dimensional (4D) navigation

Four-dimensional (4D) navigation is a key air traffic control function performed by the flight management system. In 4D navigation, the aircraft's latitude, longitude, altitude and required

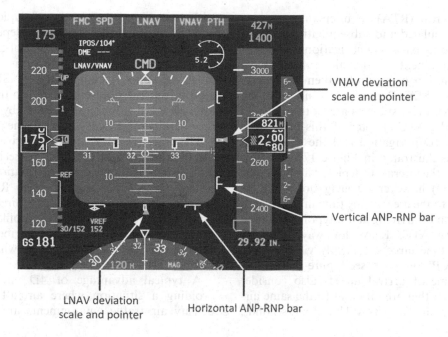

Figure 17.14   PFD with lateral and vertical guidance

Figure 17.15   Integrated multifunction display with interactive features

time of arrival (RTA) requirements are pre-planned, calculated and subsequently predicted on an on-going basis. 4D navigation is a fundamental component of the air traffic control (ATC) evolution being undertaken via several programmes (SESAR in Europe and NextGen in the USA); these subjects are both discussed further in the ATC chapter of this book. The concept of 4D navigation and the subsequent trajectory is illustrated in Figure 17.16. In this illustration, the aircraft is depicted at the top of descent (T/D), however 4D navigation and RTA is available to the crew at any time in flight, with the intention of arriving at a waypoint at a time specified by ATC. RTAs for waypoints are entered as time targets, typically via the FMS CDU, RTA PROG page, see Figure 17.17.

These time of arrival targets also consider other aircraft that are inbound to the same airport. Typical time targets will be +/− 30 seconds,

with separation between aircraft typically fixed at 90 seconds, or allowed to vary depending on prevailing conditions. NB the time of arrival targets will vary depending on the air navigation air navigation service provider (ANSP). Winds aloft forecasts will also be added to the descent forecast page, Figure 17.18, thereby updating the lateral and vertical control of the aircraft.

The objective is for aircraft to arrive at the 4D fix, or merging point, within a specified time, with a statistical time error distribution as illustrated. Some FMS versions allow RTA to be specified as before or after a required time of arrival. Monitoring of the flight profile is via the progress page as previously described. If the RTA is not achievable, a message will be given via the CDU.

A typical advantage of 4D navigation is avoiding a situation where aircraft in high-density air traffic environments are "rushed"

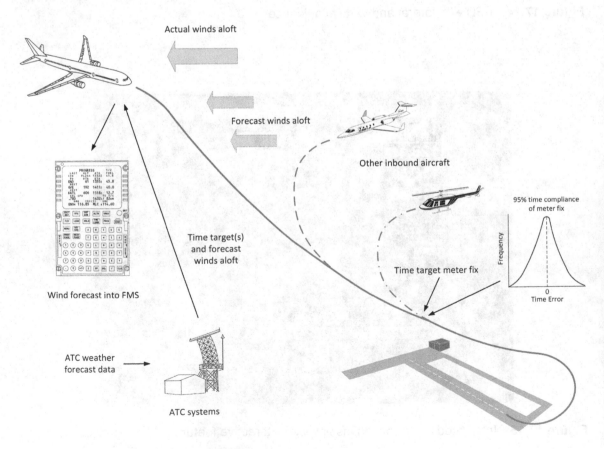

Figure 17.16 4D navigation concept

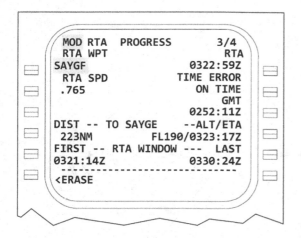

Figure 17.17 Progress page-RTA

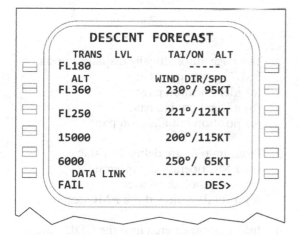

Figure 17.18 Descent forecast page – winds aloft

into arrival holding patterns. RTA flights will be more predictable enabling increase efficiency in the terminal areas.

17.7 Automatic DME tuning

The FMS will select VOR and DME navigation aids from its data base (based on the present position). DME-DME position is determined by the intersection of two DME arcs as previously described in the area navigation chapter

of this book. This position is used in the FMC position calculations. The automatic DME scanning function, Figure 17.19 will search nearby stations using the following criteria:

- The best DME geometry
- Distance to the DME station
- The navigation aid's service volume (signal strength)

The identifiers of DME stations currently providing updates to the FMS's navigation solution are displayed on the CDU. Depending on the system installed, this can be on the selected navigation aids page, Figure 17.20, or position reference page on other systems.

The FMS will use GPS as the primary navigation sensor, however if GPS data becomes unavailable due to an outage or spoofing, it will use VHF radio navigation or IRS (if installed).

17.8 FMS summary

As we have seen, the FMCS performs all the calculations and predictions required to determine the most economical flight profile, either for minimum fuel, or minimum time (or indeed some point in between depending on the operator's financial and commercial models).

Test your understanding 17.4

What are the four types of altitude constraints?

Test your understanding 17.5

Explain the principles of DME scanning, and the criteria used for updating the FMC position calculations.

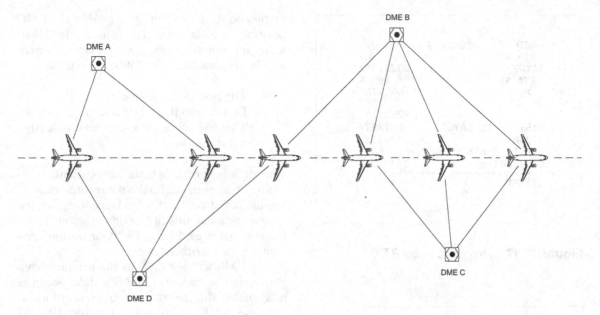

Figure 17.19   DME-DME position updates

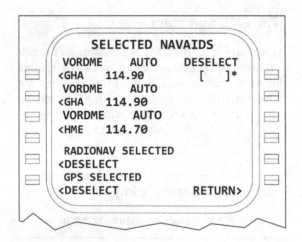

Figure 17.20   FMS selected navigation aids

17.9 Multiple choice questions

1. To define the destination airport on the FMC route page requires entry of the airfield's:
    (a) three-character identifier
    (b) four-character identifier
    (c) latitude and longitude.

2. The page automatically displayed upon FMC power-up is the:
    (a) identification page
    (b) navigation database
    (c) position initialisation page.

3. Program pins are defined by the:
    (a) FMC operational program
    (b) navigation database
    (c) aircraft wiring at the FMC connector.

4. Information entered into the CDU scratch pad is displayed in the:
    (a) lowest section of the display
    (b) box prompts
    (c) dash prompts.

5. Minimum flight time would be achieved with a cost-index of:
    (a) zero
    (b) 100
    (c) 50.

6. Aircraft and engine type can be confirmed on the:
    (a) progress page
    (b) identification page
    (c) position initialisation page.

7. The use of metric/imperial units is
   determined by:
   (a) the part number of the FMC
   (b) program pins
   (c) dashed line entries.

8. Required entries into the CDU are
   indicated by:
   (a) box prompts
   (b) dashed lines
   (c) highlighted text.

9. 'Not in database' is an example of an:
   (a) alert message
   (b) advisory message
   (c) active waypoint.

10. Display of the identification page after
    power-up confirms the:
    (a) IRS is aligned
    (b) navigation sources in use
    (c) FMC has passed its BITE check.

11. The FMC recognises specific aircraft
    types by:
    (a) CDU entry
    (b) program pins
    (c) the navigation database

12. SIDs in the navigation database refer to:
    (a) arrivals
    (b) enroute navigation
    (c) departures.

13. The EXEC key lights up when:
    (a) data is entered for initialisation/changes
    (b) advisory messages are displayed
    (c) incorrect data has been entered.

14. Alerting messages require attention from
    the crew:
    (a) before guided flight can be continued
    (b) when time is available
    (c) at the completion of the flight.

15. The highlighted waypoint on the progress
    page is the:
    (a) previous waypoint
    (b) active waypoint
    (c) destination.

Chapter 18 — Weather radar

Weather radar was introduced onto passenger aircraft during the 1950s for pilots to identify weather conditions and subsequently reroute for the safety and comfort of passengers. Extreme weather conditions are a major threat to the safe operation of an aircraft; flight crews need to be aware of these conditions and understand the consequences. There are three main technologies typically used in aircraft to detect weather conditions: on-board radar, lightning detection, and datalink services. These systems enable pilots to identify weather conditions and subsequently re-route for the safety and comfort of passengers. The majority of commercial transport aircraft are equipped with weather radar, and this represents the majority of this chapter. General Aviation aircraft that are not equipped with radar are typically equipped with an alternative technology, i.e. lightning detection, and/or datalink services.

18.1 System overview

The word radar is derived from <u>ra</u>dio <u>d</u>etection <u>a</u>nd <u>r</u>anging; the initial use of radar was to locate aircraft and display their range and bearing on a monitor (either ground based or in another aircraft). This type of radar is termed **primary radar**; energy is directed via an antenna to a 'target'; this target could be an aircraft, the ground or specific weather conditions. In the case of weather radar, we want to detect the energy reflected back from the contents of a cloud, or from **precipitation**, see Figure 18.1. The latter may defined as the result of water vapour condensing in the atmosphere that subsequently falls to the earth's surface. Precipitation can occur in many different forms including: rain, freezing rain, snow, sleet, and hail. Weather radar operates either in the C-band (4–8GHz) or X-band (8–12.5 GHz); these two bands have their advantages and disadvantages for use in weather radar

applications. C-band **microwave** energy pulses can penetrate through heavy precipitation, thereby providing **weather detection**, enabling the pilot to determine more details of the weather pattern. X-band microwave energy pulses can provide good resolution of images;

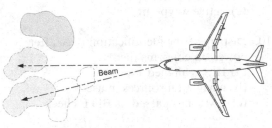

(a) Microwave energy directed via radar beam

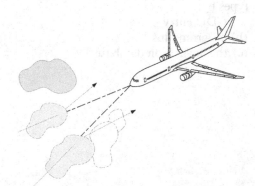

(b) Reflected energy from the contents of a cloud

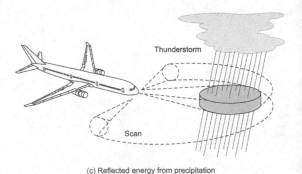

(c) Reflected energy from precipitation

Figure 18.1 Weather radar principles

DOI: 10.1201/9781003411932-18

however, this means that they can only be used for **weather avoidance**. Higher frequencies require a smaller antenna; for this reason, larger passenger aircraft use X-band radar.

The range of a weather radar system is typically 320 miles. Microwave energy pulses are reflected from the moisture droplets and returned to the radar antenna. The system calculates the time taken for the energy pulses to be returned; this is displayed as an image on a dedicated weather radar screen, or the image can be integrated with the electronic flight display system. The strength of the returned energy is measured and used to determine the size of the target. Higher moisture content in a cloud provides higher returned energy. The antenna is **scanned** in the lateral plane to provide directional information about the target.

18.2 Airborne equipment

The typical weather radar system in a large aircraft comprises the antenna (nose cone); two transceivers (nose cone); two control panels/displays (cockpit), see Figure 18.2(a). The transceivers are located in the nose cone, thereby minimising the length of the waveguide, and also avoiding the need for the waveguide to penetrate the pressure bulkhead. On single engine,

GA aircraft the antenna can be located on the wing, in a self-contained pod, Figure 18.2(b).

18.2.1 Antenna

Microwave signals are transmitted and received via the antenna. Early versions of the antenna were in the form of a parabolic dish; however, the current versions are the planar array flat-plate type, see Figure 18.3. The flat-plate antenna projects a more focused beam than the parabolic type; this is due to the reduction in side-lobes as illustrated in Figure 18.4. The antenna comprises a flat steerable plate with a large number of radiating slots, each equivalent to a half-wave dipole fed in phase. The antenna is mounted on the forward pressure bulkhead behind the radome; this is a streamlined piece of structure constructed of materials that have **low attenuation** of the radar signals. The mechanical condition of the radome is very important to the effectiveness of the weather radar system. The radome is designed to be transparent for radar waves at a given frequency range. Delamination will attenuate the radar signal. The radome also protects the antenna from the elements and air loads.

The antenna automatically traverses from left to right on a repetitive basis to be able to **scan** the weather patterns ahead of the aircraft.

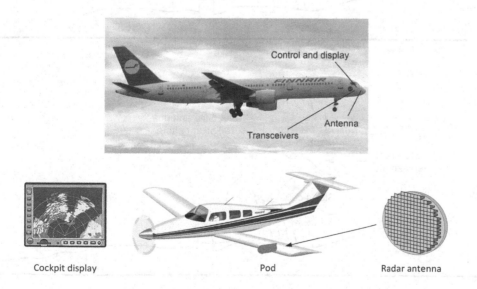

Control and display

Antenna

Transceivers

Cockpit display   Pod   Radar antenna

Figure 18.2 Location of weather radar equipment

Figure 18.3   (a) Weather radar antenna; (b) Antenna waveguide

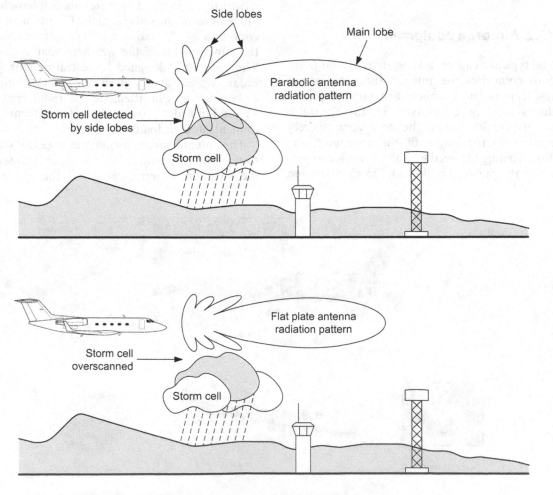

Figure 18.4   Parabolic and flat-plate antenna radiation patterns

To investigate cloud formations, the pilot can also tilt the antenna up or down to provided different viewing perspectives. The reference position is to scan the antenna so as to provide images across the horizon; inputs from the aircraft's attitude reference system are used to provide the stabilisation. Motors are used as part of a drive mechanism to traverse the antenna in azimuth and to tilt the antenna in pitch. Synchro transmitters are used to relay the various positions of the antenna back to the transceiver.

Energy pulses are carried between the antenna and transceiver via a **waveguide**, see Figure 18.3(b). This is because losses in a coaxial cable would be high at frequencies above 3 GHz, and prohibitive at frequencies above 10 GHz. Coaxial cables are also limited in terms of the peak power handling capability. Waveguides have their disadvantages; they are bulky, expensive and require more maintenance. Manufactured from aluminium alloy, in a hollow rectangular form, they have dimensions closely matched to the wavelength of the system. Chapter 2, Section 2.12 provides more details on waveguides.

Figure 18.5 Weather radar transceiver

Key point

The energy radiated from a weather radar system is hazardous and could cause injury.

18.2.2 Transceiver

The transceiver is a combined transmitte The transceiver is a combined transmitterrand receiver, Figure 18.5; antenna power output is in the order of 5–10 kW. Modern transceivers are solid-state devices, incorporating video processing for the display and stabilisation signals for the antenna. Since the energy received from a given size of water droplet varies with range, the energy returns from closer ranges will be higher than those received from droplets further away. The transceiver will automatically compensate for returns from targets that are near or far from the aircraft. This is achieved by altering the gain as a function of time from when the energy pulse is transmitted. Pulses of radar energy are transmitted on a repetitive basis; the interval between pulses depends on the range selected by the crew. Time has to be allowed for the energy pulse to be reflected from water droplets at the limit of the selected range before the next pulse is transmitted.

18.2.3 Control panel

A typical weather radar control panel is shown in Figures 18.6(a) and (b). This allows the pilot to select the left or right transceiver, select the weather radar mode, manually tilt the antenna and select the gain of the system.

18.2.4 Display

The basic display used for primary radar systems is the **plan-position-indicator** (PPI). As the beam sweeps from side to side, a radial image on the display (synchronised with each sweep) moves across the display. The image on the display depends on the amount of energy returned from the target. Original weather radar systems had dedicated monochrome displays based on a cathode ray tube (CRT); these have evolved over the years into full colour displays, often integrated with other electronic flight instruments. The full benefits of a weather radar system can be appreciated when the system is used on an aircraft with an electronic flight instrument system (EFIS) display, Figure 18.7. A symbol generator is used to provide specific weather radar

Figure 18.6   (a) Location of weather radar control panel; (b) Typical weather radar control panel

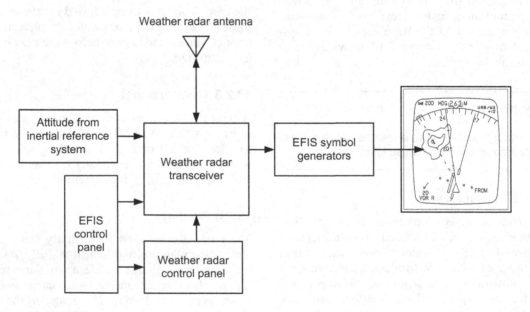

Figure 18.7   Weather radar EFIS display

images as determined by the transceiver. An electronic display control panel allows each pilot to select the range of weather radar in increments of 10, 20, 40, 80, 160 and 320 miles.

The electronic display is overlaid onto the map mode allowing the pilot to relate the aircraft's heading with the weather images. These images are colour coded to allow the pilot to assess the severity of weather conditions. Colours (ranging from black, green, yellow, red and magenta) are used to indicate rainfall rates that can be interpreted as a level of turbulence.

18.3 Precipitation and turbulence

For a more detailed understanding of weather radars, factors that affect precipitation, turbulence and the formation of clouds are now considered.

18.3.1 Cloud formation

For the comfort and safety of the passengers, we want the weather radar system to detect the turbulence resulting from precipitation that leads to severe weather conditions, i.e. thunderstorms, such that these can be avoided if possible. Precipitation may defined as the result of water vapour **condensing** in the atmosphere that subsequently falls to the earth's surface. This can occur in many different forms including: rain, freezing rain, snow, sleet, and hail. Clouds are the visible accumulation of particles of water and/or ice in the atmosphere; their formation changes on a continuous basis, often resulting in no more than a 'light shower' of rain. Under certain atmospheric conditions, clouds become **large** and **unstable** leading to hazardous flying conditions. The flight crew needs to have accurate and up to date forecasts of en route weather conditions; this includes details of cloud classifications as detailed in Table 18.1.

Within the above classification, precipitation varies with each cloud type (see Figure 18.8):

- **Altocumulus**: precipitation does not actually reach the ground
- **Altostratus**: precipitation is in the form of rain or snow
- **Nimbostratus**: the cloud base is diffuse with continuous rain or snow
- **Stratocumulus**: light rain, drizzle or snow
- **Stratus**: drizzle
- **Cumulus**: rain or snow showers
- **Cumulonimbus**: lightning, thunder, hail. Associated heavy showers of rain/snow.

It can be seen from the above that **cumulonimbus** formations present the greatest hazard to aircraft, and maximum discomfort for passengers. We need to understand the nature of **thunderstorms** that contain heavy rainfall and turbulence.

Table 18.1   Classification of clouds

Name	Base (AMSL*)	Appearance
Cirrus (Ci)	> 20,000 feet	Fibrous and detached, mainly ice crystals
Cirrocumulus (Cc)	> 20,000 feet	Thin layers or patches without shading
Cirrostratus (Cs)	> 20,000 feet	Transparent, whitish veil that can cover the sky
Altocumulus (Ac)	> 6,500 feet	Patchy groups of white/grey layers
Altostratus (As)	> 6,500 feet	Greyish/blue fibrous sheets
Nimbostratus (Ns)	< 6,500 feet	Dark grey layers, covering the sky, hiding the sun
Stratocumulus (Sc)	< 6,500 feet	Grey/white patches with dark rounded features
Stratus (St)	< 6,500 feet	Grey layers, uniform base
Cumulus (Cu)	< 6,500 feet	Dense, vertical shapes developing in mounds
Cumulonimbus (Cb)	< 6,500 feet	Heavy, dense and towering. Upper portion fibrous

* Above mean sea level

18.3.2 Thunderstorms

Three conditions are needed to create thunderstorms: instability within the atmosphere, high moisture content and a catalyst to start the air rising. Air can be forced to rise in the atmosphere from a number of causes as illustrated in Figure 18.9:

- **Frontal**: when opposing warm and cold air masses combine
- **Convective**: the ground being heated by the sun
- **Orographic**: movement of air over the terrain.

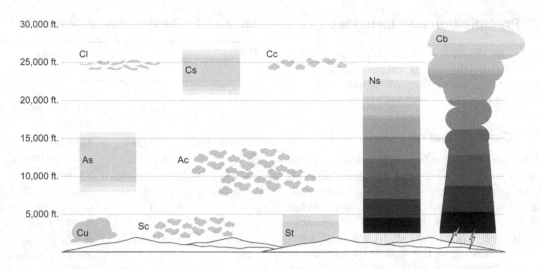

Figure 18.8   Classification of clouds

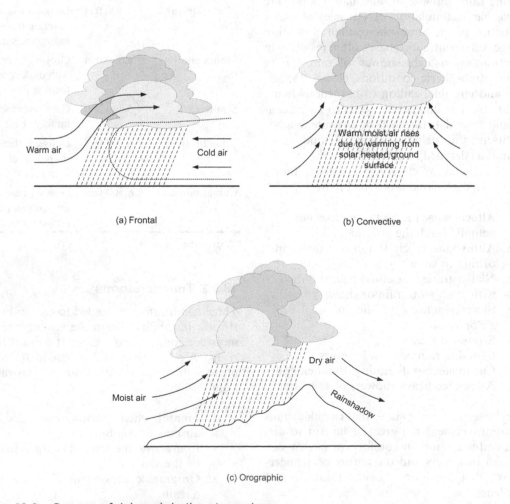

Figure 18.9   Causes of rising air in the atmosphere

Referring to Figure 18.10, the life cycle of a thunderstorm develops in three stages. During the first towering **cumulus stage**, warm, moist air containing water vapour rises up to higher altitudes. When the dew point is reached it cools down and the moisture content condenses into water droplets thereby creating clouds. As a result of the condensation process, latent heat is released causing the air to become warmer and drier and thereby less dense. This air rises as an updraft over one or two miles diameter due to convection. At this stage, air is drawn into the cell horizontally at all levels causing the updraft to become stronger with altitude. Water vapour is carried up to higher altitudes where it combines to form larger water droplets. This first stage of a thunderstorm develops over approximately 20 minutes.

When the water droplets are sufficiently large enough, they are too heavy to be supported by the updraft, and are released as rainfall. As the water droplets fall, they draw in the surrounding air causing a downdraft. The air temperature of the downdraft is cold compared with the updraft; heavy rain falls from the base of the cloud. The updraft continues to carry the remaining water droplets up to very high altitude; this can be up to 50,000 feet in the tropics (well into the troposphere). At these altitudes, strong winds in the upper atmosphere (jet streams) carry the top of the cloud away to form the characteristic 'anvil' shape. This is the **mature stage** of the thunderstorm, and can last up to 40 minutes.

These two air masses are now moving in opposite directions forming localised cells of air movement; the relative movement is known as **wind shear**, and is the basis of **turbulence**. This wind shear is characterised by the relative speed and direction of the two air movements. The amount of increase in temperature and size of air mass being heated determines the rate at which wind shear, and hence turbulence, occurs. These factors will determine the nature of the boundary between the air masses. When these boundaries are very distinct, this is when the rainfall and turbulence is most severe. The size of droplets in thunderstorms and other severe weather is large compared with general rainfall; this produces much larger radar returns as described in Section 18.3.3.

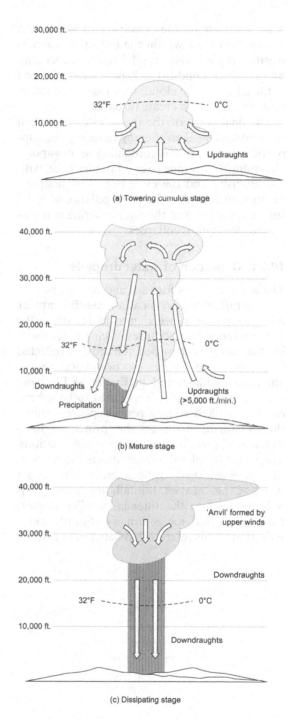

Figure 18.10 Stages in the life cycle of a thunderstorm

In addition to the upward and downward movement of air within the cell of a thunderstorm, droplets also travel laterally depending on prevailing conditions. New cells are formed at the edges of the cloud; they may not contain rain but they often create turbulence.

The **final stage** of the thunderstorm is when the updrafts weaken thereby reducing the supply of warm, moist air containing water vapour. Downdrafts continue over a broader base, with less intensity and the cycle begins to dissipate. Temperatures within the cloud balance with the surrounding air, and the once towering cumulonimbus formation collapses.

18.3.3 Detection of water droplets

The antenna scans forward and to each side of the aircraft with a conical, or **pencil** beam of microwave energy, see Figure 18.11. When the weather radar energy pulses reach a water droplet, the energy is absorbed, refracted or reflected from the front convex portion back to the antenna, as illustrated in Figure 18.12. Note that the water droplet is assumed to be spherical; it is unlikely to be a perfect sphere within a thunderstorm; however, the principles illustrated in Figure 18.12 are good approximations. The water droplet diameter affects the amount of energy returned to the antenna. With larger droplets, i.e. heavier rainfall, more energy is reflected back to the antenna. Smaller droplets from cloud and fog return significantly lower reflections. This relationship between individual

size, rainfall rate and reflected energy is the basis of detecting the severity of the storm. Weather radar wavelengths (25 mm at 12 GHz) are larger than any water droplet sizes (refer to Table 18.2). The scattering of electromagnetic radiation by particles such as raindrops, i.e. particles smaller than the wavelength of the radiated energy is characterised by a phenomenon called **Rayleigh scattering**. This is named after John William Strutt (Lord Rayleigh 1842–1919), an English physicist. The intensity of the returns is given by the Rayleigh scattering equation; in this equation, the intensity of the returns varies as the sixth power of the droplet diameter. This means that returns from water vapour within fog or clouds is small compared with raindrops.

18.3.4 Hailstones

Hailstones are formed in cumulonimbus clouds and, in addition to the turbulence, they will cause physical damage to the aircraft. They are formed in strong thunderstorms with significant updrafts as irregular lumps (or pellets) of ice; this occurs when **super-cooled water** makes contact with an object such as dust or ice particles. Super-cooled water exists at temperatures below freezing point; this is because water needs a nuclei to form ice crystals. The water freezes on the surface of the nuclei, and then grows in size by forming layers. When the hailstone is too heavy to be supported in the updraft of the storm it will fall out of the cloud. Referring to Table 18.2, the sizes of hailstones are starting to

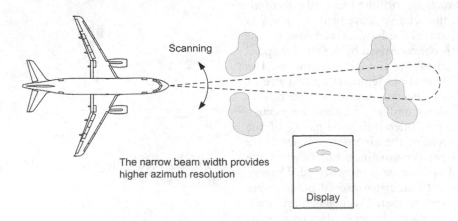

Figure 18.11　Pencil beam and azimuth resolution

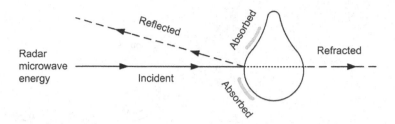

Figure 18.12 Effect of microwave energy on a water droplet

Table 18.2 Weather radar targets and approximate dimensions

Target	Diameter
Hailstones (large)	>20 mm
Hailstones (medium)	>7.5 mm
Snowflakes (large)	>10 mm
Snowflakes (medium)	>5 mm
Rain (droplets)	<5 mm
Rain (drizzle)	0.5 mm

approach the wavelength of the radar energy (25 mm at 12 GHz); hailstones have been recorded up to 60 mm in diameter. When the wavelength of the radar energy is smaller than the target diameter, the hailstone acts as a lens, focussing energy onto the concave internal surface and reflecting higher levels (compared with water droplets) of energy back to the antenna.

Key point

Microwave energy pulses from the weather radar are reflected from the moisture droplets and returned to the radar antenna.

Key point

X-band (8–12.5 GHz) energy pulses provide good resolution of images for the purposes of weather avoidance; they require a smaller antenna compared with C-band radar (4 to 8 GHz).

Key point

The weather radar antenna automatically traverses from left to right on a repetitive basis to be able to scan the weather patterns ahead of the aircraft.

18.3.5 Turbulence

Turbulence can be inferred from the measurement of precipitation, both in terms of the type of precipitation (rain or hail), droplet size and precipitation rate. Approximate dimensions (given as their diameter) of various weather targets are given in Table 18.2.

18.3.6 Predictive wind shear

Wind shear can occur in both the vertical and horizontal directions; this is particularly hazardous to aircraft during take-off and landing. Specific weather conditions known as **microbursts** cause short-lived, rapid air movements from clouds towards the ground. When the air from the microburst reaches the ground it spreads in all directions, this has an effect on the aircraft depending on its relative position to the microburst. Referring to figure 18.13, when approaching the microburst, it creates an **increase** in headwind causing a temporary increase of airspeed and lift for an aircraft approaching the cloud; if the pilot were unaware of the condition creating the increased airspeed, the normal reaction would be to reduce power. When flying through the microburst, the aircraft is subjected to a downdraft. As the aircraft exits the microburst, the downdraft now becomes a tailwind, thereby **reducing** airspeed and lift. This complete sequence of events happens very

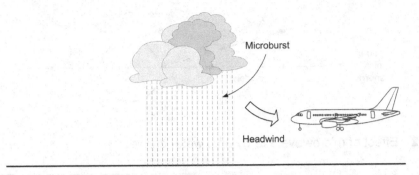

(a) An aircraft entering a microburst encounters headwinds that increase airspeed and lift

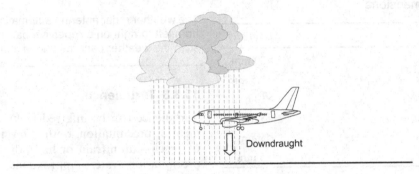

(b) The aircraft flies through the headwind and encounters a downdraught

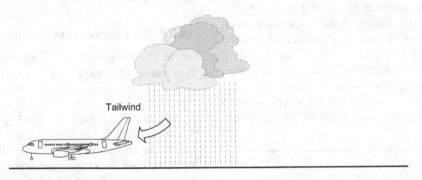

(c) The aircraft exits the microburst and encounters a tailwind, reducing airspeed and lift

Figure 18.13   Microburst and associated air movements

quickly, and could lead to a sudden loss of airspeed and altitude. In the takeoff and climb-out phase of flight, an aircraft is flying just above stall speed; wind shear is a severe threat. During approach and landing, engine thrust will be low; if a microburst is encountered, the crew will have to react very quickly to recognise and compensate for these conditions.

Modern weather radar systems are able to detect the **horizontal** movement of droplets using Doppler shift techniques. Doppler is usually associated with self-contained navigation systems, and this subject is described in a separate chapter. The Doppler effect can be summarised here as: '…the frequency of a wave apparently changes as its source moves closer to, or farther away from an observer'. This feature allows wind shear created by microbursts to be detected. Referring to Figure 18.14, the microwave energy pulses from the antenna are reflected by the

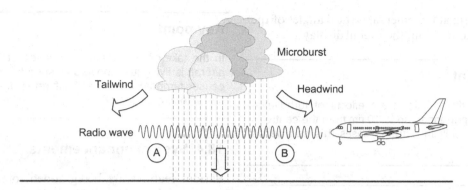

Figure 18.14   Doppler frequency shift between A and B indicates a wind shear condition

water droplets as in the conventional weather radar system. Using the Doppler shift principle, the frequency of energy pulse returned by droplets (B) moving toward the aircraft will be at a higher frequency than the transmitted frequency. The frequency of energy pulse returned by droplets moving away from the aircraft (A) will be at a lower frequency than the transmitted frequency. These Doppler shifts in frequency are used to determine the direction and velocity of the air movement resulting from a microburst.

Visual and audible warnings of wind shear conditions are provided to the crew. The visual warnings are given on the weather radar and navigation displays using a wind shear icon and message together with warning lights on the glare shield. Audible warnings are provided as computer generated voice alerts over the cockpit speakers, typically 'wind shear ahead'. The system automatically configures itself for the phase of flight; it is normally inhibited below 50 feet radio altitude during take-off and landing. During an approach, the system is activated below 2500 feet radio altitude.

Test your understanding 18.1

In which bands of radar frequencies does weather radar operate?

Test your understanding 18.2

Explain what happens when radar energy reaches a water droplet.

18.3.7 Terrain mapping

A secondary use of the weather radar system is for terrain mapping, e.g. identifying rivers, coastlines and mountains. This mode of operation is selected by the crew on the control panel. Returns from the various ground features are different, just as they are for precipitation. These variations are interpreted by the system and displayed using various colours. Since the energy of the return signal depends on the reflectivity of the terrain and angle at which the beam meets with the terrain, the gain control is used to provide the optimum display. If the gain is too low or too high, the images will be unclear.

The weather radar pencil beam is not suitable for terrain mapping since it does not cover a sufficiently large area. A fan-shaped beam (Figure 18.15) provides optimum coverage of the terrain along the intended track of the aircraft. The system can have the facility to reshape the beam depending on the selected mode. Modern systems achieve terrain mapping by sweeping the

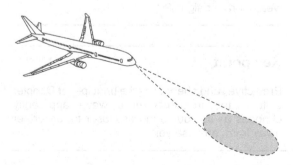

Figure 18.15   Terrain mapping

pencil beam at incremental vertical angles of the antenna to build up the overall display.

Key point

During a thunderstorm the effects of turbulence can be experienced up to 10 nm from the centre of the storm cell; in extreme cases this can be up to 20 nm.

Key point

Thunderstorms develop within cumulonimbus cloud formations and present severe hazards to aircraft:

* Reduced visibility from ground level up to 50,000 feet
* Turbulence and wind shear causing handling problems
* Hailstones up to 60 mm in diameter causing structural damage
* Lightning strikes causing, inter alia, structural damage
* Interference with navigation and communications equipment.

Key point

The temperature increase and size of air mass being heated determines the rate at which wind shear, and hence turbulence, occurs.

Test your understanding 18.3

What effect will radome delamination have on the weather radar signal?

Key point

Predictive wind shear uses the principle of Doppler shift; '...the frequency of a wave apparently changes as its source moves closer to, or farther away from, an observer'.

Key point

In the take-off and climb-out phases of flight, an aircraft is flying just above stall speed; wind shear causes temporary changes in lift and is therefore a threat to aircraft safety.

18.4 System enhancements

Various features are being added to the basic weather radar systems to provide many benefits including enhanced displays, improved turbulence detection, and integration with other systems.

The weather radar system described so far is two dimensional, i.e. it provides range and directional information. **Three-dimensional** weather radar provides volumetric information relating to the weather pattern. The antenna scans 90 degrees either side of the aircraft centreline, from ground level up to 60,000 feet and up to 320 miles ahead of the aircraft. Energy returns are stored in a computer and used to build a **volumetric model** of the airspace ahead of the aircraft. Unwanted ground returns (**clutter**) are filtered out in the computer using an on-board **terrain database**; this provides a clearer image of weather patterns. Some systems refine this de-cluttering by taking the earth's curvature into account. Weather radar software has been developed with the ability to predict storm cloud formations by analysing cloud growth characteristics, thereby providing an increased notification of turbulence. The three-dimensional image of the weather pattern is presented on the display both in terms of forward and side views. The rate of thunderstorm development can also be modelled within the weather radar system providing crews with increased notice of anticipated turbulence.

Knowledge of how thunderstorms develop over land versus oceans can be built into the system to modify the weather radar display. Different geographical areas of the world generate different types of storm with their associated turbulence; these characteristics can be built into the system's software model. These models are used to characterise the different type of storms that develop in the northern and southern hemispheres, taking into account the variations that occur with latitude. Turbulence warnings can also be customised for specific aircraft types and the various phases of flight.

18.5 Lightning detection

A complementary technology used for the detection of storms is lightning detectors. These are inexpensive and lightweight, making them very attractive for general aviation (especially for single-engine aircraft, where there is no space for a radome). Lightning detection system comprises an antenna, processor and display; the system weight is approximately 5 kg versus 15 kg for a weather radar system. The system monitors electrical activity within a storm (whereas weather radar detects precipitation as described in this chapter).

When clouds are developing as described earlier, lightning is not produced in the early stages; weather radar will indicate a developing storm before a lightning detector can. Weather radar can experience attenuation, i.e. where nearby precipitation can mask precipitation further away. Lightning detectors provide confirmation when a cloud has developed into a thunderstorm.

Lightning might occur in areas outside of the precipitation area detected by radar. Electrical activity can originate in the anvil of the thundercloud or on the outside edges of the precipitation area. Transport aircraft use weather radar in preference to lightning detectors because not all clouds develop into thunderstorms; weather radar detects smaller storms, i.e. without lightning, that also cause turbulence.

A typical general aviation lightning detection system comprises an antenna and processor and display, see Figure 18.16. This provides a simple provides a simple and intuitive colour-contoured display with instantaneous picture of surrounding weather. The system provides real time, tactical, colour weather avoidance solution. It accurately detects and clearly displays electrical discharges (the primary indicator of thunderstorms) in real time, regardless of aircraft manoeuvring. The system provides tactical weather avoidance in the critical 0 – 25nm range,

Figure 18.16 Lightning detection display

complementing to the strategic benefits of satellite-based datalink weather. Storm cells and their intensities are displayed in colour contours, providing an image of a thunderstorm's electrical activity over time. This allows the pilot to determine the areas of greatest intensity, and clearly identify areas with potential for convective wind shear, turbulence, microbursts, hail and icing.

Each cell grid is displayed as a hexagonal array of cell weights (colours). Colour-filled hexagons highlight the most intense regions of thunderstorm activity, presenting a visually contoured colour display with dynamic sectors. This helps the pilot to quickly identify the most intense regions of thunderstorm activity. In Strike Mode, up to 1024 strikes are displayed for up to three minutes. Each transmitted strike includes range, bearing, and weight (colour). Regional activity, not age, determines the strike colour. Strikes that are more recent but further away cannot "drown out" flashes that are closer and less recent. This provides a level of comfort to pilots transitioning from less capable systems. Colour enhancement helps to identify areas of high intensity while still providing a view of the discrete events.

18.6 Datalink weather

Systems have been developed to provide a datalink graphical weather system for General

Aviation operations outside the Continental United States (CONUS). Utilising the Iridium satellite communication constellation, the two-way datalink transmission technology includes high-resolution radar imagery, Meteorological Aerodrome Reports (METAR), and Terminal Aerodrome Forecasts (TAF). These are displayed in full colour on a suitable multi-function display (MFD). An aerodrome TAF is a forecast, that gives the predicted weather conditions expected at an aerodrome, usually for a 9 or 24 hour period. A METAR is a report giving the actual weather conditions at an aerodrome at the time of the report. METARs are typically issued every 3060 minutes. The system delivers satellite infrared (SAT IR) imagery for selected geographic regions; this imagery shows the emission from the infrared portion of the solar spectrum. Bright areas represent cold high cloud tops; dark areas represent warm ground and ocean surfaces.

Test your understanding 18.4

What is the significance of the black, green, yellow, red and magenta images on a weather radar display?

Test your understanding 18.5

How is wind shear created and how are microbursts detected?

18.7 Phased array radar

The weather radar antennas previously described contain moving parts, with the associated cost of serving and maintenance etc. A solid-state weather radar antenna based on phased array, or electronically steered, antenna technology. When two or more radiating elements are operated in close proximity, their interaction sets up a pattern of radiation, see Figure 18.17. By controlling the phase angle of individual elements electronically, a specific pattern of radiation can be produced.

Phased array radar antennas comprise many small elements on a flat panel. The radar beam can be "steered" electronically by changing the phase of the signal emitted from each individual antenna element. The signals from each element interact, a phenomenon known as interference, see Figure 18.18. When two waves are in phase, they create positive, or constructive interference, resulting in a wave of increased (or reinforced) amplitude. When 180° out of phase (or anti-phase), they create destructive interference, resulting in a wave of zero (or cancelled)

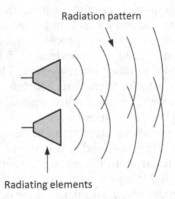

Figure 18.17 Radiating elements in close proximity

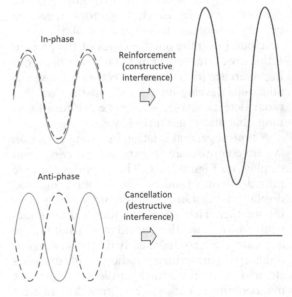

Figure 18.18 Waveform interference

amplitude. This solid-state electronically steered antenna (ESA) technology means that motors and moving parts for the radar's scanning and tilting functions are eliminated.

The phased array radar system consists of antenna elements powered by a transmitter (T_x), this system controls the radar's beam direction, see Figure 18.19. The supply current phase angle (θ) for each element is controlled by a computer. The individual wavefronts are spherical, but they combine (superpose) in front of the antenna to create a plane wave, a beam of radio waves travelling in a specific direction. Each antenna element transmits its wavefront later than the element adjacent it. This causes the resulting plane wave to be directed at an angle θ to the antenna's axis. Changing the phase angle of each wave changes the directional angle θ of the beam.

Referring to Figure 18.20, the effective radar beam is transmitted via the antenna elements. Most phased arrays have two-dimensional arrays of antennas so that the beam can be steered in two dimensions, equivalent to tilting up/down and scanning left/right.

The antenna is formed in a matrix construction, each discrete horn element has its own phase shifter, controlled by the system's computer, see Figure 18.21. The antenna is located directly in the nose cone on most aircraft types, Figure 18.22.

Aside from no moving parts and reduced weight, phased array weather radar systems incorporate several advanced features, e.g.,

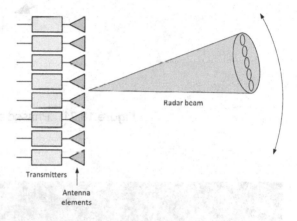

Figure 18.20   Effective radar beam

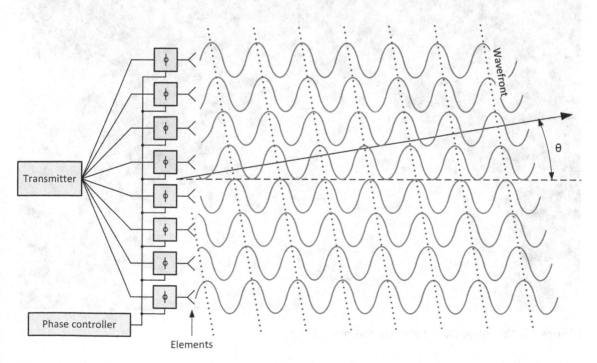

Figure 18.19   Electronically steered radar beam

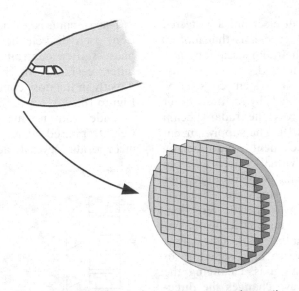

Figure 18.21   Phased array antenna schematic

Figure 18.22   Antenna location in the nose cone

faster scanning, automatic threat analysis that adjusts the antenna's "scan and tilt" patterns to accurately profile weather cells, predictive wind-shear etc.

18.8 Multiple choice questions

1. Weather radar operates in which bands of radar frequencies?
   (a) C- and X-band
   (b) L-band
   (c) HF.

2. Larger passenger aircraft use X-band radar because it can:
   (a) determine more details of the weather pattern
   (b) penetrate through heavy precipitation
   (c) provide good resolution of images and requires a smaller antenna.

3. What effect will radome delamination have on the weather radar signal?
   (a) none
   (b) decreased attenuation
   (c) increased attenuation.

4. Air can be forced to rise in the atmosphere from a number of causes including when opposing air masses combine; this is called:
   (a) convection
   (b) frontal
   (c) orographic.

5. What effect does increased water droplet diameter have on the amount of energy returned to the antenna?
   (a) increased
   (b) no effect
   (c) decreased.

6. The most severe weather radar images are colour coded:
   (a) black
   (b) magenta
   (c) green.

7. Predictive wind shear uses the principle of:
   (a) Doppler shift
   (b) detecting rain drop size
   (c) detecting rain drop shape.

8. Weather radar energy pulses are transmitted at rates that vary with selected:
   (a) range
   (b) mode
   (c) antenna tilt angle.

9. Compared with the transmitted frequency, energy pulses returned by droplets moving away from the aircraft will be at:
   (a) lower frequencies
   (b) higher frequencies
   (c) the same frequency.

10. As the aircraft travels away from a microburst, the downdraft affects airspeed and lift by:
    (a) reducing airspeed and reducing lift
    (b) increasing airspeed and reducing lift
    (c) reducing airspeed and increasing lift.

11. Weather radar energy pulses returned from a water droplet have been:
    (a) absorbed
    (b) refracted
    (c) reflected.

Chapter 19

Air traffic control systems

Air traffic control (ATC) systems enable ground controllers to maintain safe separation of aircraft, both on the ground and in the air. The system is based on secondary surveillance radar (SSR), now enhanced with high integrity digital datalinks, including ATC Mode S and automatic dependent surveillance-broadcast (ADS-B). A key organization within the ATC infrastructure is the air navigation service provider (ANSP); this organisation provides the service of managing the aircraft during en route and the terminal area. This chapter describes the various methods of communications, navigation and surveillance (CNS) used for air traffic management (ATM).

19.1 ATC overview

We have seen examples of primary and secondary radar systems in previous chapters. To reiterate; with primary radar, high energy is directed via an antenna to **illuminate** a 'target'; this target could be an aircraft, the ground or water droplets in a cloud. In the case of ATC primary radar, the energy is reflected from the aircraft's body to provide range and azimuth measurements. ATC's primary radar system places the target(s) on a **plan position indicator** (PPI). Primary surveillance radar (PSR), see Figure 19.1, has its disadvantages; one of which is that the amount of energy being transmitted is very large compared with the amount of energy reflected from the target. Small targets, or those with poor relecting surfaces, could further reduce the reflected energy. Natural and man-made obstacles such as mountains and wind farms also shield the radar signals. **Secondary surveillance radar** (SSR) transmits a specific low energy signal (the interrogation) to a known target. This signal is analysed by a **transponder** and a new (or secondary) signal, i.e. not a reflected signal, is sent back (the reply) to the origin, see Figure 19.2. Secondary radar was developed during the Second World War to differentiate between friendly aircraft and ships: this system was called Identification Friend or Foe (IFF). In the air traffic control system, the primary and secondary radar antennas are mounted on the same rotating assembly, thereby providing a coordinated system. The complete system is illustrated in Figure 19.3.

The ATC system operates on two frequencies within the L-band of radar:

- **Interrogation** codes on a 1030 MHz carrier wave
- **Reply** codes on a 1090 MHz carrier wave.

The primary radar system provides a single icon per aircraft on the ATC controller's display; this means that each icon will look similar, depending upon the amount of reflected energy. As a consequence, an aircraft would have to change direction in order for it to be uniquely identified. By implementing the SSR transponder system, each icon can be identified via a unique **four-digit code** (allocated by ATC for each flight). Using SSR also means that the effects of clutter (from trees, buildings and hills, etc.) are not displayed on the controller's screen. With an uncluttered screen, and each aircraft readily identified, more aircraft can be allowed into the controlled airspace. The combined PSR/SSR system is illustrated in Figure 19.4. Developments of the ATC transponder system have provided additional functionality allowing details such as flight number and altitude to be displayed on the controller's screen. Emergency codes can be sent in the event of radio failure or hijacking. The reader will appreciate that it is essential for an aircraft operating in controlled airspace to be equipped with an ATC transponder.

DOI: 10.1201/9781003411932-19

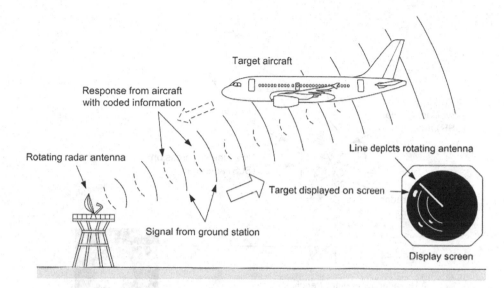

Figure 19.1 Primary surveillance radar (PSR)

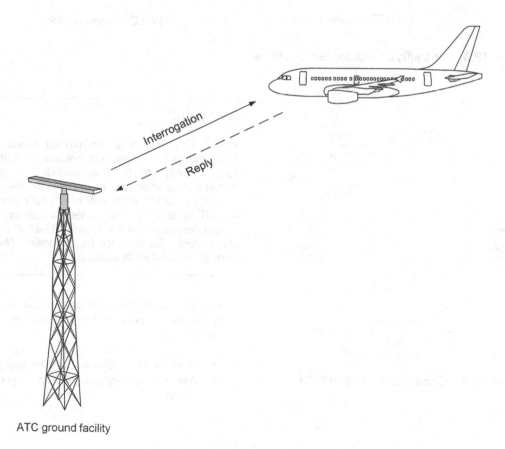

Figure 19.2 Secondary surveillance radar (SSR)

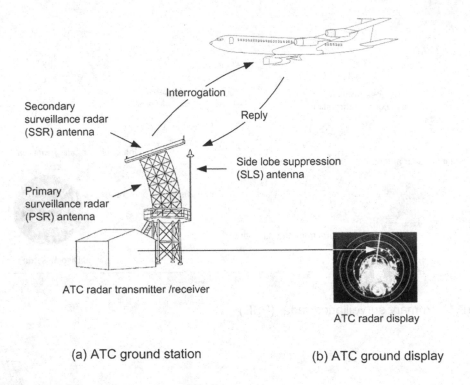

(a) ATC ground station   (b) ATC ground display

Figure 19.3   Air traffic control system overview

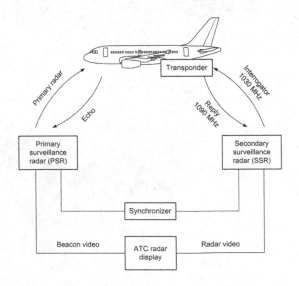

Figure 19.4   Combined PSR and SSR

Key point

ADS-B OUT refers to an aircraft broadcasting "own-ship" information, e.g. present position, altitude, velocity etc. to other aircraft and ground-based surveillance systems. ADS-B IN refers to an aircraft's ability to receive ADS-B information, e.g. ADS-B messages from other aircraft or Traffic Information Services-Broadcast (TIS-B), Automatic Dependent Surveillance-Rebroadcast (ADS-R) from the ground infrastructure.

The air navigation service provider (ANSP) supplies one or more of the following services to airspace users:

- Aeronautical information services (AIS)
- Aeronautical information management (AIM)

- Air traffic management (ATM)
- Communication navigation and surveillance systems (CNS)
- Meteorological service for air navigation (MET)
- Search and rescue (SAR)

19.2 ATC transponder modes

SSR systems have been developed for both military and commercial aircraft applications; a summary of commercial aircraft modes is as follows:

Mode A: In this transponder system, the pilot selects the four-digit code on the ATC control panel prior to each flight. The SSR system confirms this aircraft's azimuth on the controller's screen with an icon confirming that the aircraft is equipped with a transponder. If the controller needs to distinguish between two aircraft in close proximity an identity code will be requested; the pilot pushes a switch on his ATC control panel, and this highlights the icon on the controller's screen. Since each aircraft is allocated with a unique code, only one icon per aircraft will be highlighted; this unique identification is referred to as a **squawk** code. Each of the four digits ranges from 0 to 7, these are then coded as octal numbers for use by the transponder. (This system is called Mode 3 for military users.)

Mode C: Azimuth is now augmented by pressure altitude; this is displayed on the controller's screen, adjacent to the aircraft icon thereby providing threedimensional information. Altitude can be taken from the pilot's altimeter from an encoder that sends parallel data (in Gillham/Gray code) to the transponder. This coded data is in 100-foot increments. Aircraft with air data computers will send altitude to the transponder in serial data form, typically ARINC 429.

Mode S (select): In addition to the basic identification and altitude information, Mode S includes a data linking capability to provide a cooperative surveillance and communication system. Aircraft equipped with Mode S transponders allow specific aircraft to be interrogated; this increases the efficiency of the ATC resources. To illustrate this point, when aircraft equipped with Mode A or C transponders are interrogated, all aircraft with this type of transponder will send replies back to the ground station. This exchange occurs each time an interrogation signal is transmitted. Imagine a room full of people; the question is asked: 'please state your name and location in the room'. The person asking the question could become overwhelmed with the replies. If the question was posed in a different way, i.e. on a **selective** basis: 'Mike, where are you?' followed by: 'David, where are you?', the replies are only given by the person being addressed. The Mode S system has a number of advantages:

- Increased traffic densities
- Higher data integrity
- Efficient use of the RF spectrum
- Reduced RF congestion
- Alleviation of Mode A and C code shortages
- Reduced workload for ground controllers
- Additional aircraft parameters available to the ground controller.

Mode S transponders only send a reply to the first interrogation signal; the ground station logs this aircraft's address code for future reference. Mode S provides additional surveillance capability into controlled airspace; this is being introduced on a progressive basis. Aircraft equipped with Mode S transponders are also able to communicate directly with the Mode S transponders fitted to other aircraft; this is the basis of the traffic alert and collision avoidance system (TCAS) and will be described in the next chapter. (Note that Modes B and D are not used by commercial aircraft.)

19.3 Airborne equipment

Commercial transport aircraft are installed with two ATC antennas, a control panel and two transponders as illustrated in Figure 19.5. Since the ATC system and distance measuring equipment (DME) operates in the same frequency range, a mutual suppression circuit is utilised to prevent simultaneous transmissions.

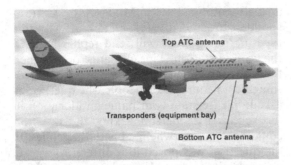

Figure 19.5 ATC airborne equipment locations

19.3.1 Control panel

This is often a combined air traffic control and traffic alert and collision avoidance system (TCAS) control panel, see Figure 19.6 (refer to the next chapter for detailed operation of TCAS). The four-digit aircraft identification code is selected by either rotary switches or push buttons, and displayed in a window. Altitude reporting for Mode C transponders can be selected on or off. When requested by ATC, a momentary make switch is pressed; this transmits the selected code for a period of approximately 15 to 20 seconds. Table 19.1 illustrates the codes that are used in emergency situations.

19.3.2 Transponder

The aircraft transponder (Figure 19.7) provides the link between the aircraft and ground stations. General aviation products have a combined panel and transponder to save space and weight. These can be Mode S capable for IFR operations.

The ground station SSR antenna is mounted on the antenna of the primary radar surveillance system, thereby rotating synchronously with the primary returns. The airborne transponder receives interrogation codes on a 1030 MHz carrier wave from the ground station via one of two antennas located on the airframe. These signals are then amplified, demodulated and decoded in the transponder. The aircraft reply is coded, amplified and modulated as an RF transmission reply code on a 1090 MHz carrier wave. If the transponder is interrogated by a TCAS II equipped aircraft, it will select the

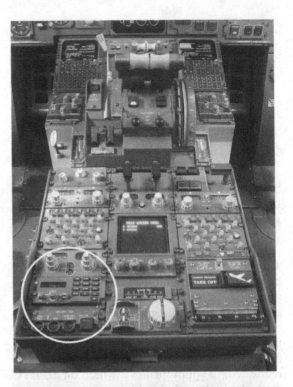

Figure 19.6 ATC control panels

Table 19.1 Emergency ATC transponder codes

Code	Meaning
7700	General air emergency
7600	Loss of radio
7500	Hijacking

Figure 19.7 ATC remote transponder

appropriate antenna to transmit the reply. This technique is called **antenna diversity**; this enhances visibility with TCAS-equipped aircraft flying above the host aircraft.

19.3.3 GA Transponder

General Aviation (GA) transponders are typically self-contained within a single panel mounted unit, see Figure 19.6(c). These are typically Mode S elementary surveillance transponders, with support for 1090 MHz Extended Squitter (ES). Some products have additional functionality, e.g. pressure altitude and GPS readout, Flight ID entry, one-touch VFR code

entry, stop watch timer, flight timer, and altitude alerter.

19.3.4 Altitude encoder

Early altitude encoders were optical-mechanical devices integrated into the barometric altimeter; the data output was a 10-bit parallel bus. Modern encoders on large aircraft are integrated into an air data computer (ADC), with serial data output, e.g. Arinc 429. General Aviation (GA) aircraft typically have a separate solid-state encoder, with options for an RS-232 serial bus and/or parallel Gillham coded bus output.

Gillham coded altitude output is based on the principle of a standard digital technique known as Gray code, NB this is not a natural binary code, see Table 19.2. The table shows the coded outputs over an altitude range of zero to 1000 feet, with each of the parallel digital bits assigned a label, A1, A2, etc. This technique ensures that only one bit of the parallel bus changes on each incremental change, i.e. no transient false values are generated as the output changes between successive values. Coded altitude outputs are in increments of 25 or 100 feet depending on the encoder specification and system requirements. A practical encoder will have an altitude range from −1000 to +30,000 feet, with some products extending up to 42,000 feet.

Table 19.2 Gillham coded altitude

ALTITUDE	A1	A2	A4	B1	B2	B4	C1	C2	C4	D1	D2	D4
0	0	0	0	0	1	1	0	1	0	0	0	0
100	0	0	0	0	1	1	1	1	0	0	0	0
200	0	0	0	0	1	1	1	0	0	0	0	0
300	0	0	0	0	1	0	1	0	0	0	0	0
400	0	0	0	0	1	0	1	1	0	0	0	0
500	0	0	0	0	1	0	0	1	0	0	0	0
600	0	0	0	0	1	0	0	1	1	0	0	0
700	0	0	0	0	1	0	0	0	1	0	0	0
800	0	0	0	1	1	0	0	0	1	0	0	0
900	0	0	0	1	1	0	0	1	1	0	0	0
1000	0	0	0	1	1	0	0	1	0	0	0	0

Key point

Failure of a single data line in a Gillham interface can generate erroneous altitude data, whilst not flagging the error; this will cause incorrect altitude reporting, with the potential for an aircraft near miss situation. Serial output encoders (that will flag an error) are therefore preferred in retrofit situations whenever practical.

Key point

In oceanic and non-radar continental airspace where it is not possible to install either radar or ADS-B ground stations, automatic dependent surveillance-contract (ADS-C) position reports are relayed via satellite communications to air traffic control (ATC); in this system, ATC specifies when the aircraft is to provide position reports, i.e. in a contract. This technology facilitates reduced separation standards.

19.4 System operation

Although SSR has many advantages over primary radar, the smaller antenna's radiation pattern contains substantial **side-lobes**. These side-lobes can generate false returns (Figure 19.8), and so a method of coding the interrogation signals via pulse techniques is employed. The solution is to superimpose an omnidirectional pattern from a second antenna onto the directional beam. Suppressing these side-lobes is discussed in the following sections on interrogations and replies.

19.4.1 Mode A and C interrogation

Interrogation is based on a three-pulse format as illustrated in Figure 19.10; each pulse is 0.8 μs wide. Two pulses (P1 and P3) are transmitted on the rotating antenna thereby producing a **directional** signal. A third pulse (P2) is transmitted on the fixed antenna that radiates an **omnidirectional** signal. The purpose of the P2 pulse is described in the Mode A reply section. Referring to Figure 19.10, P1 and P2 have a 2 μs interval; P3 is transmitted at an interval of 8 μs for Mode A and 21 μs for Mode C interrogations (see Figure 19.11). This spacing between P1 and P3 therefore determines the type of interrogation signal (Mode A or C). The pulse repetition frequency (PRF) of interrogation signals is unique to each ground station; a typical PRF is 1200 interrogation signals per second. Replies are sent by the aircraft at the same PRF.

19.4.2 Mode A reply

A given aircraft's transponder will receive maximum signal strength each time the ground station's directional beam passes, i.e. once per revolution. Since P2 is transmitted from the fixed omnidirectional antenna, it is received with constant signal strength; but with lower amplitude than P1/P3. When the aircraft's transponder receives the maximum P1/P3 signal strength, i.e. when the rotating antenna is directed at the aircraft, they are received at higher amplitude than P2. Referring to Figure 19.9, an aircraft not within the main-lobe of the directional beam would receive a P2 pulse from the omnidirectional antenna at higher amplitude than the P1/P2 pulses. The transponder

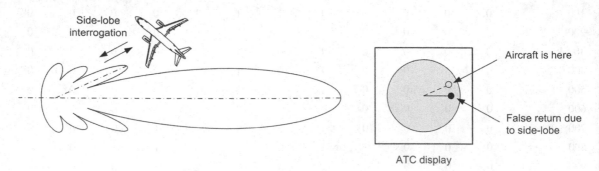

Figure 19.8   False returns from side-lobes

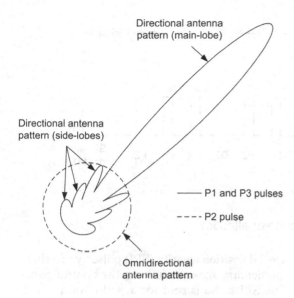

Figure 19.9   Side-lobe suppression (SLS)

recognises this as a side-lobe signal and suppresses any replies until 25 to 45 µs after P2 is received. This is called **side-lobe suppression** (SLS), a technique ensuring that only the main lobe of the rotating antenna is being replied to and not a side-lobe. The physical arrangement and antenna patterns are illustrated in Figures 19.3 and 19.9 respectively.

The Mode A reply is the ATC code allocated to that flight, formed into a series of pulses. This reply is framed between two pulses (F1 and F2) that have a time interval of 20.3 µs as illustrated in Figure 19.12. Data to be transmitted is coded by twelve pulses (plus an unused 'spare' pulse in position X) at 1.45 µs intervals within F1 and F2. The twelve pulses are grouped into four groups of three; each group represents an octal code. Each of the four groups is labelled A, B, C and D; single pulses within the group carry a

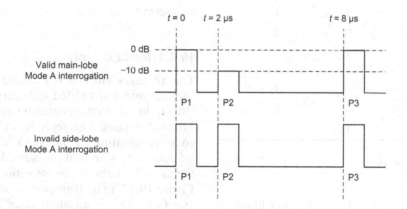

Figure 19.10   Interrogation signal validity

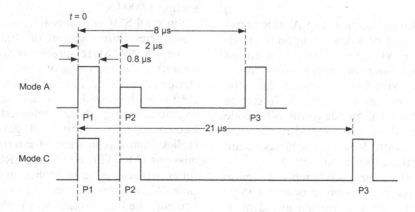

Figure 19.11   Pulse format for Mode A and Mode C interrogations

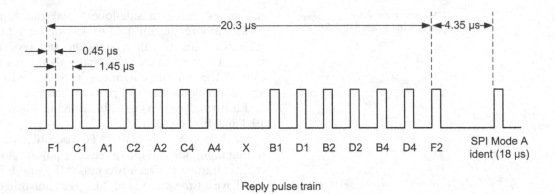

Figure 19.12   Mode A/C reply pulse train (ATC code or altitude)

Table 19.3   Illustration of Group A pulses

A1	A2	A3	Octal value
0	0	0	0
0	0	1	1
0	1	0	2
0	1	1	3
1	0	0	4
1	0	1	5
1	1	0	6
1	1	1	7

Note:
0 = no pulse transmitted
1 = pulse transmitted

numerical weighting of 1, 2 and 4. For illustration purposes, Table 19.3 shows how group A pulses represent the octal values between 0 and 7.

When a pulse occurs in group A, this represents the value 1, 2 or 4 depending on the position of the pulse. When a pulse is not transmitted in the allocated time frame, this represents a value of zero. With four groups of data, the **octal numbers** between 0000 and 77778 can be transmitted; this corresponds to the ATC code allocated to the flight and selected by the crew on their ATC control panel. (4096 codes are possible using these four octal digits.)

The final part of the aircraft reply is a pulse that is sent after F2; this pulse occurs 4.35 μs after F2, but it is only sent when an 'ident' is requested by ATC. The flight crew send this

special position identity (SPI) pulse by pressing a momentary make switch on the control panel. The SPI pulse is sent for a period of 15 to 20 seconds after the switch has been pressed; this highlights the aircraft icon on the controller's screen.

19.4.3 Mode C reply

The aircraft's altitude is encoded by the transponder and transmitted as binary coded octal (BCO) (in 100 foot increments) as described for Mode A replies. The reply will also contain a code representing the aircraft's altitude; this is referenced to standard pressure, 1013.25 mB if the aircraft is above the transition altitude, see Figure 19.12. (The transponder sends Mode A and C replies on an alternating basis.) Figure 19.13 illustrates how an ATC identification code of 2703 is combined with an altitude code representing 1,900 feet.

Since all SSR transmissions are on the same frequencies (interrogation on 1030 MHz and replies on 1090 MHz), problems can occur when aircraft are within range of two or more ground stations. Several replies could be sent by an aircraft to each ground station that sends an interrogation signal; these undesired replies are known as non-synchronised garble, or **false replies from unsynchronised interrogator transmissions** (FRUIT). Note that **FRUIT** is sometimes written as false replies uncorrelated in time. When interrogations are received simultaneously, the transponder will reply to as many ground stations as possible.

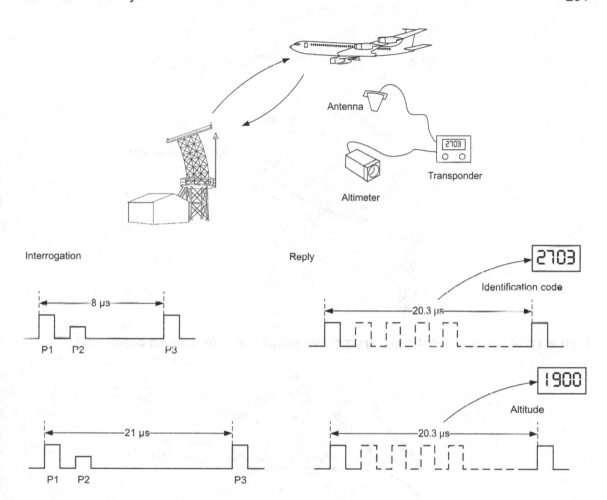

Figure 19.13 Mode C overview

If two or more aircraft are in close proximity, e.g. in a holding pattern, and within the ground station's directional antenna beamwidth, it is possible that their individual replies overlap at the ground station's computer, see Figure 19.14. The situation where replies are received from two or more interrogators answering the same interrogation is referred to as **synchronized garbling**. To resolve this, the controller can request the flight crew on a specific aircraft to provide an ident pulse. The problem is that ATCRBS for Modes A/C requires many interrogations to determine the position (**range and azimuth**) of an aircraft; this requires increased capacity of the ATCRBS infrastructure. The increasing density of aircraft within a given air space leads to false replies as the ground station saturates with garbling conditions. The solution to this is the Mode S (select) system.

19.4.4 Mode S operation

Although Mode S communication is very different to that of Modes A/C, both types of equipment operate on the same frequencies. The system is twoway compatible in that aircraft equipped with Mode A/C transponders will respond with ident and altitude data if interrogated by a Mode S ground station. The principles of Mode S are illustrated in Figure 19.15. Individual interrogations are sent to specific aircraft; only the transponder on this aircraft sends a reply. This reply contains additional information, e.g. selected altitude and flight number.

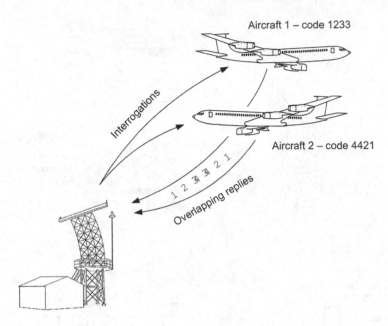

Figure 19.14   Synchronized garbling (from two aircraft within the antenna beamwidth)

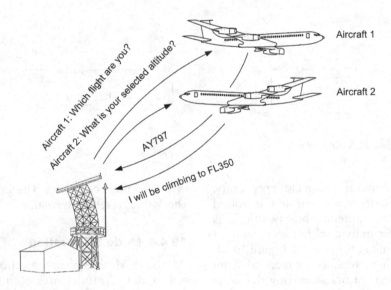

Figure 19.15   Principles of Mode S

Directional and omnidirectional beam patterns are transmitted as illustrated in Figure 19.16. Unlike ATCRBS, Mode S uses a **monopulse** SSR; this reduces the number of interrogations required to track a target. In theory, monopulse radar only requires one reply to determine the target's azimuth (direction and range).

Two interrogation uplink formats (UF) are transmitted, as shown in Figure 19.17; these are the **all-call** and **roll-call** interrogations. The two interrogations are transmitted on an alternating basis and differentiated by the width of a P4 pulse; this is either 0.8 µs or 1.6 µs. The shorter pulse is used to solicit replies from Mode A/C

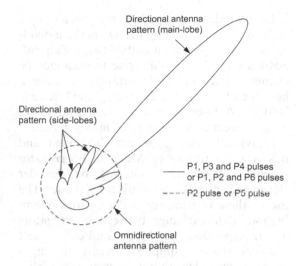

Figure 19.16   Mode S antenna pattern

transponders; they reply with their ATC code and altitude as before. Mode S transponders will not reply to this interrogation. When the P4 pulse is 1.6 µs, Mode S equipped aircraft will reply with their unique address. These replies are stored by the Mode S system as unique identifiers for each specific aircraft, see Figure 19.18(b).

The Mode S discrete addressed interrogation uplink format (UF) is shown in Figure 19.17. Pulse P1 and P2 both have the same amplitude and are part of the directional antenna's main-lobe. This pair appears as suppression pulses to Mode A/C transponders, so they do not reply. Mode S transponders then seek the start of the P6 data pulses; this is formed by a pattern of phase reversals that form a series of logic 1/0. **Phase-shift keying (PSK)** is a modulation

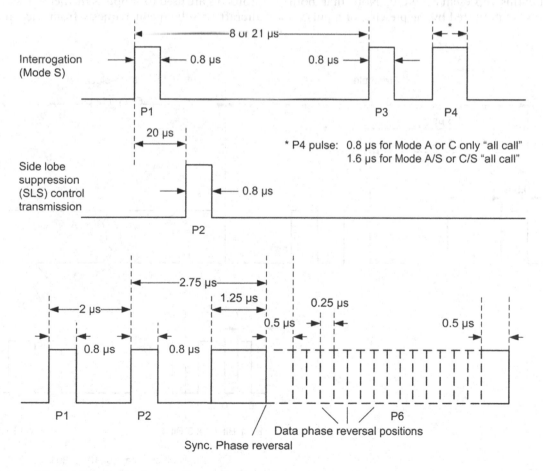

Figure 19.17   Mode S interrogation data format

technique that shifts the phase by minus 90 degrees for a logic one, and +90 degrees for a logic zero. Each data pulse's duration is 0.25 µs; the pulse's phase is sampled at these intervals. A reference pulse of 1.25 µs duration is used to indicate the start of the data word. The word length of P6 (56 or 112 bits) depends on the transponder type.

The Mode S reply is sent via the 1090 MHz carrier wave, as illustrated in Figure 19.18(a); this contains a four-pulse preamble, starting with two pairs of synchronising pulses followed by a block of data pulses (either 56 or 112 bit blocks). Using **pulse position modulation** (PPM), each data bit is allocated a 1 µs time interval, divided into two halves. If the first half of this interval contains a pulse, this represents logic 1; if the second half of the interval contains a pulse this represents logic 0. Note that both states are indicated by the presence of a pulse.

Each Mode S-equipped aircraft has a unique address allocated to it by ICAO via the individual national registration authorities; the **aircraft address** (AA) is a 24-bit code that cannot be changed. Each national authority allocates a header code within the 24 bits, e.g. the UK code is 01000. A 24-bit code of all zeros is not valid; all ones are used for the all-call interrogation.

Individual interrogators are also coded, and this is a key feature of Mode S. **Interrogator codes** (IC) comprise 15 interrogator identifier (II) and 63 surveillance identifier (SI) codes. The use of these interrogator codes ensures unambiguous data exchange between interrogators and transponders. The Mode S all-call request acquires Mode S equipped aircraft entering a given airspace. The aircraft transponder replies with its unique aircraft address (AA). Lock-out protocols are used to suppress further replies by aircraft to subsequent requests from the same

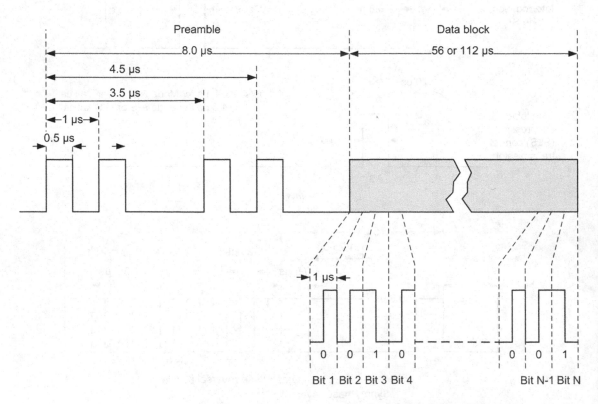

Figure 19.18   Mode S reply data format

interrogator. Transponders are thereby conditioned for replying to a specific interrogator identified by the interrogator codes (IC); the transponder will subsequently ignore requests from other interrogations. Following the all-call lock-out, the interrogator will address individual aircraft transponders on a **selective** basis. Only the selected aircraft transponder will reply;

interrogators will ignore replies not intended for them.

Mode S is being introduced on a progressive basis via a transitional phase of equipment standards. The two standards are: **elementary surveillance** (ELS) and **enhanced surveillance** (EHS). Data sent by each of these two standards is shown in Tables 19.4 and 19.5 respectively.

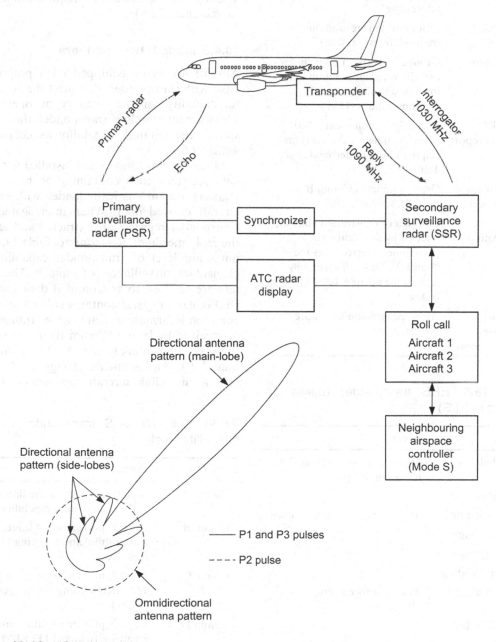

Figure 19.18   (Continued)

Table 19.4   ELS transponder replies

Data	Notes
24-bit aircraft address	Allocated to individual aircraft by ICAO via the national registration authority
SSR Mode 3/A	Range and azimuth measurements; selective addressing
SSR Mode C	25 foot altitude resolution (reduced from 100 foot)
Flight status	Ground/airborne. This includes the squawk ident function in the form of downlink aircraft communications (DAP)
Datalink capability report	This information is extracted when the transponder ID is first acquired by the interrogator, see Table 19.5
Common usage GICB report	Ground initiated Comm B (GICB), see Table 19.5
Aircraft identification	Call sign or registration number; selected by the flight crew (sometimes referred to as the flight ID). This will eventually replace the existing 4096 ATC codes
ACAS active resolution advisory report	Airborne collision avoidance system

Table 19.5   EHS transponder replies (in addition to ELS)

Data	Notes
Selected altitude	Typically from the autopilot mode control panel
Roll angle	
Track angle rate	True airspeed as an alternative
True track angle	
Ground speed	
Magnetic heading	
Indicated airspeed	Mach number as an alternative
Rate of climb/ descent	

The two standards of Mode S are applied in accordance with specified criteria: airspace requirements, Maximum Take off Weight (MTOW) and maximum True Airspeed (TAS) aircraft. With a MTOW above 5700 kg or maximum cruising TAS above 250 knots, and flying Instrument Flight Rules (IFR), enhanced Mode S surveillance (EHS) is required. Aircraft not meeting these criteria only require elementary surveillance (ELS).

19.4.5 Mode S transponders

Aircraft are being equipped on a progressive basis with transponders that meet the necessary functionality for the category of operations. There are four levels of transponder; these levels specify the datalink capability as detailed in Table 19.6.

Mode S transponders are installed with 255 data registers, each containing 56 bits. These registers are automatically loaded with specific aircraft derived data. When interrogated, the transponder registers are extracted and sent as the reply messages. Referring to Table 19.6, the minimum level of transponder capability for elementary surveillance is Comm B. The registers are referred to as Comm B data selectors (BDS); each register contains either specific or common information. Data can be transmitted or received by level 2 (Comm B) transponders; this is via a downlink format (DF) or uplink format (UF). Alphanumeric strings of data are sent as **downlink aircraft parameters** (DAP).

Table 19.6   Mode S transponder datalink capability levels

Level Type	Detail
1 Comm A	Mode A, C and S surveillance, without a datalink capability
2 Comm B	Level 1, plus standard length transmitting and receiving 112-bit messages
3 Comm C	Level 2, plus receiving of 16 linked 112 extended length messages (ELM)
4 Comm D	Level 3, plus transmitting and receiving 16 linked 112 ELM

Certain BDS registers in the transponder are used for specific parameters, other registers have common usage. These registers are checked for timely updates. A report on the status of these updates is provided when requested by the ground station: ground initiated Comm B (GICB).

Test your understanding 19.1

On which frequencies are ATC interrogations and transmitted and replies?

Test your understanding 19.2

Explain the term 'FRUIT' in the context of air traffic control.

Test your understanding 19.3

Explain the principles of pulse position modulation.

Test your understanding 19.4

What are the three emergency ATC codes?

Test your understanding 19.5

What are the differences between ATC transponder Modes A, C and S?

Key point

SSR has the following advantages compared with primary radar:
- Low power transmitter
- Superior returns from the target in terms of signal strength and integrity
- Transmissions and returns can be coded to include data
- Smaller antennas.

Key point

Altimeters are used in some aircraft to provide an encoded digital output (barometric altitude to the transponder) in Gillham code. This is a modified form of Gray code, where two successive values differ in only one digit. This code prevents spurious outputs from the analogue encoder. To illustrate how this code is used, a four bit parallel output would count from zero to seven as follows: 0000, 0001, 0011, 0010, 0110, 0111, 0101, 0100.

Key point

The 100 foot resolution used in many Mode C transponders is being updated to 25 foot resolution for Mode S; serial data from the encoder will be required to achieve this.

19.5 Automatic dependent surveillance–broadcast (ADS-B)

19.5.1 Introduction

The development of Mode S has led onto two related systems: traffic alert and collision avoidance systems (TCAS) and automatic dependent surveillance-broadcast (ADS-B). Both systems can exchange data directly between aircraft. TCAS is a **surveillance** system that provides warnings directly to the crew when other navigation systems (including ATC) have failed to maintain safe separation of aircraft. ADS-B is the technology for **air traffic management** (ATM) that is intended to replace conventional secondary surveillance radar (SSR), see Figure 19.19. TCAS is addressed in the Chapter 20; ADS-B is described below.

In addition to areas with existing SSR, ADS-B provides surveillance in remote areas where ground radar coverage is not possible, e.g. over oceans. ADS-B forms part of the FAA's next generation air transport system (NGATS). It will revolutionise how pilots obtain traffic and weather information. The intention is to increase air navigation safety by providing crews with realtime information about other traffic; this makes it possible for the crew to be responsible

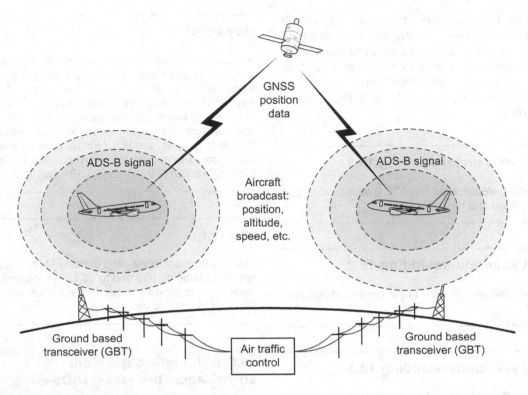

GNSS
position
data

ADS-B signal

ADS-B signal

Aircraft
broadcast:
position,
altitude,
speed, etc.

Ground based
transceiver (GBT)

Air traffic
control

Ground based
transceiver (GBT)

Figure 19.19   ADS-B overview

for their own aircraft's separation and collision avoidance. The system is **automatic** in that no interrogation is required to initiate a transponder broadcast from the aircraft; this type of unsolicited transmission is known as a **squitter**.

ADS-B utilises conventional global navigation satellite system (GNSS) and onboard equipment for communication making it **dependent**. Air traffic coordination is thereby provided though **surveillance** between aircraft; the system has a range of approximately 150 nm.

A significant benefit of ADS-B is the estimated 90% cost saving compared with replacing ageing SSR system infrastructures. Other benefits include greater access to optimum routes and altitudes; this leads to reduced fuel consumption and greater utilisation of aircraft. The system provides **real-time data** for both flight crews and air traffic controllers. Data is exchanged between aircraft and can be independent of ground equipment. Since SSR is based on range and azimuth measurements, the accuracy of determining an aircraft's position reduces as a function of range from the radar

antenna see Figure 19.21. Two aircraft in close proximity, but some range from an SSR ground station, can exchange data via ADS-B and calculate their relative positions more accurately. Referring to Figure 19.22, with the actual positions of two aircraft, and a given separation. The two positions determined by radar alone will have inaccuracies at greater ranges, and so their separation has to be increased. Secondary radars typically take up to ten seconds to update an aircraft's position; ADS-B equipment provides ATC with updated aircraft information within seconds, i.e., an order of magnitude faster; this is critical in congested airspace. This increased accuracy and faster response enables controllers to identify and resolve potentially hazardous situations rapidly and effectively. When two aircraft are both transmitting their positions via ADS-B, the position accuracy is greatly increased, and their separation can be reduced. If one aircraft is being tracked by radar, and the other is transmitting via ADS-B, then their separation would need to increase, albeit less than the radar only scenario. ADS-B

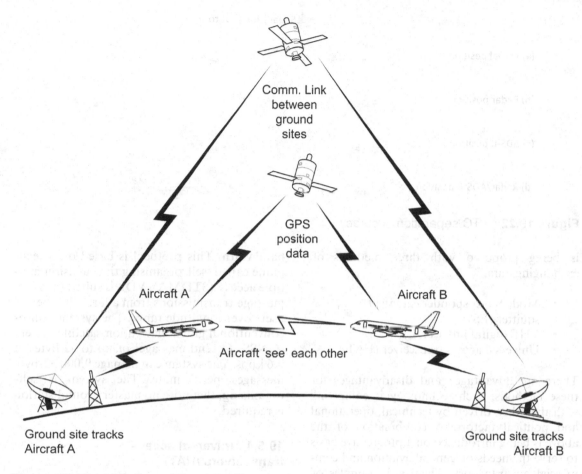

Figure 19.20   ADS-B used for air traffic management (ATM)

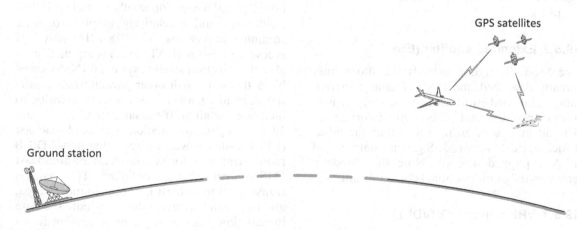

Figure 19.21   ADS-B and secondary surveillance radar

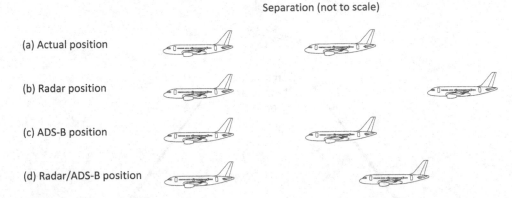

Separation (not to scale)

(a) Actual position

(b) Radar position

(c) ADS-B position

(d) Radar/ADS-B position

Figure 19.22 ATC separation accuracy

is being proposed with three methods of exchanging data:

- Mode S transponder extended squitter (ES)
- VHF digital link (VDL)
- Universal access transceiver (UAT).

There are advantages and disadvantages for these methods; each technology is competing with the other, driven by technical, operational and political factors. A combination of the above is being introduced on a progressive basis to serve the needs of general aviation and commercial air transport. There are examples of where Mode S and VDL have been integrated into single ground stations. The reader is encouraged to monitor developments via the industry press.

19.5.2 Extended squitter (ES)

The Mode S method extends the information already described above for enhanced surveillance. Extended squitter (ES) messages include aircraft position and other status information. The advantage of using ES is that the infrastructure exists via Mode S ground stations and TCAS-equipped aircraft. Note that Mode S provides only unidirectional communications.

19.5.3 VHF digital link (VDL)

VDL utilises the existing aeronautical VHF frequencies to provide bi-directional communications; digital data is within a 25 kHz bandwidth. This protocol is based on a technique called 'self organising time division multiple access' (STDMA). VDL is suited for short message transmissions from a large number of users over longitude range. The system utilises conventional global navigation satellite system (GNSS) to send messages of up to 32 bytes at 9.6 kbps. The system can manage 9,000 32-byte messages per minute. The system is self-organising, therefore no master ground station is required.

19.5.4 Universal access transceivers (UAT)

In order to illustrate the principles of ADS-B, the universal access transceiver (UAT) is described in more detail. UAT uses conventional global navigation satellite system (GNSS) technology and a relatively simple broadcast communications link. The 978 MHz universal access transceiver (UAT) receives inputs from a global navigation satellite system (GNSS), combines this data with other parameters, e.g. airspeed, heading, altitude and aircraft identity, to facilitate the **air traffic management**, see Figure 19.20. Flight information services-broadcast (FIS-B), such as weather and other non-ADS-B radar traffic information services-broadcast (TIS-B), can also be uplinked. This data is transmitted to aircraft in the surrounding area, and to ground receivers that distribute the data in real time via existing communication infrastructures. The system thereby allows operations in remote and/or mountainous areas not covered by ground radar.

Key point

ADS-B's position reporting is more accurate than from ground-based radars. This enables reduced traffic separation, making better use of the airspace.

Key point

ADS-B transmits reports to any station capable of receiving them. These transmissions are automatic and continuous, i.e., not in response to requests from air traffic ground stations.

19.5.5 ADS-B Mandates

ADS-B is being mandated throughout the world, via the regulatory authorities, e.g. the EASA and FAA. The reader is encouraged to monitor the industry announcements and rule making process for details, e.g. aircraft and airspace applicability.

ADS-B in Europe will use 1090 MHz extended quitter (ES) as the preferred initial technology, thereby using transponders that are already installed via Mode S and TCAS mandates for European airspace. 1090 MHz ES is fully compatible with other ADS-B technology, ensuring global interoperability for commercial aircraft. The European mandate applies to aircraft >5700 kg/ 12,500lbs or max cruise >250kts TAS.

The FAA mandate supports two technologies for ADS-B, 1090 MHz ES and Universal Access Transceiver (UAT). Since there is a difference in frequencies between the two technologies, UAT receivers cannot receive 1090-ES ADS-B transmissions and 1090-ES receivers cannot receive UAT transmissions.

The 1090-ES link is required for aircraft that fly above FL180. The ADS-B ground stations will re-transmit the information on the opposite link so that all aircraft can be seen on cockpit traffic displays; this is known as ADS-R, R for rebroadcast.

Key point

"ADS-B IN" is the aircraft's reception of ground station broadcast data, e.g. FIS-B and TIS-B. It also includes other ADS-B data, i.e. communication from nearby aircraft. "ADS-B OUT" periodically broadcasts information about the aircraft, e.g. identification, current position, altitude, and velocity.

19.5.6 Automatic Dependent Surveillance-Contract (ADS-C)

This is a development of ADS-B, where the data exchanges are via an open contract between the air navigation service provider (ANSP) and an aircraft. ADS-C provides a two-way communications system that provides comprehensive information critical to flight safety, see Figure 19.23. A contract is established between the aircraft and ANSP for regular position reporting; the two-way procedure transmits position reports of the aircraft on an agreed specified frequency, typically between 10 to 14 minutes. This frequency could be reduced to shorter intervals, typically 2–3 minutes, to further reduce separation depending on the air traffic environment.

ADS-C data exchanges are transmitted over the ACARS network via satellite communications (SATCOM) and can operate either with controller pilot data link communications (CPDLC) or autonomously. Typical ADS-C data exchanges include the aircraft's:

- Identification
- Position
- Altitude
- Speed
- Heading
- Rate of climb or descent
- Next waypoints (with estimated time of arrival – ETA and intended altitude)
- Weather information

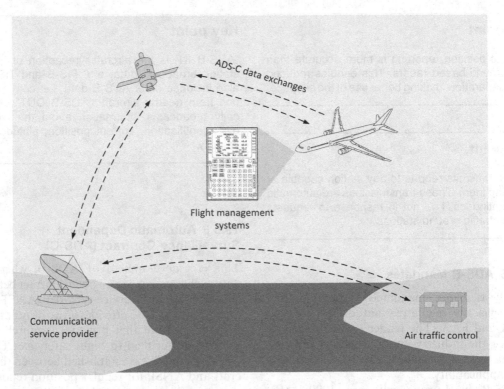

Figure 19.23 Automatic Dependent Surveillance-Contract (ADS-C)

Key point

ADS-C is a prior agreement between the aircraft and a specific ground station for data exchange, i.e., the data exchange is not broadcast to all stations. The contract is managed by the ANSP that determines the scope of the data exchange, and when this will occur.

19.6 Communications, navigation and surveillance/air traffic management (CNS/ATM)

19.6.1 Introduction

This subject is derived from the numerous disciplines and technologies required to enable aircraft to navigate, and air traffic control to manage, airspace. In many parts of the world, dense traffic flows are currently being managed; other parts of the world are seeing continuing increases in air traffic. Today's ATC infrastructure, including operating methods and equipment, cannot

possibly manage the predicted demands of air traffic management. It is vital that global standards are developed and implemented for the delivery of a safe, efficient and economic air navigation service provision.

19.6.2 Infrastructure

Navigation is not always about flying great circle routes for the shortest distance between two points, e.g. tailwinds should be exploited and headwinds avoided. This requires real-time weather information for pilots and controllers. Close cooperation is also required with airports to ensure efficient arrivals and departures thereby minimising delays. Air traffic management addresses traffic flow at the optimum speed, height and route to minimise fuel consumption. Numerous enabling factors for CNS/ATM will lead to higher navigation accuracy at lower cost (not just the cost of fuel, but also the impact of air travel on the environment). Area navigation already provides a flexible and efficient means of en route and terminal area operations in place of airway routings.

19.6.3 CPDLC

A critical factor in CNS/ATM is the saturation and congestion of voice communication channels. Controller pilot datalink communications (CPDLC) helps alleviate this situation by providing an additional communications medium for aircrews and controllers. CPDLC allows air traffic controllers to communicate with pilots over a datalink system, e.g. ATN, FANS, or VDL Mode 2. CPDLC is being globally implemented, and is in different implementation stages; European airspace will require CPDLC for operations above FL285.

CPDLC is an air/ground datalink application which enables the exchange of datalink messages between controllers and pilots, thereby complementing traditional voice communications. The objective of CPDLC is to improve the safety and efficiency of air traffic management, enabling routine, non time-critical communications between pilots and controllers. Examples of CPDLC message include:

- changes SSR code/squawk ident
- change of communications frequency
- ATC clearances, e.g. level changes, vectoring, direct routing speed control, etc.

Aircrew are also able to reply to ATC instructions via CPDLC, and request alternative routings. Urgent messages can still be via traditional voice communications if required. CPDLC will supplement company or engineering information currently handled by the lower capacity ACARS datalink system. The phased introduction of CPDLC will eventually see existing voice communications for air traffic control purposes only used as a backup. Typical CPDLC messages are displayed on the FMS CDU, see Figure 19.24.

Key point

ADS-B provides real-time data for both flight crews and air traffic controllers; data is exchanged between aircraft and is independent of ground equipment. The system allows operations in remote and/or mountainous areas not covered by ground radar.

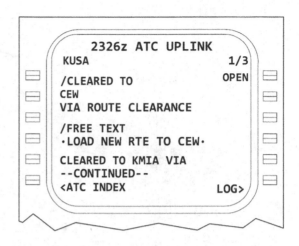

Figure 19.24 Typical CPDLC messages

Key point

Side-lobe suppression is achieved by transmitting directional and omnidirectional pulses.

19.6.4 Remote airfield control towers

Deployment continues with remote airfield control towers, enabling air traffic controllers to manage flights without the need for staff to be physically present in the airfield control tower. This will not affect the on-board aircraft systems, e.g., VHF communications, transponders etc. and so the ground-based technology is not described in any detail.

Initially developed for airports with low traffic movements, remote towers can also be considered when existing (staffed) towers are to be decommissioned. In 2021, the UK's National Air Traffic Services (NATS) announced that London City Airport (UK) became the first major international airport in the world to be fully controlled by a remote digital air traffic control tower. This airport is a major civil aerospace hub, with dense traffic in a complex airspace environment.

Flights are remotely guided to land or take off from the airfield by air traffic controllers at a control centre using an 'enhanced reality' view supplied by a digital control tower, Figure 19.25. High-definition cameras and sensors mounted on the digital control tower capture a 360-degree

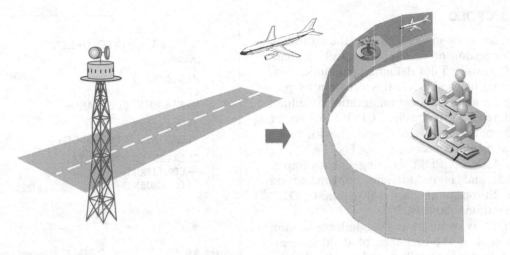

Figure 19.25 Remote airfield control tower

view of the airfield. This data is relayed through a super-fast fibre connection to an air traffic control centre.

Key point

Mode S eliminates synchronous garbling, increases the capacity of a given air space and improves surveillance accuracy.

Key point

Enhanced Mode S provides data for the current state of motion of the aircraft together with the aircraft's vertical intention, i.e. selected altitude.

Key point

ATC and DME transponders operate in the same frequency band; their transmissions have to be coordinated.

Key point

The phased introduction of CPDLC will see existing voice communications for air traffic control purposes only used as a backup.

Key point

The demands on ATC resources are further reduced by Mode S since replies are only sent during the initial 'all-call' broadcast, and thereafter only when specifically requested.

Key point

The Mode S reply is in a different form to that of Modes A or C, and is only sent to a Mode S interrogation.

Test your understanding 19.6

Who allocates the following: ATC identity codes, 24-bit aircraft address codes?

Test your understanding 19.7

What is the difference between elementary and enhanced surveillance?

Test your understanding 21.8

Explain each of the following terms: Automatic Dependent Surveillance–broadcast

19.7 Single European Sky

With increasing air traffic movements in the late 20th century, European airspace was traditionally fragmented, with limited capacity across national airspace regions. This often led to delays and many other inefficiencies. The Single European Sky (SES) initiative was created to address these challenges whilst reducing the cost of air traffic service provision and increasing Europe's capacity to meet forecast growth in demand for air traffic.

The Single European Sky ATM Research programme (SESAR) is a public–private crossindustry initiative. It brings together the aviation industry to develop new technologies and solutions that will improve the way Europe's airspace is managed and oversee their implementation. The Single European Sky ATM Research (SESAR) project was set up in 2004 as the technological pillar of the Single European Sky (SES), aiming to modernise Europe's Air Traffic Management (ATM) system. The high-level goals of SESAR are: tripling capacity, reducing costs per flight by 50%; reducing emissions by 10% and improving safety by a factor of 10.

SESAR affects all aspects of air traffic management, i.e. not just airborne equipment. The latter includes the Communications, Navigation and Surveillance (CNS) systems described in this book, e.g. ADS-B, P-RNAV, CPDLC, etc. New technologies are being developed to meet the communication capacity and performance requirements of air traffic management in the future. This includes the introduction of multipurpose communications equipment capable of fulfilling conventional CNS functions using generic computing platforms and software. CNS systems are not always used simultaneously in all airspace; dynamically reconfigurable CNS systems operate specific functions only when required. This flexibility reduces the number of separate hardware components carried on board the aircraft.

Other technologies that are included within the scope of SESAR, and already described in this book include: P-RNAV, FMS, 4D navigation, ADS-B and CPDLC. Variations in technologies already described in this book include increased second glide slope (ISGS), with a second, steeper angle, glide slope signal. Another concept is the second runway aiming point (SRAP), enabling aircraft to land on a second runway aiming point, further along the runway. ISGS and SRAP could be combined with increased glide slope angle to the second runway aiming point. Other technologies including enhanced vision systems (EVS), synthetic vision systems (SVS) head up displays (HUD) are described in another book title in the series *Aircraft flight instruments and guidance systems* (AFIGS).

Air traffic management for operations above FL550 do not yet exist on a large scale in European airspace. Flights in this region of the atmosphere are no longer exclusive to military and space operations. Typical flights now include commercial space operations, high altitude platforms, e.g., balloons and pseudo satellites. It is anticipated that operations will be significantly expanded via new business opportunities. Operational and regulatory requirements in higher airspace is being coordinated via the European concept for higher airspace operation (ECHO) SESAR 2020 project.

19.8 FAA Next Generation

The FAA's Next Generation Air Transportation System (NextGen) is a programme for the systematic transformation of the National Airspace System (NAS) of the United States, planned in stages between 2012 and 2025. NextGen represents a progression from a ground-based system of air traffic control to a satellite-based system of air traffic management. This will be achieved through the development of aviation-specific applications for existing, widely-used technologies, such as the global positioning system (GPS) and technological innovation in areas such as weather forecasting, data networking and digital communications. The objective of NextGen is to facilitate higher air traffic density on more direct routes, thereby reducing delays, carbon emissions, fuel consumption and noise.

Progression to NextGen is being facilitated by the incorporation of satellite-based and digital technologies, together with new and updated procedures that will be integrated to achieve the objectives. NextGen airborne technologies already described in this book include: P-RNAV, FMS, 4D navigation, ADS-B and CPDLC.

19.9 Future Air Navigation Systems (FANS)

Future Air Navigation System (FANS) collectively refers to the systems and procedures required to operate in specific countries and oceanic areas around the globe. This allows more aircraft to safely and efficiently utilise a given volume of airspace that is not under direct, i.e. radar surveillance. FANS is currently used primarily in oceanic regions, utilising both satellite communications and navigation systems to effectively create a virtual radar environment for aircraft surveillance.

FANS provides direct datalink communication between the pilot and Air Traffic Control (ATC), based on CPDLC and ADS-C described in this book. Satellite communications (SATCOM) and VHF data link (VDL) technology is used to enable digital transmission of messages between the aircraft and ground stations.

There are two technology approaches to FANS, developed by Boeing and Airbus.

The Boeing approach (referred to as FANS-1) is based mainly on updating the software in existing systems, e.g., the flight management system (FMS) and primary flight display (PFD). The FMS then provides additional functions such as CPDLC. The PFD provides lateral and vertical guidance when in LNAV and VNAV modes. Existing satellite communications systems are used for ACARS.

The Airbus approach (referred to as FANS-A) is based on introducing new items of equipment. The Air Traffic Services Unit (ATSU) includes the functions of ACARS, with additional air traffic messaging; the data link control display unit allows the flight crew to respond to CPDLC messages; the digital VHF datalink (VDL).

Where there are common features between the two FANS technology approaches, e.g., ADS-C, these are collectively referred to FANS 1/A+ (where the plus sign designates minor enhancements).

Figure 19.26 Urban air mobility (UAM) concept

19.10 Drones

The term "drone" is a generic term for unmanned aerial vehicle/systems (UAV/UAS). Drones cover a wide range of products, from recreational items that are available to the general public, through to aerial vehicles and systems that require certification. In the latter case, there is an increasing use of drones for applications in agriculture, surveying, delivery of food, medicines etc, see Figure 19.26.

Automatic systems incorporate a remotely located pilot that can take control of the drone, e.g., to intervene in unforeseen events for which the drone has not been programmed. Autonomous systems incorporate artificial intelligence (AI), i.e., there is no intervention from a pilot.

Work continues with integrating drones into national airspace systems. Along with commercial aircraft, drones need to provide tracking and identification whilst in flight. Urban air mobility (UAM), also referred to as advanced air mobility (AAM), is an air transportation concept for passengers and cargo in urban environments. A European system called "U-space" is being developed to manage the integration of drones into the air traffic management system.

19.11 Multiple choice questions

1. Side-lobes are inherent in which part of the SSR interrogation?
   (a) Omnidirectional antenna
   (b) Rotating antenna
   (c) Transponder reply.

2. The transponder always suppresses any reply when:
   (a) P2 amplitude is =P1/P3
   (b) P2 amplitude is ≤P1/P3
   (c) P2 amplitude is ≥P1/P3.

3. Compared with primary radar, the transmission power used by secondary surveillance radar is:
   (a) higher
   (b) the same
   (c) lower.

4. ATCRBS ident codes are formatted in which numbering system?

(a) Binary
(b) Octal
(c) Decimal.

5. Directional and omnidirectional pulses are transmitted by the SSR for:
   (a) ident and altitude data
   (b) DME suppression
   (c) side-lobe suppression.

6. Mode C (altitude information) is derived from an:
   (a) altimeter or air data computer
   (b) ATC transponder
   (c) ATC control panel.

7. The special position ident (SPI) pulse is transmitted for a period of:
   (a) indefinite time
   (b) 15 to 20 seconds
   (c) 8 to 21 μs.

8. Side-lobe suppression is achieved by transmitting:
   (a) directional (P1/P3) and omnidirectional (P2) pulses
   (b) directional (P2) and omnidirectional (P1/3) pulses
   (c) directional (P1) and omnidirectional (P2/3) pulses.

9. The transponder code of 7700 is used for:
   (a) general air emergency
   (b) loss of radio
   (c) hijacking.

10. ATC interrogations and replies are transmitted on the following frequencies:
    (a) interrogation on 1030 MHz, replies on 1090 MHz
    (b) interrogation on 1090 MHz, replies on 1030 MHz
    (c) interrogation on 1030 MHz, replies on 1030 MHz.

11. The 'Mode S all-call' interrogation will cause Mode A and C transponders to reply with their:
    (a) ident and altitude data
    (b) aircraft address code
    (c) special position ident (SPI) pulse.

Chapter 20

Traffic alert and collision avoidance systems

Several airborne collision avoidance system (ACAS) technologies have been developed since the 1950s; this chapter focuses on the traffic alert and collision avoidance system (TCAS). TCAS is an automatic surveillance system that helps aircrews and ATC maintain safe separation of aircraft. It is an airborne system based on secondary radar that interrogates and replies directly between aircraft via a high-integrity datalink. The system is functionally independent of ground stations, and alerts the crew if another aircraft comes within a predetermined time to a potential collision. TCAS is used on larger transport aircraft; smaller general aviation aircraft use a range of technologies.

20.1 Airborne collision avoidance systems (ACAS)

There are various collision avoidance technologies in use, or being planned:

20.1.1 Passive receivers

These units are intended for general aviation use; they monitor ATC transponder signals in the immediate area, and provide visual or audible signals to warn of nearby traffic. They have a range of approximately six miles and monitor 2,500 feet above or below the host aircraft. The receiver is normally located on the aircraft's glare-shield. It has an internal antenna, which can lead to intermittent coverage depending on how and where the unit is positioned. Passive systems provide approximations of where another aircraft is, and its position relative to the host aircraft. Some passive devices provide vertical trend information, e.g. indicating if the other aircraft is climbing or descending.

20.1.2 Traffic information system (TIS)

This uses the host aircraft's Mode S transponder (refer to Chapter 19) to communicate with the ground-based secondary surveillance radar (SSR) network. Traffic information can be obtained within a five-mile radius, 1,200 feet above or below the host aircraft. This traffic information is provided on a (near) real-time basis. The attraction of TIS is that aircraft hardware and software are minimal since the system 'feeds' off ground station computations. TIS is unavailable outside of areas covered by SSR.

20.1.3 Image-based systems

One major factor to consider with an integrated air traffic airspace environment is the "non-cooperative traffic" scenario i.e. where there no conventional exchange of information (e.g., TIS, Mode-S, ADS-B etc) from drones, other nearby aircraft or flocks of birds.

Collision avoidance systems for unmanned aerial vehicles (UAV), or drones, have been developed using range of sensors, e.g., cameras, radar etc, together with artificial intelligence (AI). The drones incorporate automatic or autonomous detection and avoidance features with a "sense and avoid" capability. Depending on the type and purpose of the drone, this can be a centralised protocol where the collision avoidance manoeuvres for all drones in a designated airspace are planned simultaneously. Alternatively, individual drones will have their own independent collision avoidance protocols, with the manoeuvring sequence(s) programmed in accordance with a pre-designated priority.

Although drone collision avoidance technology is outside the scope of this book, there are applications for image-based systems in general aviation (GA). The optical sensors and associated

DOI: 10.1201/9781003411932-20

AI can be used to perceive the surrounding flight environment, thereby enhancing the pilot's situational awareness. The image-based system will continuously scan the surrounding environment for non-cooperative traffic. The optical sensors can be specified with a visual range greater than that of the pilot's eyes; the system is able to provide continuous scanning of the sky for traffic and other airborne hazards. Image-based systems with associated AI could also be used to identify suitable landing areas.

20.1.4 Traffic advisory system (TAS)

The host aircraft's TAS actively monitors the airspace seeking nearby transponder-equipped aircraft and provides relevant traffic information via a display and audio warning. TAS uses active interrogation of nearby transponders to determine another aircraft's position and movement. A typical system can track up to 30 aircraft with a range of up to 21 nm, 10,000 feet above or below the host aircraft.

20.1.5 Traffic alert and collision avoidance system (TCAS)

This is the industry standard system mandated for use by commercial transport aircraft, and the main subject of this chapter. Two types of TCAS are in operation, TCAS I and II. Both systems provide warnings known as 'advisories' to alert the crew of a potential collision. TCAS I assists the crew in visually locating and identifying an intruder aircraft by issuing a traffic advisory (TA) warning. TCAS II is a collision avoidance system and, in addition to traffic advisories, provides vertical flight manoeuvre guidance to the crew. This is in the form of a resolution advisory (RA) for threat traffic. A resolution advisory will either increase or maintain the existing vertical separation from an intruder aircraft. If two aircraft in close proximity are equipped with TCAS II, the flight manoeuvre guidance is coordinated between both aircraft. A third type of system (TCAS III) was intended to provide lateral guidance to the crew; however, this has been superseded by a automatic dependence surveillance broadcast (ADS-B) as described in the air traffic control chapter of this book.

20.2 TCAS overview

Secondary surveillance radar (SSR) transmits a specific low energy signal (the interrogation) to a known target. This signal is analysed and a new (or secondary) signal, i.e. not a reflected signal, is sent back (the reply) to the origin, see Figure 20.1. In the TCAS application, interrogations and replies are sent directly between the on-board ATC transponders, see Figure 20.2. The TCAS computer interfaces with the ATC transponder and calculates the time to a potential collision known as the **closest point of approach** (CPA).

TCAS creates a **protected volume of airspace** around the host aircraft, see Figure 20.3; this is based on altitude separation and a calculated time to the CPA. The Greek letter **tau** (T) is the symbol used for the approximate time (in seconds) to the CPA, or for the other aircraft reaching the same altitude. This protected volume of airspace is determined as a function of time (tau) for both range and vertical separation:

$$\text{Rangetau} = \frac{3,600 \times \text{slant range (nm)}}{\text{closing speed (knots)}}$$

$$\text{Verticaltau} = \frac{\text{altitude separation (feet)} \times 60}{\text{combined vertical speed (fpm)}}$$

TCAS interrogates other aircraft within this protected airspace and obtains their flight path details, i.e. range, altitude and bearing. This data is analysed along with the host aircraft's flight path. If there is a potential conflict between flight paths, a visual and audible warning is given to the crew. This warning depends on the type of equipment installed in the host and other traffic, as shown in Figure 20.4.

Tau is programmed for varying sensitivity levels determined by altitude bands as illustrated in Table 20.1. For each altitude band, there is a different sensitivity level and corresponding value of tau for traffic and resolution advisories (TA and RA respectively). Higher sensitivity levels provide a larger protected volume of airspace. If closure rates are low, modifying the range boundaries required to trigger a TA or RA

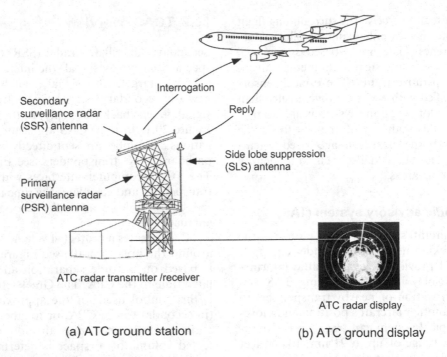

(a) ATC ground station   (b) ATC ground display

Figure 20.1 Secondary surveillance radar

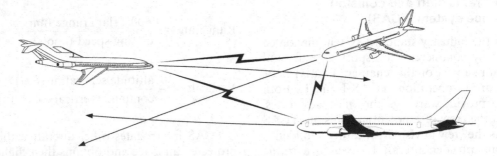

Figure 20.2 Airborne equipment—datalink communication

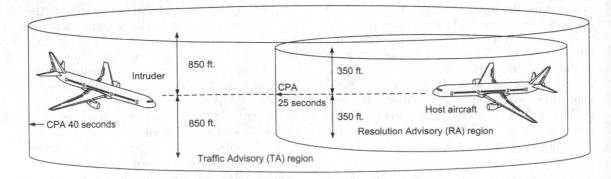

Figure 20.3 Protected airspace volume

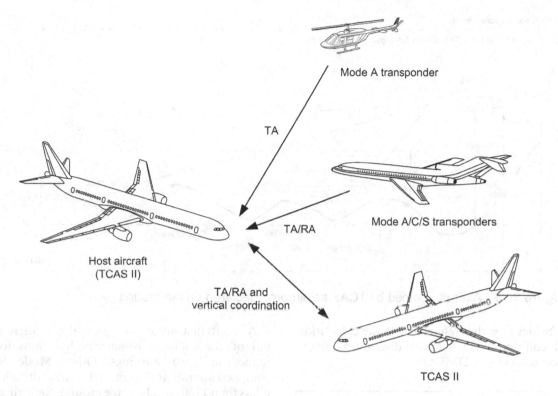

Figure 20.4 Variations of warning given to host aircraft with TCAS II

Table 20.1 TA/RA sensitivity levels

Altitude (feet)*	Sensitivity levels	Tau (seconds)		DMOD (nm)		Altitude Threshold (feet)	
		TA	RA	TA	RA	TA	RA
0 to 1,000	2	20	None	0.30	None	850	n/a
1,000 to 2,350	3	25	15	0.33	0.20	850	300
2,350 to 5,000	4	30	20	0.48	0.35	850	300
5,000 to 10,000	5	40	25	0.75	0.55	850	350
10,000 to 20,000	6	45	30	1.00	0.80	850	400
20,000 to 42,000	7	48	35	1.30	1.10	850	600
>42,000	7	48	35	1.30	1.10	1200	700

* Radio altitude for sensitivity levels 2/3, thereafter pressure altitude from the host aircraft's barometric altimeter (Data source: ARINC)

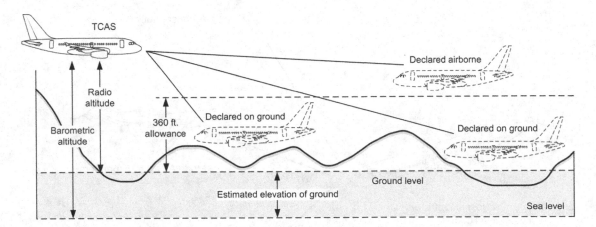

Figure 20.5   Aircraft deemed by TCAS as airborne/declared on the ground

provides a further refinement to the calculation of collision avoidance; this distance modification is known as DMOD.

Key point

TCAS is an airborne system based on secondary radar that interrogates and replies to other aircraft; the system utilises the aircraft's Mode S transponders, and is functionally independent of the aircraft navigation systems and ground stations.

Key point

The closest point of approach is derived as a function of time, referred to as tau.

Test your understanding 20.1

What is the difference between TCAS I and TCAS II?

Test your understanding 20.2

What do the abbreviations TA and RA stand for?

Aircraft that are on the ground are filtered out of the collision avoidance algorithms to reduce nuisance warnings. Other Mode S equipped aircraft are monitored if their altitude is less than 1750 feet above the ground. Referring to Figure 20.5, by using a combination of the host aircraft's barometric and radio altitude, together with the barometric altitude of the target aircraft, any target below 360 feet is deemed to be on the ground.

20.3 TCAS equipment

The system consists of one TCAS computer, two directional antennas, a control panel and two displays, see Figure 20.6. TCAS operates in conjunction with the Mode S surveillance system which includes two transponders, a control panel and antennas, as described in Chapter 19. Visual warnings can be displayed on the instantaneous vertical speed indicator (IVSI) or electronic flight instrument system (EFIS).

20.3.1 Antennas

These are located on the top and bottom of the fuselage as illustrated in Figure 20.7; this provides **antenna diversity**, a technique in which the datalink signal is transmitted along different propagation paths. The upper antenna is

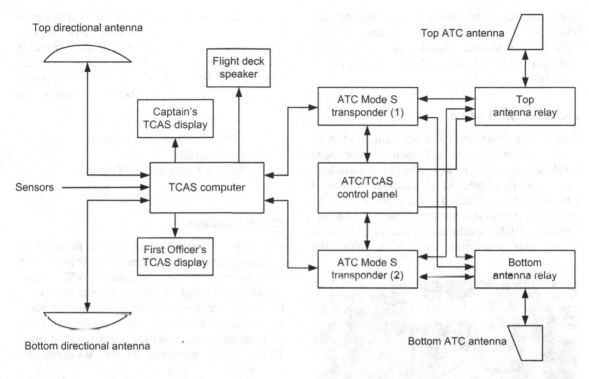

Figure 20.6 TCAS airborne equipment

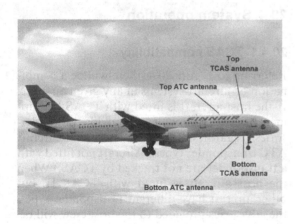

Figure   20.7 Location   of   ATC/TCAS
antennas

directional and is used for tracking targets above
the host aircraft; the bottom antenna can either
be **omnidirectional**, or **directional** as an operator
specified option. Interrogation codes are transmitted via the Mode S transponder on a 1030
MHz carrier wave; reply codes are transmitted
on a 1090 MHz carrier wave. The phase array

directional antennas are electronically steerable
and transmit in four lateral segments at varying
power levels. Note that two Mode S transponder
antennas (Figure 20.7) are also required for
TCAS operation. The latter is suppressed when
the Mode S transponder is transmitting so that
TCAS does not track the host aircraft.

20.3.2 Computer

This is a combined transmitter, receiver and
processor that performs a number of functions
including:

- **monitoring** of the surveillance airspace
  volume for aircraft
- **tracking** other aircraft
- **monitoring** its own aircraft altitude
- issuing **warnings** for potential flight path
  conflicts
- providing recommended **manoeuvres** to
  avoid potential flight path conflicts.

Inputs to the computer include the host aircraft's heading, altitude and maximum airspeed.

Other configuration inputs include landing gear lever position and weight on wheels sensor. Collision avoidance algorithms are used to interpret data from the host aircraft and proximate traffic.

20.3.3 Control panel

This is a combined ATC/TCAS item, see Figure 20.8. The four-digit aircraft identification code is selected by either rotary switches or push buttons, and displayed in a window. (Refer to Chapter 19 for detailed ATC operation.) The mode select switch is used to disable TCAS surveillance, enable traffic advisories only, or both traffic and resolution advisories. With the

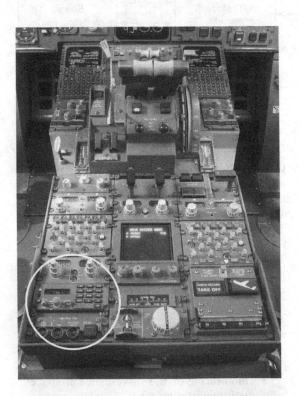

Figure 20.8 ATC/TCAS control panel

system in standby, the TCAS transponder is powered but it will not interrogate or reply to interrogations; all surveillance and tracking functions are disabled. The above/below switch (ABV-N-BLW) allows the crew to select three bands of surveillance above or below the aircraft:

- ABV +7000 feet/−2700 feet
- N ±2700 feet
- BLW +2700 feet/−7000 feet.

20.3.4 Displays

The displays used for TCAS advisories vary between aircraft types. These include the instantaneous vertical speed indicator (IVSI) and/or the electronic flight instrument system (EFIS). In either case, the advisory warnings are based on the same icons. Details of both IVSI and EFIS displays are provided in the description of TCAS system operation that follows in the next section.

20.4 System operation

20.4.1 TCAS compatibility

Referring to Figure 20.4, there will be a combination of aircraft systems in any given airspace. If the host aircraft is fitted with TCAS I equipment, then its computer will provide traffic advisories, regardless of the surrounding aircraft ATC transponder types. (Aircraft not fitted with a transponder are not tracked by TCAS.) When the host aircraft is fitted with TCAS II, but other aircraft have different transponder types, the advisories provided to the host crew will be, as shown in Figure 20.4.

Test your understanding 20.3

TCAS requires that the aircraft is equipped with a Mode S ATC transponder; the computers in TCAS II equipped aircraft will coordinate their guidance commands such that they provide **complementary** manoeuvres. In the latest versions of TCAS, aircraft performance is taken into account when providing these commands.

20.4.2 Advisory warnings

Traffic icons are shown relative to the host aircraft; these are colour coded to depict their threat level, as shown in Figure 20.9. These icons are supplemented by altitude information for the other aircraft: relative altitude (± to depict if the other aircraft is above or below the host aircraft) and vertical manoeuvre (climbing or descending indicated by an arrow). These icons are displayed on the IVSI (Figure 20.10) or EHSI (Figure 20.11).

Referring to Figure 20.11, the EHSI also has a display area for TCAS system indications (system off, self testing, etc.), traffic icons and TA/RA indications:

- **Off-scale**: intruder aircraft is out of display range
- **Traffic**: TCAS has detected intruder within the protected airspace

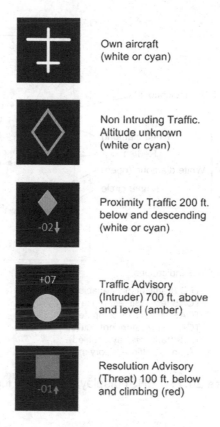

Figure 20.9 Traffic warning icons

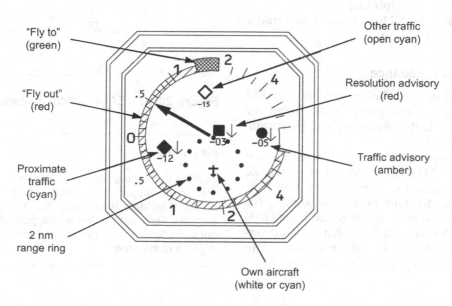

Figure 20.10 IVSI display with TCAS icons

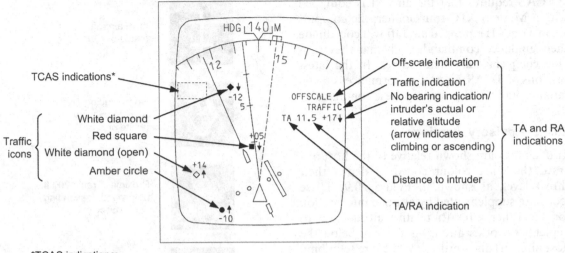

*TCAS indications:

    TFC (traffic display enabled)
    TCAS test (self test mode)
    TCAS fail (TCAS computer)
    TCAS off (system not active)
    TFC (traffic display enabled)
    TA only (traffic advisory mode)

Figure 20.11 EHSI display with TCAS icons and messages

- **No bearing**: TCAS cannot determine the bearing of an intruder
- **TA/RA**: TCAS has identified an intruder or threat aircraft.

20.4.3 TCAS guidance

In the event of a resolution advisory, the IVSI will indicate red and green bands around the display to guide the pilot into a safe flight path, see Figure 20.10.

Aircraft with an electronic attitude direction indicator (EADI) have 'fly-out-of' guidance (Figure 20.12) in the form of a red boundary. The pilot has to climb or descend, keeping the aircraft outside of these calculated boundaries until the RA is cleared. Note that the latest versions of TCAS take aircraft performance into account when issuing vertical guidance.

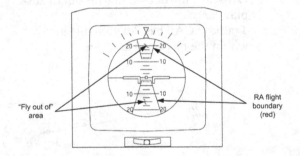

Figure 20.12 EADI vertical guidance

Key point

TCAS interrogates aircraft within a surveillance volume of airspace and obtains their flight path details. This data is analysed along with the host aircraft's flight path. If there is a potential conflict between flight paths, a visual and audible warning is given to the crew.

Key point

TCAS traffic and resolution advisories are displayed on either the IVSI or EFIS. If the host aircraft is fitted with TCAS II, but other traffic has different TCAS equipment or ATC transponders, the information provided to the host aircraft will vary.

20.4.4 TCAS aural annunciations

Aural voice alerts and warnings are produced via dedicated TCAS speakers in the cockpit, or through the aircraft's audio system so that they are heard in the pilots' headsets. Traffic advisories (TA) are announced by the words 'traffic, traffic', stated once for each TA. Resolution advisories (RA) are announced, as shown in Table 20.2. The table covers TCAS II Versions 6.04a, 7.0 and 7.1, thereby depicting the progression of TCAS development through lessons learned. Aural warnings are inhibited at altitudes below 500 feet (+/- 100 feet) above ground level (AGL).

20.4.5 TCAS communication links

Referring to Figure 20.13, there is an interrogation/identification communication link between aircraft equipped with TCAS/Mode S transponders. The TCAS on aircraft A monitors the Mode S transponder broadcast transmission (squitters) of surrounding traffic, including aircraft B. The squitter returns from all traffic includes the unique Mode S address of aircraft B. Upon receipt of a valid squitter message, the identification of aircraft B is added to a list that

Table 20.2 TCAS aural annunciations

TCAS Advisory	Version 7.1 Annunciation	Version 7.0 Annunciation	6.04a Annunciation
Traffic Advisory		Traffic, Traffic	
Climb RA	Climb, Climb		Climb, Climb, Climb
Descend RA	Descend, Descend		Descend, Descend, Descend
Altitude Crossing Climb RA		Climb, Crossing Climb;	Climb, Crossing Climb
Altitude Crossing Descend RA		Descend, Crossing Descend;	Descend, Crossing Descend
Reduce Climb RA	Level Off, Level Off	Adjust Vertical Speed, Adjust	Reduce Climb, Reduce Climb
Reduce Descent RA	Level Off, Level Off	Adjust Vertical Speed, Adjust	Reduce Descent, Reduce Descent
RA Reversal to Climb RA		Climb, Climb NOW;	Climb, Climb NOW
RA Reversal to Descend RA		Descent, Descent NOW;	Descent, Descent NOW
Increase Climb RA		Increase Climb, Increase Climb	
Increase Descent RA		Increase Descent, Increase Descent	
Maintain Rate RA	Maintain Vertical Speed, Maintain		Monitor Vertical Speed
Altitude Crossing, Maintain Rate RA (Climb and Descend)	Maintain Vertical Speed, Crossing Maintain		Monitor Vertical Speed
Weakening of RA	Level Off, Level Off	Adjust Vertical Speed, Adjust	Monitor Vertical Speed
Preventive RA (no change in vertical speed required)	Monitor Vertical Speed		Monitor Vertical Speed, Monitor Vertical Speed
RA Removed		Clear of Conflict	

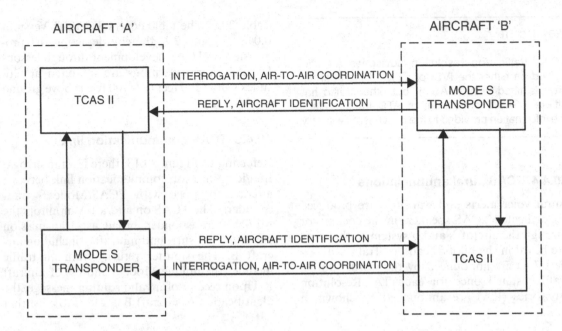

Figure 20.13 TCAS communication links

aircraft A will subsequently interrogate. The resulting exchange of squitter data allows the TCAS in aircraft A to determine the range and the altitude of aircraft B. Note that there is a corresponding exchange of interrogation/ identification communications from aircraft B to surrounding traffic, including aircraft A.

20.4.6 Whisper-Shout

A technique known as "whisper-shout" and directional transmissions can be used to reduce the number of transponders that reply to a single interrogation. Whisper-shout varies the power level of interrogations on a progressive basis; aircraft that are close to the host aircraft send their replies. A low power level is used for the first interrogation step; in the second step, a suppression pulse is first transmitted at a slightly lower level than the first interrogation, followed 21 µs later by an interrogation at a slightly higher power level than the first interrogation, see Figure 20.14. The next interrogation suppresses the transponders that have already replied, but seeks replies from aircraft that did not reply to the first interrogation. This process is repeated 24 times to ensure that all Mode C transponders in the given airspace provide a

reply. This procedure reduces the possibility of garble by suppressing most of the transponders that had replied to the previous interrogation, but eliciting replies from an additional group of transponders that did not reply to the previous interrogation. Using directional signals in the synchronous garble area, as illustrated in Figure 20.15, further reduces the number of overlapping replies.

Key point

TCAS aural annunciations are integrated with other systems on the aircraft in accordance with a priority scheme; TCAS aural annunciations will be inhibited during any terrain awareness systems (TAWS) alerts, including windshear. (TAWS is covered in another book title in the series: Aircraft electrical and electronic systems – AEES)

Test your understanding 20.4

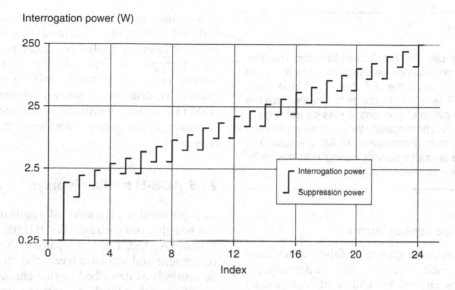

Figure 20.14   Whisper-shout interrogation

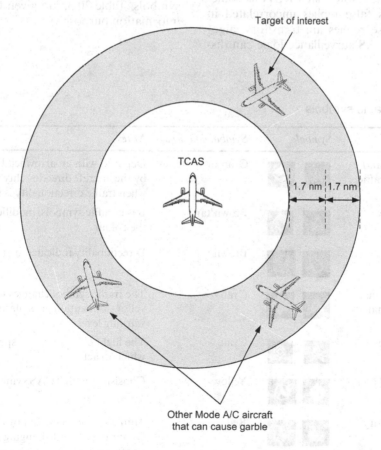

Figure 20.15   Synchronous garble area

Key point

RAs can be classified as preventive or corrective, depending on whether own aircraft is, or is not, in conformance with the RA target altitude rate. Corrective RAs are instructions to deviate from the current vertical rate; preventive RAs do not require a change in vertical speed, they are an instruction to avoid certain manoeuvre. TCAS will take into account the aircraft's service ceiling; RAs are inhibited near its maximum altitude.

20.4.7 False replies/errors

Non-synchronised garble or false replies from unsynchronised interrogator transmissions (FRUIT) is caused by undesired transponder replies that were generated in response to interrogations from ground stations or TCAS equipped aircraft. (Note that FRUIT is sometimes written as false replies uncorrelated in time.) Since these replies are transitory, algorithms in the TCAS surveillance logic can discard them.

Another consideration in TCAS surveillance is the effect of multipath errors; these are caused by more than one reply being received for a single interrogation. This is a reflected reply, and usually occurs over flat terrain. A technique known as dynamic minimum triggering level (DMTL) is used within the computer to discriminate against delayed and lower power level replies.

20.5 ADS-B traffic displays

Cockpit display of traffic information (CDTI) can be enhanced with an ADS-B traffic advisory system (ATAS). Depending upon the installed equipment and software levels, the "basic" traffic symbols as described in this chapter can be modified with ADS-B, to provide status information, such as on-ground, selected, designated, and alerted. (The following examples of traffic symbols, Table 20.3, are given for training and information purposes.)

Table 20.3   Traffic symbols

Purpose	Symbol	Symbol colour(s)	Notes
Basic directional and non-directional traffic symbol		Cyan or white	Depicted with an arrowhead shape oriented by the aircraft directionality, or diamond when traffic directionality is invalid
On-ground traffic		Brown/tan	Basic traffic symbols, modified by changing the colour
Surface vehicles		Brown/tan	Directionality indicated by triangular shape
Proximate directional and non-directional traffic		Cyan	The traffic symbol changes to (a) amber/yellow for caution level alerts and (b) red for warning level alerts.
Designated traffic		Cyan	The flight crew know the specific traffic upon which to act.
Traffic Advisory (TA)		Yellow	Consistent with TCAS symbol convention
Resolution Advisory (RA)		Red	Traffic symbol modified by changing the colour to red, and changing the shape to a square.

20.6 Traffic advisory system (TAS)

Traffic Advisory Systems are based on the technology originally developed for air-transport category Traffic Alert and Collision Avoidance Systems (TCAS). These systems have been available for General Aviation (GA) aircraft for several years, but have been cost-prohibitive for many smaller, privately owned aircraft. Traffic advisory systems give the pilot visual and audible guidance on the proximity of other aircraft, Figure 20.16. This is via traffic icons on a display screen, Figure 20.17, and messages on the audio system, e.g., "Traffic! twelve o'clock high, one mile!". The system actively interrogates other aircraft, providing timely audible and visual alerts with the precise location of conflicting traffic. An optional heading input permits rapid repositioning of targets during high-rate turns, providing optimal performance for helicopter operations.

Figure 20.17 Multi-function display with traffic advisories

Test your understanding 20.5

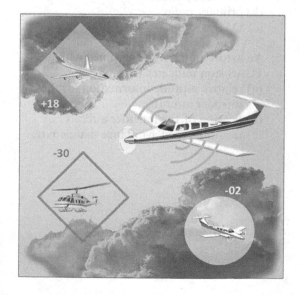

Figure 20.16 TAS overview

20.7 Multiple choice questions

1. The closest point of approach is determined as a function of:
   (a) range
   (b) time
   (c) closing speed.

2. TCAS interrogation codes are transmitted and received on:
   (a) 1030 MHz and 1090 MHz carrier waves
   (b) 1090 MHz and 1030 MHz carrier waves
   (c) 1030 MHz or 1090 MHz carrier waves.

3. What colour and shape of symbol is used for proximity traffic?
   (a) Solid red square
   (b) Solid cyan diamond
   (c) Solid orange circle.

4. What TCAS equipment is required on host aircraft and threat traffic in order to provide coordinated manoeuvres?
   (a) TCAS I
   (b) Mode C transponder
   (c) TCAS II.

5. TCAS II requires which type of ATC transponder:
   (a) Mode S
   (b) Mode A
   (c) Mode C.

6. The version of TCAS that provides vertical flight manoeuvre guidance to the crew is:
   (a) TAS
   (b) TCAS II
   (c) TCAS I.

7. The directional TCAS antennas transmit in:
   (a) four vertical segments at varying power levels
   (b) four lateral segments at fixed power levels
   (c) four lateral segments at varying power levels.

8. The whisper-shout technique:
   (a) varies the power level of interrogations on a decreasing basis
   (b) varies the power level of interrogations on an increasing basis
   (c) maintains the power level of interrogations on a progressive basis.

9. Recommended manoeuvres needed to increase or maintain vertical separation are provided by what type of TCAS warning?
   (a) Resolution advisory
   (b) Traffic advisory
   (c) Non-threat traffic.

10. Traffic advisories (TA):
    (a) assist the crew in visually searching and identifying an intruder
    (b) provide recommended manoeuvres needed to maintain vertical separation
    (c) provide lateral guidance to the crew.

11. TCAS interrogations and replies are sent:
    (a) directly between the onboard ATC transponders
    (b) via a ground link
    (c) directly to ATC ground controllers.

12. TCAS warning icons are shown relative to the:
    (a) host aircraft
    (b) intruder aircraft
    (c) proximate traffic.

13. Range tau is based on the:
    (a) altitude separation and closing speed of traffic
    (b) slant range and closing speed of traffic
    (c) altitude separation and combined vertical speed.

14. TCAS is an airborne system based on:
    (a) secondary radar
    (b) primary radar
    (c) DME.

15. An arrow and ± sign combined with a TCAS icon indicates the:
    (a) relative altitude information for the intruder aircraft
    (b) bearing of the intruder aircraft
    (c) recommended avoidance manoeuvre.

Chapter 21

Electromagnetic compatibility (EMC)

It is an unfortunate fact of life that the operation of virtually all items of electrical and electronic equipment can potentially disturb the operation of other nearby items of electronic equipment. In the aerospace world the increasing incidence of interference has prompted the introduction of legislation that sets strict standards for the design, construction, operation, and maintenance of avionic equipment.

The name given to this type of disturbance is electromagnetic interference (EMI) and the property of an electrical or electronic device or system (in terms of its immunity to the effects of EMI generated by other equipment and its ability to generate and radiate its own EMI) is referred to as its electromagnetic compatibility (EMC). Given the proliferation of avionics in modern aircraft it's perhaps not surprising that EMC and EMI have become a very important consideration for the designers, suppliers, and maintainers of avionic equipment.

21.1 The avionic electromagnetic environment

When considering EMC in relation to aircraft avionics it is essential to be fully aware of the environment in which the equipment operates. Modern aircraft use complex and diverse electrical and electronic systems. They may also carry several dozen antennas operating at different wavelengths. Some of these antennas may be radiating at appreciable power levels whilst others need to simultaneously receive signals of only a few microvolts per metre (see Figure 21.1).

The systems that are most impacted by, and most impact on, the electromagnetic environment in a modern aircraft include:

- Aircraft weather radar operating with peak radiated power levels in excess of 10 kW when compared with an isotropic radiator

- HF and VHF communication systems as well as VOR/LOC systems that must respond to incoming signals of as little as 1 µV
- C-band radio altimeters operating in a band immediately adjacent to recently introduced 5G mobile radio services
- Low-frequency ADF loop antennas that operate at field strengths of as low as 25 µV/m
- Cabled and wireless networks operating at fast data rates. Unless properly designed and maintained, the associated cabling can radiate a significant amount of radio frequency energy over a frequency range that extends from several hundred kHz to well over 100 MHz
- Switching and other systems for aircraft lighting and power control.

Interference can also be an unintentional consequence of the operation of faulty or poorly designed electronic equipment. To put this into a practical context, Figure 21.1 shows the frequency spectrum and waterfall display of radio signals in the range 500 kHz to 1.75 MHz resulting from a defective switched mode power supply (SMPS). The upper spectrum display shows signal amplitude varies with frequency while the lower waterfall display shows how signal amplitude varies with time. The noise floor has been raised from less than −110 dB to around −84 dB with peaks of noise at integer multiples of the switching frequency. MF radio reception would be severely compromised with the 26 dB, or so, increase in noise prevalent when the SMPS is running.

The waveform at the output of the EMI-producing 5V SMPS is shown in Figure 21.2. This reveals a switching transient with an amplitude of 100 mV and a repetition frequency of around 20 kHz (the power supply's switching frequency). Additional low-pass filtering, screening and grounding (see later) is needed to

DOI: 10.1201/9781003411932-21

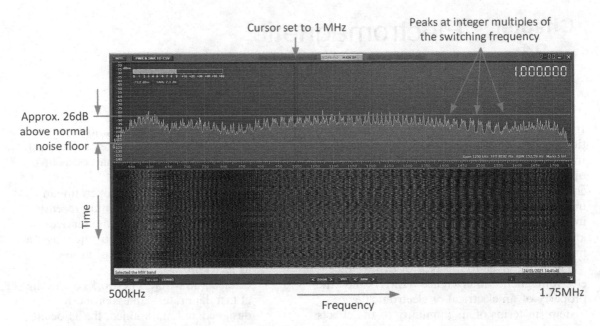

Figure 21.1 Radio frequency spectral analysis showing the effect of SMPS noise in the frequency range 500 kHz to 1.75 MHz

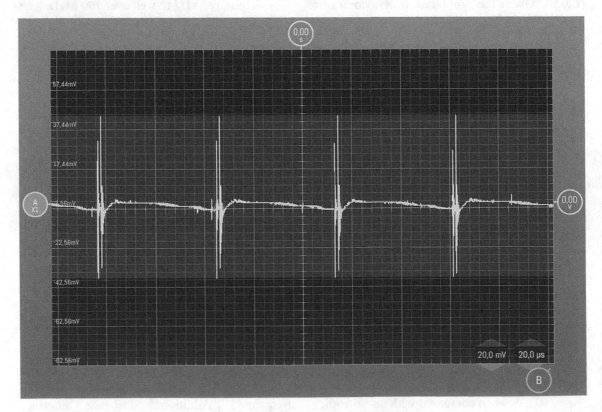

Figure 21.2 Waveform of the output of the EMI-producing SMPS

improve performance and comply with the appropriate EMC standards.

21.2 Effects of EMI

EMI can be defined as the presence of unwanted voltages or currents which can adversely affect the performance of an avionic system. The effects of EMI include errors in instrument readings (both above and below true values), heterodyne whistles present on audio signals, herring-bone patterns superimposed on video displays, repetitive pulse noise (buzz) on inter-com and cabin phone systems, desensitising or 'blocking' of radio and radar receivers, false indication in radar and distance-measuring equipment, unwanted triggering of alarms, and so on. Note that, several of these effects may be simultaneously present and may impact on the correct operation of multiple avionic systems (see the Case Study in Section 21.9).

21.3 Sources of EMI

Some items of avionic and electrical equipment are well-known as sources of EMI. Such equipment includes display screens (but not CRT), radio and radar transmitters, power lines, window heat controllers, induction motors, switching and light-dimming circuits, microprocessors and associated digital circuitry, pulsed high-frequency circuits, bus cables (but not fibreoptic cables), static discharge and lightning. The energy generated by these sources can be conducted and/or radiated as an electromagnetic field.

Unless adequate precautions are taken to eliminate the generation of interference at source and/or to reduce the equipment's susceptibility to EMI, radiated or conducted energy can become coupled into other circuits. Conduction is the process by which energy is transmitted through electrically conductive paths such as circuit wiring or metal parts of an aircraft structure. In electromagnetic field radiation, energy is transmitted through electrically non-conductive paths, such as air, plastic, or non-conductive composite materials.

Systems which may be susceptible to electro-magnetic interference include radio and radar receivers, microprocessors and other microelec-tronic systems, electronic instruments, control systems, audio and inflight entertainment systems (IFE).

Whether a system will have an adverse response to electromagnetic interference depends on the type and amount of emitted energy in conjunction with the susceptibility threshold of the receiving system. The threshold of suscepti-bility is the minimum interference signal level (conducted or radiated) that results in equip-ment performance that is indistinguishable from the normal response.

If the threshold is exceeded, then the perfor-mance of the equipment will become degraded. Note that when the susceptibility threshold level is greater than the levels of conducted or radi-ated emissions, electromagnetic interference problems do not exist. Systems to which this applies are said to be electromagnetically com-patible. In other words, the systems will operate as intended and any EMI generated is at such a level that it does affect normal operation.

21.4 Classification of EMI

EMI can be classified by bandwidth, amplitude, waveform and occurrence. The bandwidth of interference is the frequency range in which the interference exists. The interference bandwidth can be narrow or wide/broadband.

Narrowband interference can be caused by such items as AC power rails, microprocessor clocks (and their harmonics), radio transmitters and receivers. These items of equipment all con-tain sources (e.g., clock oscillators) that work on specific frequencies. These signals (along with unwanted harmonics) can be radiated at low levels from the equipment.

Broadband interference is caused by devices generating random frequencies and noise which may be repetitive but is not confined to a single frequency or range of frequencies. Examples of this type of interference are power supplies, LCD and AMLCD (by virtue of their use of high frequency AC supplies), switched-mode power supplies, switching power controllers and microprocessor bus systems.

Interference amplitude is the strength of the signal received by the susceptible system. The amplitude can be constant or can vary predictably with time or can be totally random. For example, a 115 V AC power line can induce a stable sinusoidal waveform on adjacent 28 V DC power or signal lines. The amplitude of the interference will depend on the load current in the AC power line (recall that the magnetic field produced around a conductor is directly proportional to the current flowing in the conductor). Examples of random interference are environmental noise and inductive switching transients. Environmental noise is the aggregate of all electromagnetic emissions present in a particular space or area of concern at any one time. This is usually measured over a defined spectrum (e.g., 30 kHz to 30 MHz).

It is important to be aware that there is no one specific waveform that produces electromagnetic interference. Instead, it is the change from one signal level to another in conjunction with the rate at which it changes that determines the amount of electromagnetic energy released. More energy is released when the change in signal level and rate is increased.

In terms of the irregularity of its occurrence, EMI can be categorised as being either periodic (continuously repetitive), aperiodic (predictable but not continuous), or random (totally unpredictable).

21.5 Some examples of EMI

Figure 21.3 shows normal VHF traffic present in a 2 MHz wide band between 119.5 and 121.5 MHz. The upper half of the display shows amplitude (in dBm) plotted against frequency. The lower half of the screen shows how signal amplitude varies over a short period of time

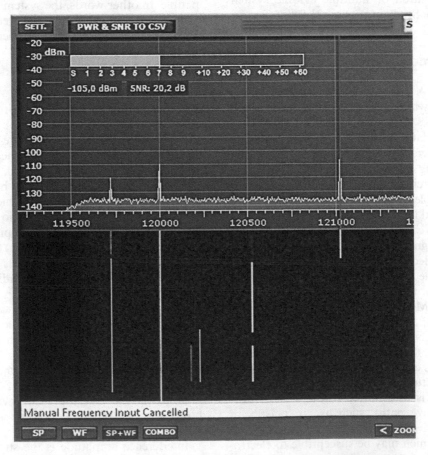

Figure 21.3 Spectral analysis showing normal VHF traffic. The cursor line is set to 121.025 MHz (25 kHz below the emergency guard channel

(approximately 30 seconds). The colours present in the lower screen area provide an indication of relative signal amplitude. The display shows several exchanges between aircraft and ground controllers with the receiver tuned to 121.025 Hz (as indicated by the cursor where an aircraft is currently replying to a ground controller). The signal level indicator (green bar) shows the show the signal level at −105 dBm and a received signal-to-noise ratio of 20.2 dB (adequate for voice communication). Note that the noise floor is around −133 dBm.

21.5.1 Intermittent broadband noise

Leakage and radiation from data cables is a common cause of broadband noise, as shown in Figure 21.4 which shows intermittent bursts of

interference from a data source filtered above 120.3 MHz. Note how the noise floor increases to −128 dBm but the filter is effective in reducing interference to services above 120.35 MHz and also to the guard channel at 121.05 MHz.

21.5.2 Continuous wideband noise

Figure 21.5 shows the appearance of continuous noise over the complete measurement spectrum from 119.5 MHz to 121.5 MHz. This makes it difficult to receive signals with levels below about −120 dBm.

21.5.3 Instability and drift

Figure 21.6 shows the appearance of a noisy and unstable signal produced by faulty VHF

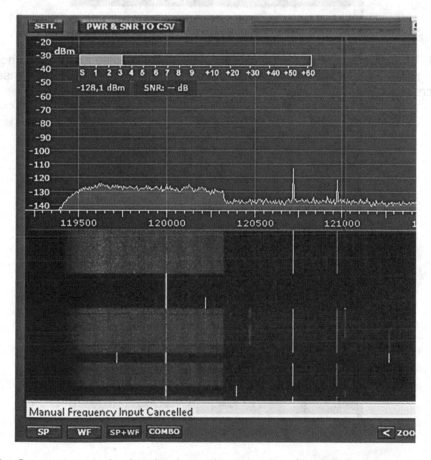

Figure 21.4 Spectral analysis showing intermittent bursts of broadband noise caused by radiation from a data cable (filtered above 120.3 MHz). Note that there is negligible interference to services above 120.35 MHz

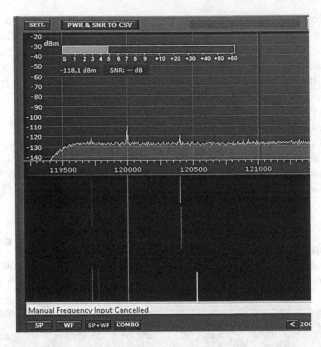

Figure 21.5 Spectral analysis showing continuous wideband noise and interference produced by a faulty lighting panel. Note the large reduction in signal-to-noise ratio that arises from the apparent increase in noise floor

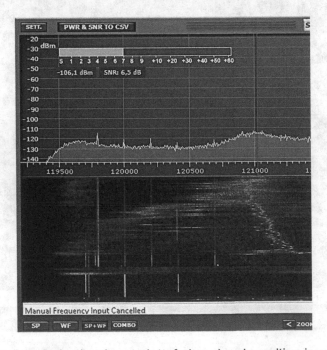

Figure 21.6 Spectral analysis showing an interfering signal resulting in a 20 dBm increase in noise floor. Note how the interference appears to be drifting slowly with a centre that moves from around 121.025 MHz to 120.950 MHz

transceiver. The display shows an interfering signal drifting across the measurement spectrum. Note that there is an approximate 20 dBm increase in noise level that degrades all VHF signals within the wanted band.

Test your understanding 21.1

Figure 21.7 shows the frequency spectrum between 113.500 and 114.350 MHz with the receiver (cursor line) tuned to the Midhurst (UK) VOR beacon transmitting on 114.000 MHz. The display shows interference from a strong drifting carrier.

1. What is the received signal level of the Midhurst VOR beacon?
2. What is the received signal to noise ratio of the Midhurst VOR beacon?
3. What is the signal level of the interfering carrier?
4. At what frequency does the interfering carrier start before drifting?
5. At what frequency does the interfering carrier settle?

21.6 EMI reduction

Planning for electromagnetic compatibility must be initiated in the design phase of a device or system (as discussed in Section 21.2). If this is not satisfactorily achieved, interference problems may arise. The three factors necessary to produce an EMI problem are a noise source, a means of coupling (by conduction or radiation) and a susceptible receiver. To reduce the effects of EMI, at least one of these factors must be addressed. The following lists techniques for EMI reduction under these three headings (note that some techniques address more than one factor).

1. Suppress interference at source.
   * Enclose interference source in a screened metal enclosure and then ensure that the enclosure is adequately grounded.
   * Use transient suppression on switching devices and contactors.
   * Twist and/or shield bus wires and data bus connections.
   * Use screened (i.e., coaxial) cables for audio and radio frequency signals.
   * Keep pulse rise times as slow and long as possible.
   * Check that enclosures, racks and other supporting structures are grounded effectively.

2. Reduce noise coupling.
   * Separate power leads from interconnecting signal wires.
   * Twist and/or shield noisy wires and data bus connections.
   * Fit an optical fibre data bus where possible.
   * Use screened (i.e., coaxial) cables for audio and radio frequency signals.
   * Keep ground leads as short as possible.
   * Break interference ground loops by incorporating isolation transformers, differential amplifiers, and balanced circuits.
   * Ensure that supply connections are filtered adequately (see Figure 21.8).
   * Physically relocate receivers and sensitive equipment away from interference source.

3. Increase susceptibility thresholds.
   * Limit bandwidth to only that which is strictly necessary.
   * Limit gain and sensitivity to only that which is strictly necessary.
   * Ensure that enclosures are grounded and that internal screens are fitted.
   * Fit components that are inherently less susceptible to the effect of stray radiated fields.

Key point

Avionic equipment incorporates low-pass filters in order to reject noise and spurious harmonic signals that might otherwise be conveyed and/or radiated from supply voltage connections.

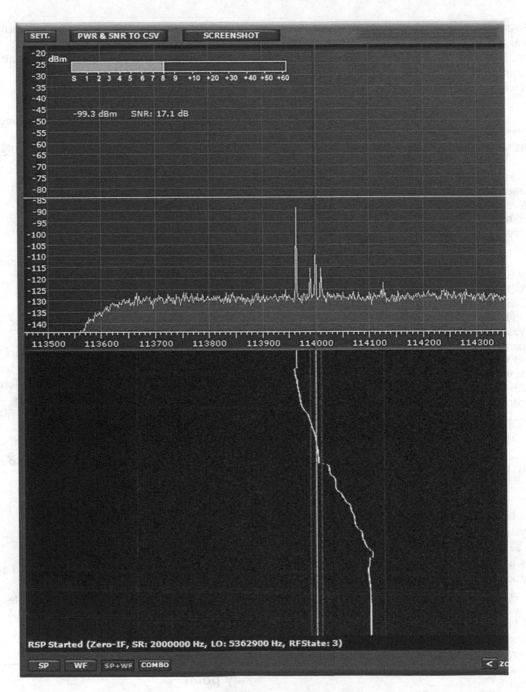

Figure 21.7   See Test your understanding 21.1

Test your understanding 21.2

Sketch the circuit of a low-pass EMI filter suitable for
incorporation in the connection to a 400 Hz AC supply.

21.7 Aircraft wiring and cabling

When many potential sources of EMI are present in a confined space, aircraft wiring and
cabling has a crucial role to play in maintaining

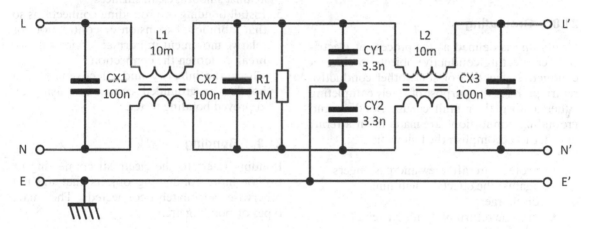

Figure 21.8 A typical supply line low-pass EMI filter. The filter provides a rejection of more than 60 dB at 1 MHz. Values shown are for 50/60 Hz operation. Correspondingly smaller values are used at 400 Hz.

electromagnetic compatibility. The following points should be observed:

1. Adequate wire separation should be maintained between noise source wiring and susceptible wiring (for example, ADF wiring should be strategically routed away from radiating sources in the aircraft to ensure a high level of EMC).
2. Any changes to the routing of this wiring could have an adverse effect on the system. In addition, the wire separation requirements for all wire categories must be maintained.
3. Wire lengths should be kept as short as possible to keep coupling at a minimum.

Where wire shielding is incorporated for lightning protection, it is important that the shield grounds (pigtails) be kept to their designed length. An inch or two added to the length will result in degraded lightning protection.

4. Equipment grounds must not be lengthened beyond design specification. A circuit ground with too much impedance may no longer be a true ground.
5. With the aid of the technical manuals, grounding and bonding integrity must be maintained. This includes proper preparation of the surfaces where electrical bonding is made.

21.8 Grounding and bonding

The electrical integrity of the aircraft structure is extremely important as a means of reducing EMI as well as protecting the aircraft, its passengers, crew and systems from the effects of lightning strikes and static discharge. Grounding and bonding are specific techniques that are used to achieve this (see Figure 21.9). Grounding and bonding can also be instrumental in minimising the effects of high-intensity radio frequency fields (HIRF) emanating from high power radio transmitters and radar equipment. Grounding and bonding resistances of less than 0.001–0.003 Ω are usually required.

21.8.1 Grounding

Grounding is defined as the process of electrically connecting conductive objects to either a conductive structure or some other conductive return path for the purpose of safely completing either a normal or fault circuit. Bonding and grounding connections are made in an aircraft in order to accomplish the following:

- protect aircraft, crew and passengers against the effects of lightning discharge;
- provide return paths for current;
- prevent the development of RF voltages and currents;
- protect personnel from shock hazards;

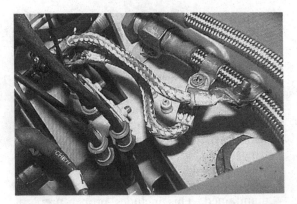

Figure 21.9   Bonding straps

- maintain an effective radio transmission and reception capability;
- prevent accumulation of static charge.

The following general procedures and precautions apply when making bonding or grounding connections:

1. Bond or ground parts to the primary aircraft structure where possible
2. Make bonding or grounding connections so that no part of the aircraft structure is weakened.
3. Bond parts individually if feasible
4. Install bonding or grounding connections against smooth, clean surfaces
5. Install bonding or grounding connections so that vibration, expansion or contraction, or relative movement in normal service will not break or loosen the connection
6. Check the integrity and effectiveness of a bonded or grounded connection using an approved bonding tester.

21.8.2 Bonding

Bonding refers to the electrical connecting of two or more conducting objects that are not otherwise adequately connected. The main types of bonding are:

1. Equipment Bonding. Low impedance paths to the aircraft structure are generally required for electronic equipment to provide radio frequency return circuits and to facilitate reduction in EMI.
2. Metallic Surface Bonding. All conducting objects located on the exterior of the airframe should be electrically connected to the airframe through mechanical joints, conductive hinges, or bond straps, which can conduct static charges and lightning strikes.
3. Static Bonds. All isolated conducting paths inside and outside the aircraft with an area greater than 3 in² and a linear dimension over 3 inches that are subjected to electrostatic charging should have a mechanically secure electrical connection to the aircraft structure of adequate conductivity to dissipate possible static charges.

Key point

Initial control of EMI is achieved in modern aircraft by careful design and rigorous testing. Routine maintenance helps to ensure that the aircraft retains electromagnetic compatibility, thereby keeping EMI problems to a minimum.

Key point

Effective grounding and bonding provide a means of ensuring the electrical integrity of the aircraft structure as well as minimising the effects of HIRF fields and the hazards associated with lightning and static discharge.

21.9 Case study – The 5G Problem

The introduction of 5G services in many countries has raised concerns about the risk of interference to C-band radio/radar altimeter equipment operating in the frequency range 4.2 to 4.4 GHz (see Figure 21.10).

The results of various studies have indicated that the risk of 5G interference is widespread

and is most likely to impact on aircraft when on landing approach, particularly in conditions of restricted visibility. Suggested short terms solutions to this problem include:

*   Replacing outdated altimeters with modern equipment
*   Retrofitting sharp cut-off antenna filters
*   Use of software algorithms to improve the integrity and validity of altimeter data
*   Establishing 5G "buffer zones" around at-risk airports (determined by assessment of risk factors such as traffic patterns, geographical location, and the prevalent number of low-visibility days)

In response to initial concerns, the Federal Aviation Authority (FAA) issued airworthiness directives (AD) warning that interference from 5G ground stations could affect radio altimeters. The directives applied to all Boeing 737 aircraft with the exception of the 200 and 200-c variants. The AD cautioned:

This AD was prompted by a determination that radio altimeters cannot be relied upon to perform their intended function if they

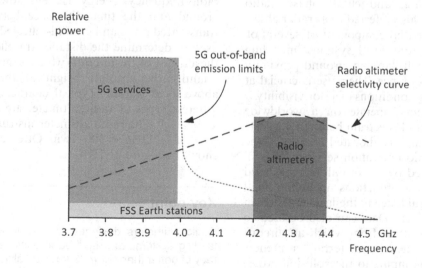

Figure 21.10   C-band radio spectrum showing allocations for 5G and radio altimeter equipment (see also Table 21.2)

experience interference from wireless broadband operations in the 3.7–3.98 GHz frequency band (5G C-Band), and a recent determination that, during approach, landings, and go-arounds, as a result of this interference, certain airplane systems may not properly function, resulting in increased flight crew workload while on approach with the flight director, auto-throttle, or autopilot engaged, which could result in reduced ability of the flight crew to maintain safe flight and landing of the airplane. This AD requires revising the limitations and operating procedures sections of the existing airplane flight manual (AFM) to incorporate specific operating procedures for instrument landing system (ILS) approaches, speed brake deployment, go-arounds, and missed approaches, when in the presence of 5G C-Band interference as identified by Notices to Air Missions (NOTAMs). The FAA is issuing this AD to address the unsafe condition on these products.

– FAA

21.9.1 Radio altimeters

Providing accurate and reliable indications of an aircraft's height above ground during the approach, landing, and climb phase, radio altimeters (often also referred to as radar altimeters) are an essential component of several of an aircraft's safety-critical systems, including precision approach, landing, ground proximity and collision avoidance. Their use is crucial at night and during conditions of poor visibility.

Radio altimeters operate on a worldwide basis in the C-band spectrum between 4.2 GHz and 4.4 GHz which is allocated for use by the aeronautical radio navigation service (ARNS). The data returned by a radio altimeter is used for several crucial functions including final approach flare guidance in the last stages of an auto-land approach. Data can also be used to determine the altitude from which an aircraft can safely land, generating terrain awareness warnings, and as inputs to the collision avoidance and weather radar (predictive wind shear) systems.

It is important to be aware that radio altimeter systems are designed to operate for the entire life of any aircraft in which they are installed. Because installed life can exceed 30 years, radio altimeter equipment usually exhibits a wide range of equipment age, performance and tolerance.

21.9.2 Types of radio altimeter

Several different types of radio altimeter are in common use. They can be based on either LFMCW (linear frequency modulated continuous wave) or pulsed modulation (PM) and also be based on either analogue or digital technology. In an LFMCW radio altimeter, the frequency of the C-band transmitter is continuously swept up and down between two limits at a constant rate. The ground reflected signal returned to the aircraft has the same frequency modulation imparted on it however, because the signal has had to make a return journey between the aircraft and the ground, at any instant of time the frequency received will no longer be the same as the frequency transmitted. The difference between the two frequencies (determined by mixing the transmitted and received frequencies) can then be used to determine the aircraft's altitude above the local terrain.

In a PM radio altimeter, a short pulse of radio-frequency energy is sent towards the ground and the time difference between the transmitted pulse and ground-reflected return is used to determine the distance travelled (twice the altitude of the aircraft) which is then used to quantify the absolute height of the aircraft above local terrain. Typical characteristics of different types of radio altimeter are shown in Table 21.1. The radio altimeter instrument display of a DHC-6-300 Twin Otter aircraft is shown in Figure 21.11.

Key point

Radio altimeters designed for use in automated landing systems are required to achieve an accuracy of better than 0.9 m (3 ft.) at an altitude of less than 46 m (150 ft.).

Table 21.1 Summary of typical characteristics of some different types of radio altimeter

Characteristic	LFMCW		Pulsed		Units
	Analogue	Digital	Analogue	Digital	
Operating frequency (nominal)	4.3 GHz	4.3 GHz	4.3 GHz	4.3 GHz	GHz
Transmitted power	0.6 W	0.4 W	40 W	5 W	W
Modulation	LFMCW	LFMCW	Pulsed	Pulsed	
Bandwidth	133 MHz	150 MHz	n/a	n/a	MHz
Reported altitude range	−6 to 2500(−20 to 8200)	−6 to 1676 (−20 to +5700)	1500(5000)	−6 to 2500(−20 to 8200)	m(ft.)
Pulse repetition frequency	n/a	n/a	5000–10000	5000–20000	Hz
Antenna gain	10	10	10	10	dBi
Beamwidth	45	45	45°	45°	°

Figure 21.11 Radio altimeter display on the flight deck of a DHC-6-300 Twin Otter aircraft

21.9.3 Safety criticality

Because of the importance of radio altimeters to the safe operation of an aircraft, they are included in the minimum equipment list on aircraft certified for passenger service. Furthermore, they must be certified at a safety criticality rating or Design Assurance Level (DAL) of A (where a software or hardware failure would cause or contribute to a catastrophic failure of the aircraft flight control systems) for all transport aircraft or B (where a software or hardware failure would cause or contribute to a hazardous/severe failure condition in the flight control systems) for business and regional aircraft.

Key point

Safety criticality is expressed in terms of Design Assurance Level (or DAL). DAL is quoted as a level extending from A to E with A and B the most safety critical and requiring the most stringent certification process.

Key point

Altimeter data is used by core aircraft sensor systems during the critical stages of flight and rotorcraft operation, particularly during landing and take-off at night and in conditions of low visibility.

Key point

Any interference that compromises the reported data from a radio altimeter can immediately affect an aircraft's safety systems.

21.9.4 Radio altimeter antennas

The antennas used in conjunction with radio altimeters typically provide around 10 dBi gain, are horizontally polarised, and exhibit a beamwidth of around 45° to cope a wide range of pitch and roll angles. The radio altimeter antenna placement for an A380 aircraft is shown in Figure 2.12.

The fact that radio altimeter antennas are directed towards the Earth's surface makes the system vulnerable to potential sources of interference that appear within the antenna's footprint on the ground during final approach (see Figure 21.13). Such potential sources of interference may be located inside or outside buildings or other structures and may also be fixed or mobile. Additionally, and unlike other antennas fitted to an aircraft, they look vertically downwards and are unable to benefit from any shielding or screening that might otherwise be offered by the aircraft's structure.

21.9.5 Potential consequences of interference to radio altimeter systems

Several consequences can occur when an aircraft loses or receives erroneous data from a radio altimeter. The severity of these will depend on many factors including aircraft type, airport

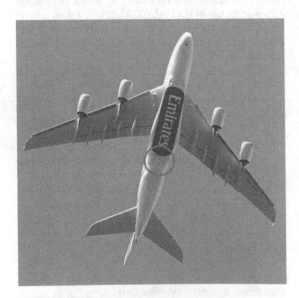

Figure 21.12 Radio altimeter placement below the fuselage of an A380 aircraft

landing requirements, weather, and other environmental conditions. Loss of radio altimeter data may disable the autopilot resulting in the pilot and co-pilot having to resort to manual flight and landing. Some airport categories or certain weather conditions may prohibit the landing of some aircraft types when there is a loss of altimeter data.

Where only one radio altimeter is operational, the altitude at which the decision to land must be increased. If visibility is poor, then the aircraft might be forced to wait until the weather improves or divert to a different airport. Finally, if the radio altimeter signal receives severe interference during the final stages of landing, then a hazardous or catastrophic situation could occur. At best, the flight crew workload increases significantly; at worst the aircraft, crew and passengers may be placed in a catastrophic situation. These concerns underpin the need for action to resolve the potential problems associated with interference from 5G masts and other ground installations.

21.9.6 5G spectrum allocation

The internationally designated C-band spectrum reserved for aircraft radio navigation service (ARNS) occupies a 200 MHz band extending from 4.2 GHz to 4.4 GHz. Within this region radio altimeters have primary status.

The frequency spectrum allocated to 5G straddles the C-band spectrum used by radio altimeters and comes particularly close to the lower band edge at 4.2 GHz. Table 21.2 shows the 5G allocations of various countries together with regulated power limits. Note the proximity of 5G to the ARNS band in several countries, notably Japan and the USA, and the variation in regulated power limits.

Due to the proximity of the 5G allocation to the band used by radio altimeters (particularly in the United States and Japan) the adjacent channel rejection (see Section 3.8) offered by a radio altimeter must roll off sharply below 4.2 GHz and above 4.4 GHz. Unfortunately, the desired rejection in not generally a characteristic of equipment that pre-dates the introduction of 5G (see Figure 21.10). This, coupled with the introduction of 5G masts and other 5G installations around airports, has been a primary cause of concern.

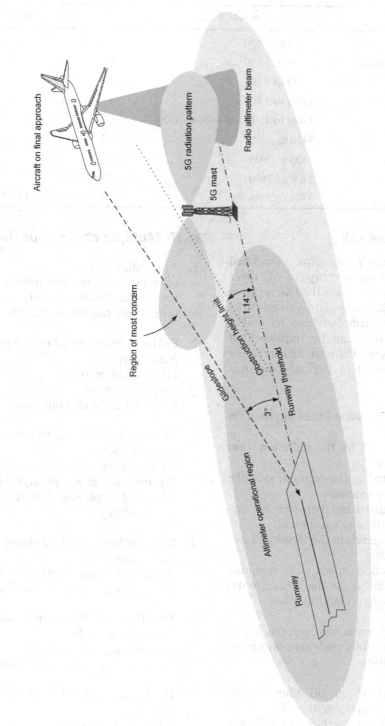

Figure 21.13 Potential worst-case scenario for 5G interference along an approach flight path

Table 21.2 5G spectrum allocation for various countries

Country	C-band allocation (GHz)	Power limit (dBm/MHz)
Canada	3.450 to 3.650	61
Denmark, Finland, Sweden	3.400 to 3.800	61
Europe	3.400 to 3.800	61
France	3.400 to 3.800	55.4 to 60.3
Japan	3.400 to 4100 and 4600 to 4700	56
New Zealand	3410 to 3580	52
Saudi Arabia	3400 to 3800	58
UK	3400 to 3800	58
United States	3700 to 3980	65.15 (rural) and 62.15 (non-rural)

21.9.7 Managing the risk

Despite obvious safety concerns associated with the rapid growth in coverage of 5G there is a need to find ways in which the two services safely coexist. This has led not only to an effort to quantify the problem but also to find a range of solution that can be implemented quickly and effectively. Current measures for 5G operators proposed in various countries include:

1. Reduce spurious emission limits imposed on 5G operators
2. Adopt downward tilt for the radiation from 5G antenna masts
3. Minimise unwanted and undesirable sidelobe radiation from 5G antennas
4. Implement a guard band between 5G services and radio altimeter equipment
5. Reduce 5G power levels in the vicinity of airports and helipads
6. Introduce protection zones for incoming flight paths as well as for the areas around heliports and rotorcraft landing pads.

Of equal importance, measures currently implemented by manufacturers and maintainers of radio altimeters include:

1. Retrofitting of sharp cut-off filters in the radio altimeter's receive signal path
2. Improved algorithms used for digital processing that will specifically recognise and counteract interference from 5G sources.

21.10 Multiple choice questions

1. Aperiodic noise is:
   (a) regular but not continuous
   (b) regular and continuous
   (c) entirely random in nature.

2. EMI can be conveyed from a source to a receiver by:
   (a) conduction only
   (b) radiation only
   (c) conduction and radiation.

3. The display produced by a spectrum analyser shows:
   (a) frequency plotted against time
   (b) time plotted against signal amplitude
   (c) signal amplitude plotted against frequency.

4. EMI can be reduced by means of:
   (a) screening only
   (b) screening and filtering
   (c) screening, bonding and filtering.

5. The effects of HIRF can be reduced by:
   (a) screening only
   (b) screening and filtering
   (c) screening, bonding and filtering.

6. The typical maximum value of bonding resistance is:
   (a) less than 0.005 Ω
   (b) between 0.005 Ω and 0.05 Ω
   (c) more than 0.05 Ω.

7. Effective protection against lightning and static discharge damage to an aircraft requires that:
   (a) all isolated conducting parts must have high resistance to ground
   (b) all parts of the metal structure of the aircraft must be bonded to ground
   (c) all power and bus cables must be well insulated.

8. Supply borne noise can be eliminated by means of:
   (a) a low-pass filter
   (b) a high-pass filter
   (c) a band-pass filter.

9. Noise generated by a switching circuit is worse when:
   (a) switching is fast and current is low
   (b) switching is slow and current is high
   (c) switching is fast and current is high.

10. Interference from 5G cellular services to C-band radio altimeters can be categorised as:
    (a) adjacent channel interference
    (b) co-channel interference
    (c) image channel interference.

Abbrev	Meaning
AA	Aircraft Address
AAIS	Advanced Aircraft Information System
AAM	Advanced air mobility
ABAS	Aircraft-based Augmentation System
ACARS	Aircraft Communications Addressing and Reporting System
ACAS	Airborne Collision Avoidance System
ACM	Aircraft Condition Monitoring
ACMS	Aircraft Condition Monitoring System
ACP	Audio Control Panel
ADAPT	Air Traffic Management Data Acquisition, Processing and Transfer
ADF	Automatic Direction Finder
ADI	Attitude Director Indicator
ADIRS	Air Data/Inertial Reference System
ADIRU	Air Data Inertial Reference Unit
ADM	Air Data Module
ADR	Air Data Reference
ADS	Air Data System
ADS	Automatic Dependent Surveillance
ADS-B	Automatic Dependent Surveillance-Broadcast
ADS-C	Automatic Dependent Surveillance-Contract
AEEC	Airlines Electronic Engineering Committee
AF	Audio Frequency
AFC	Automatic Frequency Control
AFCS	Auto Flight Control System (Autopilot)
AFDX	Avionics Full Duplex
AFIS	Airborne Flight Information Service
AFMS	Advanced Flight Management System
AFS	Automatic Flight System (Autopilot)
AGC	Automatic Gain Control
AGL	Above Ground Level
AHRS	Attitude/Heading Reference System
AHS	Attitude Heading System
AI	Airbus Industries
AI	Artificial intelligence
AIAA	American Institute of Aeronautics and Astronautics
AIDS	Aircraft Integrated Data System
AIMS	Airplane Information Management System
AIS	Aeronautical Information System
AIV	Anti-Icing Valve
ALT	Altitude
AM	Amplitude Modulation
AMD	Advisory Map Display
AMI	Airline Modifiable Information
AMLCD	Active Matrix Liquid Crystal Display
AMSL	Above Mean Sea Level
AMU	Audio Management Unit
ANSI	American National Standards Institute
ANSIR	Advanced Navigation System Inertial Reference
ANSP	Air navigation air navigation service provider
ANT	Antenna
AOA	Angle of Attack
AOC	Airline Operational Control
AP	Autopilot
APATSI	Airport Air Traffic System Interface
API	Application Programming Interface
APM	Advanced Power Management

APP	Approach	BER	Bit Error Rate
APR	Auxiliary Power Reserve	BFO	Beat Frequency Oscillator
APU	Auxiliary Power Unit	BGW	Basic Gross Weight
APV	Approach Procedures Vertical Guidance	BIOS	Basic Input/Output System
ARINC	Aeronautical Radio Incorporated	BIST	Built-In Self-Test
ARR	Arrival	BIT	Built-in Test
ARTAS	Advanced Radar Tracker and Server	BITE	Built-in Test Equipment
		BIU	Bus Interface Unit
ARTS	Automated Radar Terminal System	BLEU	Blind Landing Experimental Unit
		BPS	Bits Per Second
ASAAC	Allied Standard Avionics Architecture Council	CAA	Civil Aviation Authority
		CABLAN	Cabin Local Area Network
ASCB	Aircraft System Common Data Bus	CADC	Central Air Data Computer
		CAI	Computer Aided Instruction
ASCII	American Standard Code for Information Interchange	CAN	Controller Area Network
		CAS	Collision Avoidance System
ASI	Air Speed Indicator	CAS	Crew Alerting System
ASIC	Application Specific Integrated Circuit	CAT	Clear-Air-Turbulence
		CCA	Circuit Card Assembly
ASR	Aerodrome Surveillance Radar	CDDI	Copper Distributed Data Interface
ASTM	American Society for Testing and Materials	CDI	Course Deviation Indicator
		CDROM	Compact Disk Read-Only Memory
ATA	Actual Time of Arrival		
ATA	Air Transport Association	CDS	Common Display System
ATAS	ADS-B Traffic Advisory System	CDTI	Cockpit Display of Traffic Information
ATC	Air Traffic Control		
ATCRBS	ATC Radio Beacon System	CDU	Control Display Unit
ATE	Automatic Test Equipment	CEATS	Central European Air Traffic Service
ATFM	Air Traffic Flow Management		
ATI	Air Transport Indicator	CFDS	Central Fault Display System
ATLAS	Abbreviated Test Language for Avionics Systems	CH	Compass Heading
		CIDIN	Common ICAO Data Interchange Network
ATM	Air Traffic Management		
ATN	Aeronautical Telecommunications Network	CIDS	Cabin Intercommunication Data System
ATR	Air Transportable Racking	CIO	Carrier Insertion Oscillator
ATS	Air Traffic Services	CLB	Climb
ATS	Air Traffic System	CMC	Central Maintenance Computer
ATSU	Air Traffic Services Unit	CMOS	Complementary Metal Oxide Semiconductor
AVC	Automatic Volume Control		
AVG	Approach with Vertical Guidance	CMP	Configuration Management Plan
AVLAN	Avionics Local Area Network	CMS	Centralized Maintenance System
AWIN	Aircraft Weather Information	CMU	Communications Management Unit
AWLU	Aircraft Wireless Local Area Network Unit		
		CNS	Communications Navigation and Surveillance
B-C	Back-Course		
BCD	Binary Coded Decimal	COMPAS	Computer Orientated Metering, Planning and Advisory System
BCO	Binary Coded Octal		
BDS	Comm. B Data Selector	CONUS	Continental United States

COTS	Commercial Off-The-Shelf	DMA	Direct Memory Access
CPA	Closest Point of Approach	DME	Distance Measuring Equipment
CPDLC	Controller Pilot Datalink Communications	DMEP	Data Management Entry Panel
		DMOD	Distance Modification
CPM	Core Processing Module	DO	Design Organisation
CPU	Central Processing Unit	DP	Decimal Point
CRC	Cyclic Redundancy Check	DP	Departure Procedures
CRM	Crew Resource Management	DPM	Data Position Module
CRT	Cathode Ray Tube	DPSK	Differential Phase Shift Keying
CRZ	Cruise	DPU	Display Processor Unit
CTO	Central Technical Operations	DRAM	Dynamic Random Access Memory
CVOR	Conventional VOR		
CVR	Cockpit Voice Recorder	DS	Data Segment
CW	Continuous Wave	DSB	Double Sideband
D8PSK	Differential Eight Phase Shift Keying	DSB-SC	Double Sideband Suppressed Carrier
DA	Drift Angle	DSP	Digital Signal Processing
DAC	Digital to Analog Converter	DSP	Display Select Panel
DADC	Digital Air Data Computer	DSRTK	Desired Track
DAP	Downlink Aircraft Parameters	DTG	Distance To Go
DATAC	Digital Autonomous Terminal Access Communications System	DTG	Dynamically Tuned Gyroscope
		DTOP	Dual Threshold Operation
DBRITE	Digital Bright Radar Indicator Tower Equipment	DU	Display Units
		DUATS	Direct User Access Terminal System
DCPC	Direct Controller Pilot Communications	DVOR	Doppler VOR
		EADI	Electronic Attitude Director Indicator
DCU	Data Concentrator Unit		
DDM	Difference in Depth of Modulation	EARTS	En-route Automated Radar Tracking System
DDR	Digital Data Recorder		
DECU	Digital Engine Control Unit	EAS	Express Air System
DEOS	Digital Engine Operating System	EASA	European Union Aviation Safety Agency
DEP	Departure		
DES	Descent	EAT	Expected Approach Time
DEU	Digital Electronics Unit	EATMP	European Air Traffic Management Programme
DF	Downlink Format		
DFDAU	Digital Flight Data Acquisition Unit	EATMS	Enhanced Air Traffic Management System
DFDR	Digital Flight Data Recorder	EC	European Commission
DFGC	Digital Flight Guidance Computer	ECAM	Electronic Centralized Aircraft Monitoring
DFGS	Digital Flight Guidance System	ECB	Electronic Control Box
DFLD	Database Field Loadable Data	ECHO	European concept for higher airspace operation
DG	Directional Gyro		
DGPS	Differential Global Positioning System	ECM	Electronic Countermeasures
		ECS	Environmental Control System
DH	Decision Height	ECU	Electronic Control Unit
DIR INTC	Direct Intercept	EEC	Electronic Engine Control
DIS	Distance	EEPROM	Electrically Erasable Programmable Read-Only Memory
DITS	Digital Information Transfer System		

EFCS	Electronic Flight Control System		FCC	Federal Communications Commission
EFIS	Electronic Flight Instrument System		FCC	Flight Control Computer
EGNOS	European Geostationary Navigation Overlay Service		FCC	Federal Communications Commission
EGPWS	Enhanced Ground Proximity Warning System		FCGC	Flight Control and Guidance Computer
EHF	Extremely High Frequency		FCS	Flight Control System
EHS	Enhance Surveillance		FCU	Flight Control Unit
EHSI	Electronic Horizontal Situation Indicator		FD	Flight Director
			FDAU	Flight Data Acquisition Unit
EIA	Electronic Industries Association		FDC	Flight Director Computer
EICAS	Engine Indication and Crew Alerting Systems		FDD	Floppy Disk Drive
			FDDI	Fibre Distributed Data Interface
EIDE	Enhanced Integrated Drive Electronics		FDE	Fault Detection and Exclusion
			FDM	Frequency Division (Domain) Multiplexing
EIS	Electronic Instrument System			
EL	Elevation-Station		FDMU	Flight Data Management Unit
ELAC	Elevator and Aileron Computer		FDR	Flight Data Recorder
ELF	Extremely Low Frequency		FDS	Flight Director System
ELM	Extended Length Message		FET	Field Effect Transistor
eLORAN	Enhanced LORAN		FFS	Full Flight Simulator
ELS	Electronic Library System		FG	Flight Guidance
ELS	Elementary Surveillance		FGC	Flight Guidance Computer
ELT	Emergency Locator Transmitter		FGI	Flight guidance by digital Ground Image
EMC	Electromagnetic Compatibility			
EMI ·	Electromagnetic Interference		FGS	Flight Guidance System
EPROM	Erasable Programmable Read-Only Memory		FIR	Flight Information Region
			FIS	Flight Information System
EROPS	Extended Range Operations		FIS-B	Flight Information Services-Broadcast
ERU	Electronic Routing Unit			
ES	Extended Squitter		FL	Flight Level
ESA	Electronically steered antenna		FLIR	Forward Looking Infrared
ESA	European Space Agency		FLS	Field Loadable Software
ESD	Electrostatic Discharge		FM	Frequency Modulation
ESD	Electrostatic Sensitive Device		FMC	Flight Management Computer
ETA	Estimated Time of Arrival		FMCDU	Flight Management Control and Display Unit
ETOPS	Extended Range Twin-engine Operations			
			FMCS	Flight Management Computer System
EVS	Enhanced vision systems			
EXEC	Execute		FMGC	Flight Management Guidance Computer
FAA	Federal Aviation Administration			
FAC	Flight Augmentation Computer		FMS	Flight Management System
FADEC	Full Authority Digital Engine Control		FOG	Fibre Optic Gyroscope
			FRUIT	False Replies from Unsynchronised Interrogator Transmissions
FANS	Future Air Navigation Systems			
FAR	Federal Aviation Regulations		FSK	Frequency Shift Keying
FBL	Fly-By-Light		FSS	Fixed Satellite Service
FBW	Fly-By-Wire		G	Giga (10^9 multiplier)

GA	General Aviation	**IAS**	Indicated Air Speed
GAT	General Air Traffic	**IATA**	International Air Transport Association
GBAS	Ground-based Augmentation System	**IC**	Interrogator Codes
GBST	Ground-Based Software Tool	**ICAO**	International Civil Aviation Organization
GBT	Ground Base Transmitter		
GDOP	Geometric Dilution Of Precision	**IDENT**	Identification
GEO	Geostationary Earth Orbit	**IF**	Intermediate Frequency
GES	Ground Earth Station	**IFALPA**	International Federation of Air Line Pilots' Associations
GHz	Gigahertz (10^9 Hz)		
GICB	Ground Initiated Comm. B	**IFE**	In-Flight Entertainment
GLONASS	Global Navigation Satellite System	**IFF**	Identification, Friend or Foe
		IFOG	Interferometric Fibre Optic Gyro
GLS GNSS	Landing System	**IFPS**	International Flight Plan Processing System
GMT	Greenwich Mean Time		
GND	Ground	**IFR**	Instrument Flight Rules
GNSS	Global Navigation Satellite System	**IHF**	Integrated Human Interface Function
GPM	Ground Position Module	**IHUMS**	Integrated Health and Usage Monitoring System
GPS	Global Positioning System		
GPWS	Ground Proximity Warning System	**II**	Interrogator Identifier
		ILS	Instrument Landing System
GRI	Group Repetition Interval	**IM**	Inner Marker
GS	Glide Slope	**IMA**	Integrated Modular Avionics
GS	Ground Speed	**IMU**	Inertial Measurement Unit
GW	Gross Weight	**INS**	Inertial Navigation System
HDD	Head Down Display	**IP**	Internet Protocol
HDG	Heading	**IPC**	Instructions Per Cycle
Hex	Hexadecimal	**IPR**	Intellectual Property Right
HF	High Frequency	**IPX/SPX**	Inter-network Packet Exchange/ Sequential Packet Exchange
HFDL	High Frequency Datalink		
HFDS	Head-up Flight Display System	**IR**	Infra-Red
HGS	Head-up Guidance System	**IRMP**	Inertial Reference Mode Panel
HIRF	High-energy Radiated Field/High-intensity Radiated Field	**IRS**	Inertial Reference System
		IRU	Inertial Reference Unit
HIRF	High-Intensity Radiated Field	**ISA**	Inertial Sensor Assembly
HIRL	High-Intensity Runway Lights	**ISA**	International Standard Atmosphere
HM	Health Monitoring		
HMCDU	Hybrid Multifunction Control Display Unit	**ISAS**	Integrated Situational Awareness System
HPA	High Power Amplifier	**ISDB**	Integrated Signal Database
HSI	Horizontal Situation Indicator	**ISDU**	Inertial System Display Unit
HUD	Head-Up Display	**ISGS**	Increased second glide slope
HUGS	Head-Up Guidance System	**ISM**	Industrial, Scientific and Medical
Hz	Hertz (cycles per second)	**ISO**	International Standards Organisation
I/O	Input/Output		
IAC	Integrated Avionics Computer	**IVSI**	Instantaneous Vertical Speed Indicator
IAPS	Integrated Avionics Processing System		
		IWF	Integrated Warning Function

JAA	Joint Airworthiness Authority	**MCU**	Modular Component Unit
JAR	Joint Airworthiness Requirement	**MDA**	Minimum Descent Altitude
JEDEC	Joint Electron Device Engineering Council	**MDAU**	Maintenance Data Acquisition Unit
k	Kilo (10^3 multiplier)	**MEL**	Minimum Equipment List
kHz	Kilohertz (10^3 Hz)	**MEMS**	Micro electromechanical systems
KIAS	Indicated Airspeed in Knots	**METAR**	Meteorological Aerodrome Report
km	kilometre		
Knot	Nautical Miles/Hour	**MF**	Medium Frequency
KT	Knots	**MFD**	Multifunction Flight Display
LAAS	Local Area Augmentation System	**MFDS**	Multifunction Display System
LAN	Local Area Network	**MHRS**	Magnetic Heading Reference System
LASER	Light Amplification by Stimulated Emission of Radiation	**MHz**	Megahertz (10^6 Hz)
Lat.	Latitude	**MLS**	Microwave Landing System
LATAN	Low-Altitude Terrain-Aided Navigation	**MLW**	Maximum Landing Weight
		MM	Middle Marker
LCD	Liquid Crystal Display	**MMI**	Man Machine Interface
LDA	Localizer Directional Aid	**MMR**	Multimode Receiver
LDU	Lamp Driver Unit	**MMS**	Mission Management System
LED	Light-Emitting Diode	**MNPS**	Minimum Navigation Performance Specification
LF	Low Frequency		
LIDAR	Light Radar	**MOS**	Metal Oxide Semiconductor
LNAV	Lateral Navigation	**MOSFET**	Metal Oxide Semiconductor Field Effect Transistor
LO	Local Oscillator		
LOC	Localizer	**MRC**	Modular Radio Cabinets
Long	Longitude	**MRO**	Maintenance/Repair/Overhaul
LORADS	Long Range Radar and Display System	**MSB**	Most Significant Bit
		MSD	Most Significant Digit
LOS	Line of Sight	**MSG**	Message
LRM	Line Replaceable Module	**MSI**	Medium Scale Integration
LRNS	Long Range Navigation System	**MSK**	Minimum Shift Keying
LRRA	Low Range Radio Altimeter	**MSL**	Mean Sea Level
LRU	Line Replaceable Unit	**MSU**	Mode Select Unit
LSAP	Loadable Aircraft Software Part	**MSW**	Machine Status Word
LSB	Least Significant Bit	**MT**	Maintenance Terminal
LSB	Lower Sideband	**MTBF**	Mean Time Between Failure
LSD	Least Significant Digit	**MTBO**	Mean Time Between Overhaul
LSI	Large Scale Integration	**MTC**	Mission and Traffic Control systems
LSS	Lightning Sensor System		
LUF	Lowest Usable Frequency	**MTOW**	Maximum Take-off Weight
M	Mega (10^6 multiplier)	**MUF**	Maximum Usable Frequency
MAPt	Missed Approach Point	**NAS**	National Airspace System
MASI	Mach Airspeed Indicator	**NASA**	National Aeronautics and Space Administration
MAU	Modular Avionics Unit		
MCDU	Microprocessor Controlled Display Units	**NATS**	National Air Traffic Services
		NAVSTAR	Navigation System with Timing and Ranging
MCP	Mode Control Panel		
MCS	Master Control Station	**NCD**	No Computed Data

ND	Navigation Display	PNF	Pilot Non-Flying
NDB	Navigation Database	POS	Position
NDB	Non-Directional Beacon	Pos. init.	Position initialisation
NFF	No Fault Found	POST	Power-On Self-test
NGATS	Next Generation Air Transport System	PP	Pre-Processor
		PPI	Plan Position Indicator
NIC	Network Interface Controller	PPM	Pulse Position Modulation
nm	Nautical mile	PPOS	Present Position
NMS	Navigation Management System	PQFP	Plastic Quad Flat Package
NMU	Navigation Management Unit	PRF	Pulse Repetition Frequency
NNEW	Next Generation Network Enabled Weather	PRI	Primary
		PRN	Pseudo-random Noise
NOAA	National Oceanic and Atmospheric Administration	PROG	Progress
		PROM	Programmable Read-Only Memory
NOTAM	Notice to Airmen		
NPA	Non-Precision Approach	PSK	Phase Shift Keying
NVM	Non-Volatile Memory	PSM	Power Supply Module
NVS NAS	voice switch	PSR	Primary Surveillance Radar
OAT	Outside air temperature	Q	Quality Factor
OBI	Omni Bearing Indicator	QAM	Quadrature Amplitude Modulation
OBS	Omni Bearing Selector		
ODS	Operations Display System	QE	Quadrantal Error
OEI	One Engine Inoperative	QoS	Quality of Service
OEM	Original Equipment Manufacturer	QPSK	Quadrature Phase Shift Keying
OLDI	On-Line Data Interchange	R/T	Receiver/Transmitter
OM	Optical Marker	RA	Radio Altitude
OMS	On-board Maintenance System	RA	Resolution Advisory
OS	Operating System	Radar	Radio Direction and Ranging
OSS	Option Selectable Software	RAeS	Royal Aeronautical Society
PA	Precision Approach	RAIM	Receiver Autonomous Integrity Monitoring
PAPI	Precision Approach Path Indicator		
PBN	Performance Based Navigation	RAM	Random Access Memory
PCA	Preconditioned Air System	RDMI	Radio Distance Magnetic Indicator
PCB	Printed Circuit Board		
PCC	Purser Communication Center	RF	Radio Frequency
PCHK	Parity Check(ing)	RIMM	RAM Bus In-line Memory Module
PDB	Performance Database		
PDL	Portable Data Loader	RISC	Reduced Instruction Set Computer
PFD	Primary Flight Display		
PIC	Pilot In Command	RLG	Ring Laser Gyro
PLASI	Pulsed Light Approach Slope Indicator	RMI	Radio Magnetic Indicator
		RMP	Radio Management Panel
PLL	Phase Locked Loop	RNAV	Area Navigation
PM	Protected Mode	RNP	Required Navigation Performance
PMAT	Portable Maintenance Access Terminal	ROM	Read-Only Memory
		RPAS	Remotely Piloted Aircraft System
PMO	Program Management Organization	RTA	Required time of arrival
		RTCA	Radio Technical Commission for Aeronautics
PMS	Performance Management System		

RTE	Route	STP	Shielded Twisted Pair
RVR	Runway Visual Range	SVS	Synthetic vision systems
RVSM	Reduced Vertical	SW	Software
	Separation Minimum	SWIM	System Wide Information
RX	Receiver		Management
SA	Service Availability	SWR	Standing Wave Ratio
SAARU	Secondary Attitude/Air Data	T/C	Top of Climb
	Reference Unit	T/D	Top of Descent
SAT	Static Air Temperature	TA	Traffic Advisory
SATCOM	Satellite Communications	TACAN	Tactical Air Navigation
SBAS	Satellite-based	TAF	Terminal Aerodrome Forecast
	Augmentation System	TAS	Traffic Advisory System
SC	Suppressed Carrier	TAS	True Air Speed
SCMP	Software Configuration	TAT	Total Air Temperature
	Management Plan	Tau	Minimum time to collision
SDD	System Definition Document		threshold
SDI	Source/Destination Identifier	TAWS	Terrain Awareness
SDR	Software Defined Radio		Warning System
SEC	Secondary	TBO	Time Between Overhaul
SELCAL	Selective Calling	TBO	Trajectory Based Operations
SES	Single European Sky	TCAS	Traffic Alert and Collision
SESAR	Single European Sky ATM		Avoidance System
	Research Programme	TCVR	Transceiver
SHF	Super High Frequency	TDM	Time Division (Domain)
SI	Surveillance Indicator		Multiplexing
SID	Standard Instrument Departure	TIS-B	Traffic Information
SLS	Side Lobe Suppression		Services-Broadcast
SMART	Standard Modular Avionics	TK	Track
	Repair/Test	TKE	Track Angle Error
SNMP	Simple Network Management	TRF	Tuned Radio Frequency
	Protocol	TS	Task Switched
SPDA	Secondary Power Distribution	TTL	Transistor–Transistor Logic
	Assembly	TTP	Time Triggered Protocol
SPDT	Single Pole Double Throw	TWDL	Two-Way Datalink
SPI	Special Position Identity	TX	Transmitter
SRAM	Synchronous Random	TX/RX	Transmitter/receiver (transceiver)
	Access Memory	UAM	Urban air mobility
SRAP	Second runway aiming point	UAS	Unmanned Aircraft System
SRD	Short-range Device (radio band)	UAT	Universal Access Transceiver
SRD	System Requirement Document	UAV	Unmanned Aerial Vehicle
SROM	Serial Read Only Memory	UDP	User Datagram Protocol
SSB	Single Sideband	UF	Uplink Format
SSI	Small Scale Integration	UHF	Ultra High Frequency
SSM	Sign/Status Matrix	ULSI	Ultra Large Scale Integration
SSR	Secondary Surveillance Radar	UVPROM	Ultraviolet Programmable Read-
SSV	Standard Service Volume		Only Memory
STAR	Standard Terminal Arrival Route	UMS	User Modifiable Software
STDMA	Self-organising Time Division	USB	Universal Serial Bus
	Multiple Access	USB	Upper Sideband

USCG	United States Coast Guard	**VNAV**	Vertical Navigation
UTC	Coordinated Universal Time	**VOR**	VHF Omnirange
UTP	Unshielded Twisted Pair	**VORTAC**	VOR TACAN Navigation Aid
UV	Ultra-Violet	**VPA**	Virtual Page Address
VAC	Volts, Alternating Current	**VSI**	Vertical Speed Indicator
VAS	Virtual Address Space	**W/S**	Whisper-Shout
VCO	Voltage Controlled Oscillator	**WAAS**	Wide Area Augmentation System
VDB	VHF Data Broadcast		
VDC	Volts, Direct Current	**WACS**	Wireless Airport Communication System
VDL	Very High Frequency Datalink		
VFR	Visual Flight Rules	**WAN**	Wide Area Network
VG	Vertical Gyro	**WD**	Wind Direction
VHF	Very High Frequency	**WS**	Wind Speed
VHSIC	Very High Speed Integrated Circuit	**WX**	Weather
		WXP	Weather Radar Panel
VIA	Versatile Integrated Avionics	**WXR**	Weather Radar
VLF	Very Low Frequency	**XTAL**	Crystal
VLSI	Very Large Scale Integration	**XTK**	Cross Track
VME	Versatile Module Eurocard	**ZFW**	Zero Fuel Weight

Appendix 2 Revision papers

These revision papers are designed to provide you with practice for examinations. The questions are typical of those used in CAA and other examinations. Each paper has 20 questions and each should be completed in 25 minutes. Calculators and other electronic aids must not be used.

Revision Paper 1

1. A radio wave is said to be polarised in:
   (a) the direction of travel
   (b) the E-field direction
   (c) the H-field direction.

2. Radio waves tend to propagate mainly as line of sight signals in the:
   (a) MF band
   (b) HF band
   (c) VHF band.

3. An isotropic radiator will radiate:
   (a) only in one direction
   (b) in two main directions
   (c) uniformly in all directions.

4. A vertical quarter wave antenna will have a polar diagram in the horizontal plane which is:
   (a) unidirectional
   (b) omnidirectional
   (c) bi-directional.

5. The attenuation of an RF signal in a coaxial cable:
   (a) increases with frequency
   (b) decreases with frequency
   (c) stays the same regardless of frequency.

6. The method of modulation used for aircraft VHF voice communication is:
   (a) MSK
   (b) D8PSK
   (c) DSB AM.

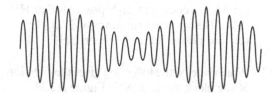

Figure A2.1   See Paper 1, Question 7

7. The type of modulation shown in Figure A2.1 is:
   (a) AM
   (b) FM
   (c) PSK.

8. The standard for ACARS is defined in:
   (a) ARINC 429
   (b) ARINC 573
   (c) ARINC 724.

9. The frequency range currently used in Europe for aircraft VHF voice communication is:
   (a) 88 MHz to 108 MHz
   (b) 108 MHz to 134 MHz
   (c) 118 MHz to 137 MHz.

10. The type of antenna shown in Figure A2.2 is
    (a) a unipole
    (b) a dipole
    (c) a Yagi.

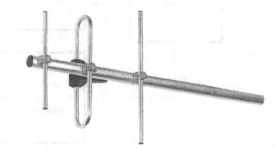

Figure A2.2   See Paper 1, Question 10

11. The typical frequency emitted by a ULB is:
    (a) 600 Hz
    (b) 3.4 kHz
    (c) 37.5 kHz.

12. ELT transmissions use:
    (a) Morse code and high-power RF at HF
    (b) pulses of acoustic waves at 37.5 kHz
    (c) low-power RF at VHF or UHF.

13. Which one of the following gives the function
    of the block marked 'X' in Figure A2.3?
    (a) power amplifier
    (b) matching unit
    (c) SWR detector.

14. The angular difference between magnetic
    north and true north is called the:
    (a) magnetic variation
    (b) great circle
    (c) prime meridian.

15. Morse code tones are used to identify
    the VOR:
    (a) identification
    (b) frequency
    (c) radial.

16. When hovering over water, the 'worst case'
    conditions for Doppler signal-to-noise
    ratios are with:
    (a) smooth sea conditions
    (b) rough sea conditions
    (c) tidal drift.

17. Once aligned, the inertial navigation
    system is always referenced to:
    (a) magnetic north
    (b) true north
    (c) latitude and longitude.

18. During prolonged periods of poor satellite
    reception, the GPS receiver:
    (a) enters into a dead reckoning mode
    (b) re-enters the acquisition mode
    (c) rejects all satellite signals.

19. What effect will radome delamination have
    on the weather radar signal?
    (a) None
    (b) Decreased attenuation
    (c) Increased attenuation.

20. The purpose of traffic advisories
    (TA) is to:
    (a) assist the crew in visually searching and
        identifying an intruder
    (b) provide recommended manoeuvres
        needed to maintain vertical
        separation
    (c) provide lateral guidance to the crew.

Revision Paper 2

1. The HF range extends from:
   (a) 300 kHz to 3 MHz
   (b) 3 MHz to 30 MHz
   (c) 30 MHz to 300 MHz.

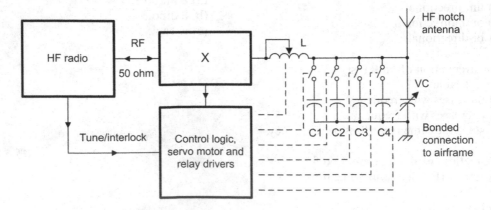

Figure A2.3   See Paper 1, Question 13

2. A transmitted radio wave will have a plane
   wavefront:
   (a) in the near field
   (b) in the far field
   (c) close to the antenna.

3. When radio waves travel in a cable
   they travel:
   (a) at the speed of light
   (b) slower than the speed of light
   (c) faster than the speed of light.

4. A radio wave at 11 MHz is most likely to
   propagate over long distances as:
   (a) a ground wave
   (b) a sky wave
   (c) a space wave.

5. The height of the E-layer is approximately:
   (a) 100 km
   (b) 200 km
   (c) 400 km.

6. The impedance measured at the input of a
   long length of correctly terminated coaxial
   cable will be:
   (a) the same as the characteristic
       impedance of the cable
   (b) zero
   (c) infinite.

7. The ATC transponder code of 7700 is
   used for:
   (a) general air emergency
   (b) loss of radio
   (c) hijacking.

8. During the alignment mode, a flashing IRS
   align light indicates:
   (a) the system is ready to navigate
   (b) the aircraft was moved during
       align mode
   (c) the present position entered agrees with
       the last known position.

9. The decision height and runway visual
   range for a Category 2 automatic
   approach are:
   (a) 100 ft and 300 m respectively
   (b) 200 ft and 550 m respectively
   (c) less than 100 ft and 200 m respectively.

10. The typical pulse rate for a ULB is:
    (a) 0.9 pulses per sec
    (b) 10 pulses per sec
    (c) 60 pulses per sec.

11. The CVR is usually located:
    (a) on the flight deck
    (b) in the avionic equipment bay
    (c) in the ceiling of the aft passenger
        cabin.

12. Transmission from an ELT is usually
    initially detected by:
    (a) low-flying aircraft
    (b) one or more ground stations
    (c) a satellite.

13. Which one of the following gives the
    approximate LOS range for an aircraft at
    an altitude of 15,000 feet?
    (a) 74 nm
    (b) 96 nm
    (c) 135 nm.

14. The function of the compressor stage in an
    aircraft VHF radio is:
    (a) to reduce the average level of
        modulation
    (b) to increase the average level of
        modulation
    (c) to produce 100% modulation at
        all times.

15. The radiation efficiency of an antenna:
    (a) increases with antenna loss
        resistance
    (b) decreases with antenna loss
        resistance
    (c) is unaffected by antenna loss
        resistance.

16. Display of the FMCS CDU identification
    page after power-up confirms the:
    (a) IRS is aligned
    (b) navigation source(s) in use
    (c) FMC has passed its BITE check.

17. DME-DME position is determined by:
    (a) three or more GPS satellites
    (b) the intersection of two DME arcs
    (c) 4D navigation

18. The characteristic impedance of a coaxial cable depends on:
    (a) the ratio of inductance to capacitance
    (b) the ratio of resistance to inductance
    (c) the sum of the resistance and reactance.

19. When a DME indicator is receiving no computed data, it will display:
    (a) dashes
    (b) zeros
    (c) eights.

20. Quadrantal error (QE) for an ADF system is associated with the:
    (a) ionosphere
    (b) physical aspects of terrain
    (c) physical aspects of the aircraft structure.

Revision Paper 3

1. The free-space path loss experienced by a radio wave:
   (a) increases with frequency but decreases with distance
   (b) decreases with frequency but increases with distance
   (c) increases with both frequency and distance.

2. For a given HF radio path, the MUF changes most rapidly at:
   (a) mid-day
   (b) mid-night
   (c) dawn and dusk.

3. In the HF band radio waves tend to propagate over long distances as:
   (a) ground waves
   (b) space waves
   (c) ionospheric waves.

4. A standing wave ratio of 1:1 indicates:
   (a) that there will be no reflected power
   (b) that the reflected power will be the same as the forward power
   (c) that only half of the transmitted power will actually be radiated.

5. An aircraft is flying on heading of 090° to intercept the selected VOR radial of 180°; the HSI will display that the aircraft is:
   (a) right of the selected course
   (b) left of the selected course
   (c) on the selected course.

6. The typical bandwidth of a DSB AM voice signal is:
   (a) 3.4 kHz
   (b) 7 kHz
   (c) 25 kHz.

7. The operational state of an ELT is tested using:
   (a) a test switch and indicator lamp
   (b) immersion in a water tank for a short period
   (c) checking battery voltage and charging current.

8. The air testing of an ELT can be carried out:
   (a) at any place or time
   (b) only after notifying the relevant authorities
   (c) only at set times using recommended procedures.

9. A Type-W ELT is activated by:
   (a) a member of the crew
   (b) immersion in water
   (c) a high G-force caused by deceleration.

10. The angle between north and the flight path of the aircraft is the:
    (a) ground track angle
    (b) drift angle
    (c) heading.

11. An RDMI provides the following information:
    (a) distance and bearing to a navigation aid
    (b) deviation from a selected course
    (c) the frequency of the selected navigation aid.

12. Marker beacon outputs are given by:
    (a) coloured lights and Morse code tones
    (b) deviations from the runway centreline
    (c) deviations from the glide slope.

13. Inertial navigation system errors are a factor of:
    (a) the aircraft's velocity
    (b) how long the system has been in the 'align' mode
    (c) how long the system has been in the 'navigation' mode.

14. The GPS navigation concept is based upon calculating satellite:
    (a) speed
    (b) altitude
    (c) range.

15. The FMC recognises specific aircraft types by:
    (a) CDU entry
    (b) program pins
    (c) the navigation database.

16. Weather radar operates in which bands of radar frequencies?
    (a) C- and X-band
    (b) L-band
    (c) HF.

17. ATC Mode C (altitude information) is derived from an:
    (a) altimeter or air data computer
    (b) ATC transponder
    (c) ATC control panel.

18. TCAS II requires which type of ATC transponder?
    (a) Mode S
    (b) Mode A
    (c) Mode C.

19. The outer marker is displayed on the primary flying display as a coloured icon that is:
    (a) yellow
    (b) white
    (c) cyan.

20. DME is based on what type of radar?
    (a) Primary
    (b) Secondary
    (c) VHF.

Revision Paper 4

1. Which one of the following gives the approximate length of a half-wave dipole for use at 300 MHz?
   (a) 50 cm
   (b) 1 m
   (c) 2 m.

2. In the horizontal plane, a vertical dipole will be:
   (a) bi-directional
   (b) omnidirectional
   (c) unidirectional.

3. The function of the HF antenna coupler is to:
   (a) reduce static noise and interference
   (b) increase the transmitter output power
   (c) match the HF antenna to the HF radio.

4. Another name for a quarter-wave vertical antenna is:
   (a) a Yagi antenna.
   (b) a dipole antenna
   (c) a Marconi antenna.

5. A full-wave dipole fed at the centre must be:
   (a) current fed
   (b) voltage fed
   (c) impedance fed.

6. The antenna shown in Figure A2.4 is used for:
   (a) ILS
   (b) GPS
   (c) VHF communications.

7. The channel spacing currently used in Europe for aircraft VHF voice communication is:
   (a) 8.33 kHz and 25 kHz
   (b) 12.5 kHz and 25 kHz
   (c) 25 kHz and 50 kHz.

8. On which frequencies do ELT operate?
   (a) 125 MHz and 250 MHz
   (b) 122.5 MHz and 406.5 MHz
   (c) 121.5 MHz and 406.025 MHz.

Figure A2.4 See Paper 4, Question 6

9. The DME interrogator is part of the:
   (a) airborne equipment
   (b) DME navigation aid
   (c) VORTAC.

10. Localizer transmitters are located:
    (a) at the threshold of the runway, adjacent to the touchdown point
    (b) at the stop end of the runway, on the centreline
    (c) at three locations on the extended centreline of the runway.

11. When two RF waves from adjacent transmitting elements are in phase, they create interference that is:
    (a) positive (destructive) resulting in a wave of decreased (or reinforced) amplitude.
    (b) negative (destructive) resulting in a wave of decreased (or reinforced) amplitude.
    (c) positive (constructive) resulting in a wave of increased (or reinforced) amplitude.

12. Autotuning of navigation aids is used by RNAV systems to:
    (a) update the navigation database
    (b) create waypoints in the CDU
    (c) select the best navigation aids for optimised area navigation.

13. When hovering directly over an object in the sea with a six-knot tide, the Doppler navigation system will indicate:
    (a) six knots, drift in the opposite direction of the tide
    (b) six knots, drift in the direction of the tide
    (c) zero drift.

14. Errors in an inertial navigation system are:
    (a) random and build up as a function of time
    (b) fixed and irrespective of time
    (c) random and irrespective of time.

15. The purpose of RAIMS in GPS is to:
    (a) identify the selected satellite
    (b) speed up the satellite acquisition process
    (c) provide error detection.

16. FMCS alerting messages require attention from the crew:
    (a) before guided flight can be continued
    (b) when time is available
    (c) at the completion of the flight.

17. The most severe weather radar images are colour coded:
    (a) black
    (b) magenta
    (c) green.

18. Mode A replies contain the following aircraft information:
    (a) identification
    (b) identification and altitude
    (c) identification, altitude and aircraft address.

19. Recommended manoeuvres needed to increase or maintain vertical separation are provided by what type of TCAS warning:
    (a) resolution advisory
    (b) traffic advisory
    (c) non-threat traffic.

20. During sunrise and sunset, ADF transmissions are affected by:
    (a) coastal refraction
    (b) static build-up in the airframe
    (c) variations in the ionosphere.

Appendix 3 Answers to multiple choice questions

Answers to review questions

Chapter 1 (Page 24)

1. b
2. c
3. a
4. a
5. c
6. c
7. c
8. a
9. c
10. b
11. a
12. c
13. c
14. c
15. b
16. c
17. c
18. c
19. c
20. a
21. a
22. c
23. a
24. b
25. b
26. a
27. c

Chapter 2 (Page 55)

1. c
2. c
3. a
4. b
5. b
6. a
7. a
8. b
9. c

10. b
11. c
12. a
13. a
14. c
15. a
16. b
17. c

Chapter 3 (Page 79)

1. c
2. a
3. a
4. c
5. b
6. b
7. b
8. c
9. c
10. b
11. c
12. a
13. a
14. b
15. b
16. c
17. c

Chapter 4 (Page 92)

1. a
2. c
3. a
4. c
5. b
6. a
7. b
8. b
9. b
10. c
11. c
12. a

Chapter 5 (Page 106)

1. a
2. a
3. a
4. c
5. b
6. a
7. b
8. c
9. a
10. c

Chapter 6 (Page 114)

1. b
2. a
3. a
4. b
5. c
6. c
7. c
8. b
9. a
10. a
11. a
12. c
13. c
14. a
15. c
16. c

Chapter 7 (Page 123)

1. c
2. b
3. c
4. a
5. c
6. c
7. c
8. c
9. b
10. b
11. b
12. a
13. b

Chapter 8 (Page 143)

1. b
2. a
3. a
4. b
5. b
6. c
7. a
8. a
9. b
10. b
11. c
12. a
13. b
14. a
15. c

Chapter 9 (Page 153)

1. b
2. a
3. b
4. a
5. c
6. a
7. a
8. b
9. b
10. c
11. a
12. c
13. b
14. a
15. c
16. b

Chapter 10 (Page 168)

1. c
2. c
3. b
4. a
5. b
6. b
7. b
8. a

9. a
10. b
11. a
12. a
13. a
14. b
15. a
16. a

Chapter 11 (Page 180)

1. b
2. c
3. a
4. a
5. a
6. a
7. b
8 a
9. a
10. a
11. b
12. b
13. a
14. a
15. b
16. c
17. a

Chapter 12 (Page 193)

1. c
2. b
3. c
4. c
5. a
6. b
7. a
8. a
9. a
10. a
11. a
12. a
13. c
14. a

Chapter 13 (Page 201)

1. a
2. b
3. c
4. c
5. b
6. b
7. a
8. a
9. c
10. c
11. a
12. b
13. b

Chapter 14 (Page 218)

1. a
2. c
3. b
4. b
5. b
6. b
7. a
8. b
9. c
10. c
11. b
12. a
13. b
14. b
15. c
16. c

Chapter 15 (Page 233)

1. b
2. c
3. a
4. b
5. c
6. a
7. a
8. a
9. b
10. a

11. c
12. c
13. a
14. a
15. b
16. a
17. c

Chapter 16 (Page 244)

1. a
2. b
3. a
4. b
5. a
6. c
7. a
8. c
9. c
10. b
11. b
12. b
13. c
14. b
15. c
16. b
17. a
18. b

Chapter 17 (Page 262)

1. b
2. c
3. a
4. b
5. c
6. a
7. a
8. a
9. b
10. a
11. c
12. c
13. a
14. a
15. b

Chapter 18 (Page 281)

1. a
2. c

3. c
4. b
5. a
6. b
7. a
8. a
9. a
10. a
11. c

Chapter 19 (Page 307)

1. b
2. c
3. c
4. c
5. c
6. a
7. b
8. a
9. a
10. a
11. a

Chapter 20 (Page 321)

1. c
2. a
3. b
4. c
5. a
6. b
7. b
8. b
9. a
10. a
11. a
12. a
13. b
14. a
15. a

Chapter 21 (Page 338)

1. a
2. c
3. c
4. c
5. c
6. a

7. b
8. a
9. c
10. a

Answers to revision papers

Revision Paper 1

1. A
2. C
3. C
4. B
5. A
6. C
7. A
8. C
9. C
10. C
11. C
12. C
13. C
14. A
15. A
16. A
17. B
18. B
19. C
20. A

Revision Paper 2

1. B
2. B
3. B
4. B
5. A
6. A
7. A
8. B
9. A
10. A
11. C
12. C
13. C
14. B
15. B
16. C
17. B

18. A
19. A
20. C

Revision Paper 3

1. C
2. C
3. C
4. A
5. A
6. B
7. A
8. C
9. B
10. C
11. A
12. A
13. C
14. C
15. B
16. A
17. A
18. A
19. C
20. B

Revision Paper 4

1. A
2. B
3. C
4. C
5. B
6. C
7. A
8. C
9. A
10. B
11. C
12. C
13. C
14. A
15. B
16. A
17. B
18. A
19. A
20. C

Appendix 4   Decibels

Decibels (dB) are a convenient means of expressing gain (amplification) and loss (attenuation) in electronic circuits. In this respect, they are used as a *relative* measure (i.e. comparing one voltage with another, one current with another, or one power with another). In conjunction with other units, decibels are sometimes also used as an *absolute* measure. Hence dBV are decibels relative to 1 V, dBm are decibels relative to 1 mW, etc.

The decibel is one-tenth of a bel which, in turn, is defined as the logarithm, to the base 10, of the ratio of output power (P_{out}) to input power (P_{in}).

Gain and loss may be expressed in terms of power, voltage and current such that:

$$A_p = \frac{P_{out}}{P_{in}} \quad A_v = \frac{V_{out}}{V_{in}} \text{ and } A_i = \frac{I_{out}}{I_{in}}$$

where A_p, A_v or A_i is the power, voltage or current gain (or loss) expressed as a ratio, P_{in} and P_{out} are the input and output powers, V_{in} and V_{out} are the input and output voltages, and Iin and $Iout$ are the input and output currents. Note, however, that the powers, voltages or currents should be expressed in the same units/multiples (e.g. P_{in} and P_{out} should both be expressed in W, mW, μW or nW). It is often more convenient to express gain in decibels (rather than as a simple ratio) using the following relationships:

$$A_p = 10\log_{10}\left(\frac{P_{out}}{P_{in}}\right) \quad A_v = 20\log_{10}\left(\frac{V_{out}}{V_{in}}\right)$$

$$\text{and } A_i = 20\log_{10}\left(\frac{I_{out}}{I_{in}}\right)$$

Note that a positive result will be obtained whenever P_{out}, V_{out}, or I_{out} is greater than P_{in}, V^{out}, or I_{out}, respectively. A negative result will be obtained whenever P_{out}, V_{out}, or I_{out} is less than P_{in}, V in, or I in.

A negative result denotes attenuation rather than amplification. A negative gain is thus equivalent to an attenuation (or loss). If desired, the formulae may be adapted to produce a positive result for attenuation simply by inverting the ratios, as shown below:

$$A_p = 10\log_{10}\left(\frac{P_{in}}{P_{out}}\right) \quad A_v = 20\log_{10}\left(\frac{V_{in}}{V_{out}}\right)$$

$$\text{and } A_i = 20\log_{10}\left(\frac{I_{in}}{I_{out}}\right)$$

where A_p, A_v, or A_i is the power, voltage or current gain (or loss) expressed in decibels, P_{in} and P_{out} are the input and output powers, V_{in} and V_{out} are the input and output voltages, and I_{in} and I_{out} are the input and output currents. Again note that the powers, voltages or currents should be expressed in the same units/multiples (e.g. P_{in} and P_{out} should both be expressed in W, mW, μW or nW).

It is worth noting that, for identical decibel values, the values of voltage and current gain can be found by taking the square root of the corresponding value of power gain. As an example, a voltage gain of 20 dB results from a voltage ratio of 10 while a power gain of 20 dB corresponds to a power ratio of 100.

Finally, it is essential to note that the formulae for voltage and current gain are only meaningful when the input and output impedances (or resistances) are identical. Voltage and current gains expressed in decibels are thus only valid for matched (constant impedance) systems. Table A4.1 gives some useful decibel values.

Example A4.1

An amplifier with matched input and output resistances provides an output voltage of 1V for an input of 25 mV. Express the voltage gain of the amplifier in decibels.

Solution

The voltage gain can be determined from the formula:

$$A_v = 20\log_{10}(V_{out} / V_{in})$$

where $V_{in} = 25mV$ and $V_{out} = 1V$
   Thus:

$$A_v = 20\log_{10}(1V / 25mV)$$
$$= 20\log_{10}(40) = 20 \times 1.6 = 32dB$$

Example A4.2

An audio amplifier provides a power gain of 33 dB. What output power will be produced if an input of 2 mW is applied?

Solution

Here we must rearrange the formula to make P_{out} the subject, as follows:

$$A_v = 10\log_{10}(P_{out} / P_{in})$$

thus:

$$A_p / 10 = \log_{10}(P_{out} / P_{in})$$

or

$$\text{antilog}_{10}(A_p / 10) = P_{out}/P_{in}$$

Hence:

$$P_{out} = P_{in} \times \text{antilog}_{10}(A_p / 10)$$

Now $P_{in} = 2mW = 20 \times 10^{-3}W$ and $A_v = 33dB$, thus:

$$P_{out} = 2 \times 10^{-3} \text{antilog}_{10}(33/10)$$
$$= 2 \times 10^{-3} \times \text{antilog}_{10}(3.3)$$
$$= 2 \times 10^{-3} \times 1.995 \times 10^{-3} = \mathbf{4\ W}$$

Example A4.3

An antenna has a gain of 7 dB relative to a reference dipole. What power should be applied to the antenna in order to maintain the same signal strength as that produced when 20 W is fed to the dipole?

Solution

The required power can be determined from the formula:

$$A_p = 10\log_{10}(P_{ant} / P_{ref})$$

from which:

$$P_{ant} = P_{ref} / \text{antilog}_{10}(A_p / 10)$$

Now $P_{ref} = 20W$ and $A_p = 7dB$, thus:

$$P_{ant} = 20 / \text{antilog}_{10}(7/10)$$
$$= 20 / \text{antilog}_{10}(0.7) = 20/5 = 4W$$

Table A4.1 Decibels and ratios of power, voltage and current

Decibels (dB)	Power gain (ratio)	Voltage gain (ratio)	Current gain (ratio)
0	1	1	1
1	1.26	1.12	1.12
2	1.58	1.26	1.26
3	2	1.41	1.41
4	2.51	1.58	1.58
5	3.16	1.78	1.78
6	3.98	2	2
7	5.01	2.24	2.24
8	6.31	2.51	2.51
9	7.94	2.82	2.82
10	10	3.16	3.16
13	19.95	3.98	3.98
16	39.81	6.31	6.31
20	100	10	10
30	1,000	31.62	31.62
40	10,000	100	100
50	100,000	316.23	316.23
60	1,000,000	1,000	1,000
70	10,000,000	3,162.3	3,162.3

Index

Pages in *italics* refer to figures and pages in **bold** refer to tables.

AAC 23
ACARS 22–3, 85, 88–90; over IP 23
ACAS 308
accelerometer 219, 223
acceptor circuit 64
ACP 20–2
actual navigation performance *see* ANP
address code 98
ADF 133, 144, *146*; antenna 147–8; bearing display 149; bearing indicator *150*; control panel 148–9; display *150*; homing 153; receiver 147
adjacent channel rejection 66, 83
ADS–B 297–*9*; mandate 301; traffic display 320
ADS–C 301–2
AES 17
AFCS mode control panel 190
AGC 63, 68
air: navigation service provider 284; traffic control 138, 282–4; traffic management *see* ATM
airborne collision avoidance system *see* ACAS
aircraft address 294; communication, addressing and reporting system *see* ACARS; earth station *see* AES; radio navigation service *see* ARNS
airline: administrative control *see* AAC; operational control 85, 98
airway 164–5, *205*
all–call interrogation 292
altitude encoder 287
AM 59
amplitude modulation *see* AM
AMU 20–2
angle of attack 12
ANP 217
antenna 27; analysis 54; coupler *105*; coupling unit 60, 103; gain 31; waveguide *266*
AoIP 23
area navigation 135, 203
ARNS 336
ATC 138; control panel 286; separation accuracy 300; transponder 285
atmosphere 5, 151; layers 151
ATM 19, 297–300
ATN router 88
atomic clock 236
attenuation 43
audio: accessory unit 109; control panel *see* ACP; management unit *see* AMU; selector panel 109–10
augmented approach 243
autoland 190

automatic DME tuning 261; approach and landing 190; dependent surveillance *see* ADS–B, ADS–C; direction finder *see* ADF
azimuth resolution 272

backscatter 198
balanced modulator 99
band–pass filter 65
bandwidth 65
baro–VNAV 242
beamwidth 33
bearing 126
beat frequency oscillator 58
beyond line of sight *see* BLOS
BLOS 22–3
bonding 332; strap 333
boom microphone 111
broadband noise 327

C–band 264–5, 334, 336
cabin interphone 107
cabling 331
capture range 71
carrier: insertion oscillator 96; signal 60
CDI 161–*2*
celestial navigation 130
Cervit glass 224
channel spacing 83
characteristic impedance 40
chart 130
closest point of approach *see* CPA
clouds 269–*70*
clutter 276
CME 16
CNS/ATM 302
coastal refraction 151–2
coaxial cable *41*–2
cockpit: speaker unit 109; voice recorder *see* CVR
communication system 19–20
compression 84–5
connector 44
continuous wave 58
control segment 237
controller pilot datalink communication 303
conventional VOR 157
Coordinated Universal Time *see* UTC
corner reflector 37
coronal mass ejection *see* CME
Cospas–Sarsat 121, 123

course deviation indicator *see* CDI
CPA 309
CPDLC 303
critical frequency 11
current–fed antenna 36
CVOR 157
CVR 107, 110, 112–4

D–layer 9–10
D8PSK 88
DAL 335
data collision 99
datalink weather 277; service provider *see* DSP
dead reckoning 129, 135
decision height 192
demodulation 59
demodulator 60
depth of modulation 84
descent forecast page 261
design assurance level *see* DAL
detector 58
difference in depth of modulation 182
differential eight phase shift keying 88
diffraction 7
digital frequency synthesis 70, 72
dipole antenna 32
direct: path 5; wave 5, 9
directional: characteristic of an antenna 34;
    gyroscope 127, 132
director 33
distance 128; measuring equipment *see* DME;
    modification 312
divide–by–*n* counter 72
DMA 133, 134, 170; antenna 174; display 175;
    equipment 173; operation 172; overview 171;
    panel 175; position updates 262; terminology
    172; transponder *204*
Doppler: beam 199; effect 196; equipment 199;
    navigation 196; shift 136, 196, 276; VOR 158, 201
double: conversion 69; sideband 95–6; superhet
    receiver 68
downlink 85, 89; aircraft parameter 296
drift 129, 329; angle 129
drone 307–8
DSB: modulation 82; suppressed carrier 99
DSP 22
dual conversion 69
ducting 7
DVOR 158

E–field 3, 50
E–layer 8–10
Earth station antenna 39
EFB 141–2
EHSI 163
electrical length 28
electromagnetic: compatibility *see* EMC;
    environment 323 *see* EMI; wave 1, 3

electronic: flight bag *see* EFB; horizontal situation
    indicator 163
elementary surveillance 295
ELT 116–22; types 117
EMC 323–35
emergency ATC code 286; locator transmitter 116–9
EMI 323–4; classification 325; effects 325; examples
    326; filter 331; reduction 330; sources 325
enroute navigation 175
enhanced surveillance 295
European GNSS 241
extended squitter 300

F–layer 8–10
FAA Next Generation 305
false reply 290, 320
FANS 98, 306
feeder 40–2
fibre optic gyroscope 225–6
flare mode 192
flight: interphone amplifier 109; management
    computer system *see* FMCS; management
    system *see* FMS
FM 59
FMCS 247; CDU 248; operation 250
FMS 247; CDU 249; control display unit 248;
    overview 247
focal plane reflector 38
four–dimensional navigation 258, 260
frequency 4; division multiplexing 98; modulation
    see FM; shift keying *see* FSK
FSK 86
Future Air Navigation Systems *see* FANS

GA transponder 287
GEO 18
geomagnetic storm 14
geostationary: earth orbit *see* GEO; satellite 18
Gillham code 287, 297
glide slope 183; antenna *184*
global: navigation satellite systems *see* GNSS; global
    positioning system *see* GPS
GNSS 235; augmentation 241; error detection 242;
    integration 240; operation 239; vulnerability 240
GPS 236; receiver 238; signal 239; space
    segment 237
great circle 128
ground: wave 5; crew call system 107; network 19,
    88; speed 196; wave 5
grounding 332
guard band 83
gyro–magnetic compass 131–2, 144
gyroscope 219

H–field 3, 50
hailstone 272
half–wave dipole 28–9
headset 111

HF: antenna 103; communication 94; datalink *see* HFDL; propagation 94; radio controller 102; radio specification 102
HFDL 19, 22, 97–8, *100–1*
high: frequency *see* HF; intensity radio frequency field *see* HIRF
HIRF 332
horizontal situation indicator *see* HSI
horn antenna 39
HSI 161
hyperbolic navigation 134

ICAO address 98
identification page 252
IFF 171
ILS 133, 136, 182, 193; airborne equipment 186; antenna 186; approach 190; controls 187; display 187–9; frequency 182; ground equipment 182; overview 182–3; receiver 186
image channel rejection 67
impedance 30; frequency plot 53
inertial: navigation 219, *222*; navigation accuracy 231, 233; navigation system *see* INS; reference mode panel 228; reference system 228; reference unit 221; signal processing 225; space 227; system alignment 230
INS 136
instability 329
instrument: approach 139; landing system *see* ILS
interference 151
intermediate frequency 63
internet protocol *see* IP
interphone 107–8
interrogation 288–9
interrogator 171; code 294
ionised layer 9
ionosphere 8; layer 10
ionospheric sounding 8; wave 6
IP 23
Iridium 17, 19; subscriber unit *see* ISU
IRS 228
IRU 221
isotropic radiator 27
ISU *77*

j-notation 52–3
jack panel 109

Kalmann filter 212
knot 128
Ku–band 19

Lambert projection 131
lateral direction 219
latitude 125
law of reciprocity 27
legs page 254–*5*
LEO 18

LFMCW 334
lightning 277; detection 277
line: of position 238; of range 81, 156–7; of sight 5, 203
linear frequency modulated continuous wave carrier *see* LFMCW
local: oscillator 63; user terminal 121
localizer: antenna *184*; beam *184*; transmitter 182
locator beacon *see* ULB
log–on request 98
longitude 125
loop: antenna 144–5; capacitance and inductance *41*
Loran–C 137
low earth orbit *see* LEO
low range radio altimeter *see* LRRA
lower: side frequency 82; sideband 95–6
lowest usable frequency *see* LUF
LRRA 139, 189; display 140
LUF 11, 94

magnetic: compass 131; north 126–7; variation 126, 231
major lobe 33
management unit 86, 88
map 130
Marconi antenna 36
marker beacon 185
maximum: take–off weight 296; usable frequency *see* MUF
Mercator projection 130–1
meridian 125
METAR 278
micro–electromechanical system *see* MEMS
microburst 273–4
microwave landing system *see* MLS
minimum navigation performance 141
minutes 125
mission control centre 121
mixer 63–4, 69
MLS 133
MEMS 224
modulation 59, 83
modulator 60
mono–pulse SSR 292
Morse code 59
MUF 10, 94
multi: hop propagation 12–3; touch screen 258; function display *259*

nautical mile 128
navigation 125, 130–1, 203, 219; aids 135; database 203, 209, 248; terminology 132
NDB 145–*6*; code 147; frequency 147
near field 4
no computed data 173
noise 327; operated squelch 85
non–directional beacon *see* NDB
normalising 53

North 126
null: point 144; position 219, 224

OBS *162*
octal number 290
omni–bearing selector 161–*2*
open wire feeder 42
outer marker 185
overmodulation 84

parabolic reflector 37–8
parallel tuned circuit 64–5
passenger address system 107
PBN 215; errors 217
performance: based navigation 215; database 248;
    initialisation 250–1
phase: locked loop *see* PLL; shift keying *see* PSK
phased array 278, *280*
physical length 28
pinpointing 129
plan position indicator *see* PPI
platform 219
PLL 70–1
PM 334
polar radiation pattern 29
position: fixing 129; initialisation 249
PPI 267, 282
precipitation 265, 269
predictive wind shear 273
press–to–talk *see* PTT
primary: constants 40, 43; radar 170, 265;
    surveillance radar *283*
prime meridian 125
program pin 252
progress page 253, 261
propagation 3, 7, 81, 94
protected 223; airspace *310*
pseudo–range 238–9
pseudorandom noise 239
PSK 98, 293
PTT 108–11
pulse position modulation 294
pulse modulation *see* PM

Q–code 165
Q–factor 65
quadrantal error 151
quadrature modulator 88
quarter–wave antenna 36

radar: antenna 265–*6*; bands 3; navigation 137;
principle 170
radial 36, 203
radiated power 31
radiation resistance 30
radio: altimeter 334–5; altitude 192; antenna 336;
    frequency amplifier 63; frequency spectrum 1–**2**;

interference 333, *337*–8; magnetic indicator *see*
    RMI; management panel *see* RMP
Rayleigh scattering 272
RDMI 175
receiver 58
reciprocity 27
reduced vertical separation 141
reference antenna 31
reflection 6
reflector 32, 37
refraction 6
rejector circuit 64
remote airfield control tower 303–4
reply code 282
reporting points 178
required navigation performance *see* RNP
rho: rho 130; –theta 130
rhumb line 127, 130
ribbon cable 42
ring laser gyro 224
rising air 271
RMI 149, 161–*2*; bearing indicator *150*; source
    control 161
RNAV 135, 203, *207*; computer 209; control display
    unit 208; equipment 208; geometry *210*, *211*; leg 206
RMP 21
RNP 215
roll–call interrogation 292
route: page 251; selection 250
Russian GNSS 241

S–band 239
Sagnac effect 225
satellite: communication *see* SATCOM; data unit
    see SDU; navigation 137, 235; orbit 18
SATCOM 16–9, 22–4
scatter 6
SDU 19
search and rescue 121
secondary: radar 170, *204*; surveillance radar 138,
    282–*3*, *310*
seconds 125
SELCAL 97
selected navigation aids 262
selective: availability 240; calling *see* SELCAL
selectivity 63, 64
sense antenna 145
sensitivity 63
separation accuracy 300
series tuned circuit 64–5
service interphone 107
side–lobe 288; suppression 289
signal routing 20
silent zone 12–3
single: European sky 305; sideband *see* SSB
skip distance 12–3
sky wave 6, 9
slant range 171

Smith chart 51–2
SMPS 323–*4*
software defined radio 73, *78*
solar flare 14–5
space: segment 235; system 236; wave 5; weather 12
special position indicator 290
spectral analysis *324*, 327–*8*
spectrum 1
speed 128
spherical geometry 227
squawk code 285
squelch 85–6
squitter 298
SSB 96
standard: instrument departure *213*, *256*; terminal
    arrival 214, 257
standing wave ratio *see* SWR
steep approach 243
steered radar beam 278
strap–down 221
superhet receiver 63–4
suppressed carrier 96
surface wave 5
switched mode power supply *see* SMPS
SWR 45, 47–9, 53, 103–*5*
synchronised garbling 291–2
synchronous garble *319*

TA/RA sensitivity **311**
TACAN 176–7
TAF 278
TAS 309, 321
TCAS 138–9, 285–6, 308–12; advisory 315; antenna
    312; aural annunciation 317; communication
    links 317–8; compatibility 314; computer 313;
    control panel 314; display 314; equipment 312–3;
    guidance 316; icon 316; operation 314;
    overview 309
terminal aerodrome forecast *see* TAF
Terrain 151; database 276; mapping 275
theta–theta 130
three–dimensional radar 276
thunderstorm 269–71, 276
time: bias error 238; division multiplexing 98
TIS 308
track 129
traffic advisory system *see* TAS
traffic alert and collision avoidance system *see*
    TCASinformation system 308; symbol 320;
    warning icon 315
transmission line 43
transmitter 58, 61
transponder 286–7, 296
transport rate 227
TRF 62
troposphere 5
tropospheric: ducting 7; scatter 7

true airspeed 296
true north 126
tuned circuit 66
tuned radio frequency receiver *see* TRF
turbulence 269, 271, 273
twin feeder *41*

ULB 113–4
underwater locator beacon *see* ULB
universal access transceiver 300
uplink 85, 89
upper: side frequency 82; sideband 95–6
user segment 235, 237
UTC 237

VDL 22, 85, 300; Mode 0 86; Mode 2 87;
    Mode A 87
vector network analysis 51, 54–5
velocity factor 44
vertical: antenna 36–7; displacement 197; half–wave
    antenna 36; quarter–wave antenna 36
very high frequency *see* VHF
VHF: channels 83; coverage 94; datalink *see* VDL;
    digital link 300; omnidirectional range 155;
    propagation 81; range 81
VNA 54–5
voltage: controlled oscillator 70–1; fed antenna 36
volume control 109
VOR 155–6, *179*, *204*; airway 164; antenna 159;
    control panel 160; display 161, 166; electronic
    displays 163; frequency 155, 182; ground station
    156; principle 155; radial 164; receiver 159–60
VORTAC 176–7

water droplet 272–3
wave 1, 3
waveform interference 278
wavefront 3
waveguide 50, *266–7*
wavelength 4
waypoint 203, 206
weather: avoidance 265; detection 265; EFIS display
    268; radar 50, 138, 265; radar control panel 267–8;
    radar display 267; radar transceiver 267
whisper–shout 318–*9*
wideband noise 327
wind shear 271
wireless telegraphy 1
wiring 331

X–band 264
X–ray chart 15

Yagi beam antenna 32–3

zero meridian 125

Printed in the United States
by Baker & Taylor Publisher Services

Printed in the United States
by Baker & Taylor Publisher Services